Elements of Mathematical Methods for Physics

Elements of Mathematical Methods for Physics provides students with an approachable and innovative introduction to key concepts of mathematical physics, accompanied by clear and concise explanations, relevant real-world examples and problems that help them to master the fundamentals of mathematical physics. The topics are presented at a basic level, for students lacking a prior mathematical background.

This book is designed to be covered in two semesters, presenting 18 chapters on topics varying from differential equations, matrix algebra and tensor analysis to Fourier transform, including special functions and dynamical systems.

Upper-level undergraduate and graduate students of physics and engineering as well as professionals will gain a better grip of the basics and a deeper insight into and appreciation for mathematical methods for physics.

Key Features:

- Reviews and presents the basic math skills needed at the undergraduate level.
- Chapters accompanied by examples and end-of-chapter problems to enhance understanding.
- Introduces dynamical systems and includes a chapter on Hilbert Space.

Francis E. Mensah is currently Chair of the Department of Natural Sciences and Associate Professor of Physics at Virginia Union University. He is also the Coordinator for the Physics and Engineering program. He holds a PhD in Atmospheric Lidar and Remote Sensing from Howard University and a D.Sc. in Theoretical Physics from the University of Abomey-Calavi in Benin. Dr. Mensah has been Lecturer in Physics and in Mathematics at Howard University in the Department of Physics and Astronomy and in the Department of Mathematics. He was also Assistant Professor of Physics at the University of the District of Columbia in Washington DC. He is a member of the American Physical Society and the National Society of Black Physicists. In 2018, Dr Mensah received the Scott & Stringfellow Outstanding Professor Award from Virginia Union University. Dr Mensah's passion is teaching, which he loved from an early age. He has used various techniques to teach, including Project-Based Learning (PBL), a project currently sponsored by the National Science Foundation (NSF).

Elements of Mathematical Methods for Physics

Francis E. Mensah

CRC Press
Taylor & Francis Group
Boca Raton London New York

CRC Press is an imprint of the
Taylor & Francis Group, an **informa** business

First edition published 2024
by CRC Press
2385 NW Executive Center Drive, Suite 320, Boca Raton FL 33431

and by CRC Press
4 Park Square, Milton Park, Abingdon, Oxon, OX14 4RN

CRC Press is an imprint of Taylor & Francis Group, LLC

© 2024 Francis E. Mensah

Library of Congress Cataloging-in-Publication Data
Names: Mensah, Francis E., author.
Title: Elements of mathematical methods for physics / Francis E. Mensah.
Description: First edition. | Boca Raton, FL: CRC Press, 2024. | Includes bibliographical references and index. |
Identifiers: LCCN 2024004769 | ISBN 9781032762227 (hbk) | ISBN 9781032765136 (pbk) | ISBN 9781003478812 (ebk)
Subjects: LCSH: Mathematical physics.
Classification: LCC QC20 .M39 2024 | DDC 530.15--dc23/eng/20240509
LC record available at https://lccn.loc.gov/2024004769

ISBN: 9781032762227 (hbk)
ISBN: 9781032765136 (pbk)
ISBN: 9781003478812 (ebk)

DOI: 10.1201/9781003478812

Typeset in Times
by Deanta Global Publishing Services, Chennai, India

Contents

Foreword

Physics is the most powerful field of science that speaks more intimately of man's speculative inquiry into the physical world. Fundamentally, physics focuses on the miracles of everyday life, such as the dynamics of atmospheric disturbances, the rising and setting of the sun, etc. At its deepest level, physics tackles the subtlest of scientific concepts such as the quantization of matter and energy, the complex interweaving of space and time, etc. A deeper understanding and hence appreciation of the seeming simplicity and yet complexities of natural phenomena observed in the physical world cannot be manifested without a mastery of the mathematical tools used in physics. It is against this background that the writing of this book was conceived.

The intent of this book is threefold: (1) to provide the tools for the senior and graduate students to build competence in many areas of mathematics used in physics and engineering; (2) to provide advanced students who have forgotten some essential topics or want to learn new mathematical tools; and (3) to be used as an effective research tool by research professionals in physical and engineering sciences who need to review their partially forgotten mathematical tools.

The book is designed based on the assumption that the senior-graduate level is at some degree of mathematical sophistication and progressing in the knowledge of advanced physics. Examples are provided to ease the learning and applications of the concepts. However, mastery or skills are gained by solving many problems.

It is with great pleasure and gratitude that I write the foreword of this book. Initially, I was asked to review this book, a task which I did not take lightly, especially having been aware of the challenges facing students who had to take mathematical methods for physicists. The experience was very humbling and quite gratifying when I was tasked with the responsibility of writing the foreword. The result was very rewarding. It is my sincere hope that this book will prove more useful in its usage.

Dr. Peter A.D. Intsiful, PhD
Professor of Physics, Howard University

Preface

Mathematical methods in physics are usually taught to physics and engineering students in their senior year where they have already taken calculus I and II. Teaching this course for many years, I have noticed that students usually start the course somehow unprepared. I have therefore decided to start writing a book that can ease the learning process of mathematics for prospective physicists. Thus, this book, despite not covering everything in mathematical methods in physics, is a good tool for senior physics majors, especially for those who don't have a strong mathematical background, but still would like to major in physics or related fields.

The purpose of this book is then to provide a useful manual for physicists and engineers. The book is written to cover the math gap that physics and engineering majors may have and help improve their confidence in solving problems and critical thinking.

Furthermore, the book covers 18 chapters each of which is followed by a set of practice problems. Topics such as ordinary differential equations, partial differential equations, vector analysis, tensor analysis, matrices algebra, sequences, functions of a complex variable, Fourier series, Fourier and Laplace Transforms, Helmholtz's equation, the special functions, the integral equations and the dynamical systems covered are important in physics and engineering.

Your feedback will be greatly appreciated and will help improve future editions. We do not pretend to have explored all aspects of mathematical physics. However, the book gives details of many interesting topics in mathematical physics. It is designated for two semesters of teaching.

Francis E.Mensah, DSc. PhD
Associate Professor of Physics
Virginia Union University

Symbols and Notations

$\Rightarrow$	implies
$\Leftarrow$	is implied by
$\Leftrightarrow$	if and only if
$\forall$	for all
$\exists$	there exists
$\exists!$	there exists one and only one
$\in$	belongs to
$\notin$	does not belong to
$\cap$	intersection
$\cup$	union
$\subset$	is contained in; is a subset of
$\perp$	orthogonal to
$\{\}$	a sequence or set
QED	quod erat demonstrandum (indicates the end of a proof)
RHS	right-hand side
LHS	left-hand side

1 Elements of Ordinary Differential Equations

INTRODUCTION

Let us start by asking the question, "What is an Ordinary Differential Equation (ODE)?" The answer is "an ODE is an equation relating an unknown function of a single variable to one or more of its derivatives." A most general ODE may be written in an implicit form as

$$F\left(t, y, y', \ldots, y^{(n)}\right) = 0 \tag{1.1}$$

where t is the independent variable with domain $[a, b]$; $y = y(t)$ is the dependent (unknown) variable; $y^{(k)}$ is the kth derivative of y with respect to t; n is the order of the differential equation, and F is a function of $n + 2$ variables.

Such relationships often occur when expressing scientific laws connecting physical quantities and their rates of change in mathematical terms. They are called mathematical models. Three main stages characterize the study of mathematical models:

- The derivation of differential equations that describe specified physical situations.
- The determination of the appropriate solutions to these differential equations (exact or approximate solutions).
- The interpretation of the solutions found in relation to the physical situation.

Here is the solution process in mathematical modeling (Figure 1.1).

Example 1: Newton's second law of motion states that the sum $\vec{F}$ of all the forces acting on a particle of mass m located at a position represented by $\vec{x}$ in space at a given instant t is equal to the product of the mass and the acceleration of the particle at that point:

$$\vec{F} = m\frac{d^2\vec{x}}{dt^2} \tag{1.2}$$

This is a second-order ODE where $\vec{x}(t)$ is the position vector of the particle, an unknown function of the single variable t.

Example 2: The equation

$$\frac{dy}{dt} = k(t)f(y) \tag{1.3}$$

describes the growth of the unknown function y when its rate of change is proportional to a function f depending on y.

DOI: 10.1201/9781003478812-1

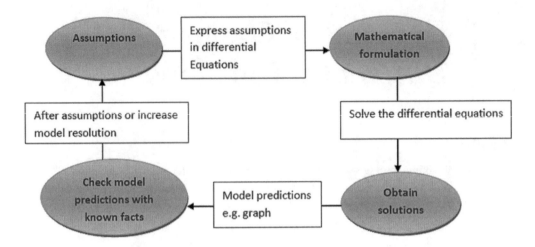

FIGURE 1.1   Solution process in mathematical modeling.

Example 3: The equation

$$\frac{d^2\theta}{dt^2} + \frac{g}{L}\sin\theta = 0 \tag{1.4}$$

describes the motion of a pendulum of length L, where θ is the angle between the pendulum and the vertical axis.

$g = 32\,\text{ft}/\text{s}^2 = 9.8\,\text{m}/\text{s}^2$ is the planet Earth's gravity acceleration.

Example 4: The second-order ODE

$$\frac{d^2Q}{dt^2} + R\frac{dQ}{dt} + \frac{Q}{C} = V(t) \tag{1.5}$$

is the LRC oscillator's equation. An LRC circuit consists of

- a resistor with a resistance of R ohms,
- an inductor with inductance of L Henries and
- a capacitor with a capacitance of C farads.

mounted in series with a source of electromotive force (a battery or a generator) that supplies a voltage of $V(t)$ volts at time t; $Q(t)$ represents the charge in coulombs on the capacitor at time t (Figure 1.2).

1.1   FIRST-ORDER ODES

1.1.1   INTEGRALS AS GENERAL AND PARTICULAR SOLUTIONS

The study of differential equations generalizes that of integral calculus as illustrated by the following two special cases of ODEs.

Special case 1: The differential equation

$$\frac{dy}{dx} = f(x) \tag{1.6}$$

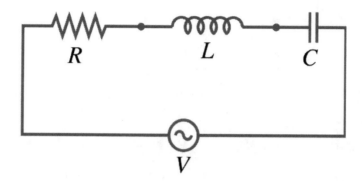

FIGURE 1.2   An LRC circuit composed of a resistor of resistance R, an inductor of self-inductance L, a capacitor of capacitance C and a generator of emf $V(t)$.

implies that

$$y(x) = \int f(x)\,dx + C = G(x) + C \tag{1.7}$$

where C is an arbitrary constant and $G(x)$ is a function of a single variable x. This is the general solution of equation (1.6). A particular solution of (1.6) is the one satisfying an initial condition of the form $y(x_0) = y_0$.

$$\text{A problem of the form } \begin{cases} \dfrac{dy}{dx} = f(x) \\ y(x_0) = y_0 \end{cases} \tag{1.8}$$

is called an initial-value problem (IVP).

Example 5: The IVP $\begin{cases} \dfrac{dy}{dx} = (x-1)^2 \\ \quad y(1) = 2 \end{cases}$. $\tag{1.9}$

has solution: $y(x) = \dfrac{(x-1)^3}{3} + 2$. $\tag{1.10}$

1.1.2 DIRECTION FIELDS AND SOLUTION CURVES

Consider the differential equation

$$\frac{dy}{dx} = f(x, y) \tag{1.11}$$

In general, one cannot find an analytic form to the solution of (1.11), that is, a solution which can be expressed using elementary functions studied in calculus. Such is the case for $\dfrac{dy}{dx} = x^2 + y^2$. One may, however, use graphical and numerical techniques to construct approximate solutions to ODEs of the form (1.11).

1.1.2.1 Slope Fields and Graphical Solutions

At a point (x, y), equation (1.11) represents the slope of the tangent line to the curve solution to the ODE. Thus, we can determine a solution curve of equation (1.11) by obtaining successive slopes of tangent lines materialized by short line segments having the correct slope $m = f(x, y)$. These line segments constitute a direction field or a slope field for $\dfrac{dy}{dx} = f(x, y)$.

Example 6

Consider the first-order differential equation $\dfrac{dy}{dx} = 2y$, where $m = f(x, y) = 2y$.

For $y = 0$, $m = 0$ so that the x-axis is a solution curve of the given ODE.

For $y = 1$, $m = 2$ so that the parallel solution curves intersect this horizontal line at an *arctan*2 angle.

Similarly, for $y = -1$, $m = -2$ so that the parallel solution curves intersect this horizontal line at an $\arctan(-2) = -arctan2$ angle.

By sketching short line segments on these lines to materialize the slopes, we obtain the slope field, as shown in Figures 1.3 and 1.4.

The slope field depicts a view of the general shapes of solutions curves.

1.1.2.2 Definition of the Isoclines

The isoclines are curves for the ODE (1.11) defined by the equation $f(x, y) = constant$, on which the slope of the solution curve is always constant. They are convenient when plotting the direction field of a differential equation and sketching its solutions graphically. Indeed, through each point, a solution curve should proceed in such a direction that its tangent line is nearly parallel to the nearby line segments of the slope field. By following these short line segments, one can depict a view of the general shapes of solution curves.

1.1.3 EXISTENCE AND UNIQUENESS OF SOLUTIONS

Consider the IVP $\begin{cases} \dfrac{dy}{dx} = f(x) \\ y(x_0) = y_0 \end{cases}$ consisting of equation (1.8) and the associated initial value. It is natural that the questions of existence and uniqueness of solution just evoked first be addressed

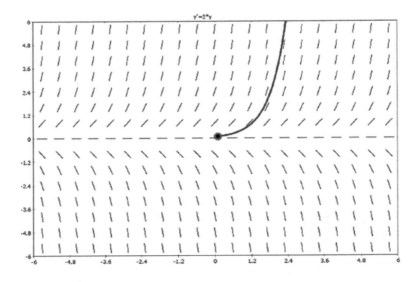

FIGURE 1.3 Slope field for $\dfrac{dy}{dx} = 2y$.

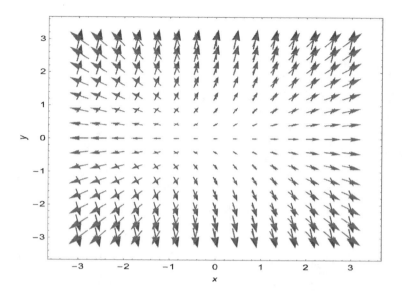

FIGURE 1.4 Slope field for $\dfrac{dy}{dx} = 2y$ shown using the VectorPlot Command with the Mathematica software.

before any attempt to solve it. Otherwise, the graphical method described above using direction fields may not always provide the desired results as illustrated by the following two examples.

Example 7: The IVP $\begin{cases} \dfrac{dy}{dx} = \dfrac{1}{x} \\ y(0) = 0 \end{cases}$ has no solution.

Example 8: The IVP $\begin{cases} \dfrac{dy}{dx} = 2\sqrt{y} \\ y(0) = 0 \end{cases}$ admits infinitely many solutions.

One can readily verify that the functions below are two such solutions:

$$y_0(x) = x^2 \text{ and } y_1(x) \begin{cases} 0 & \text{if } 0 \le x \le 1 \\ (x-1)^2 & \text{if } x \ge 1 \end{cases}$$

The following theorem provides conditions that ensure that IVP (1.8) has a unique solution.

Theorem 1: Existence and Uniqueness of Solutions

Suppose that $f(x,y)$ and $\dfrac{\partial}{\partial x} f(x,y)$ are continuous on some rectangle R in the x–y plane that contains the point (x_0, y_0) in its interior. Then, for some open interval I containing x_0,

the IVP $\begin{cases} \dfrac{dy}{dx} = f(x) \\ y(x_0) = y_0 \end{cases}$ has one and only solution that is defined on the interval I.

Observations: Looking at Examples 7 and 8 in the light of this theorem, we observe the followings:

- If $\dfrac{\partial}{\partial x} f(x,y)$ is undefined at some point (x_0, y_0), then no solution exists for which $y(x_0) = y_0$.

- If $\frac{\partial}{\partial x} f(x, y)$ and $f(x, y)$ are continuous in a rectangle about (x_0, y_0), then a unique solution to the IVP exists.
- Although we may know by Theorem 1 that IVP (1.8) has a solution, finding it, is in general, a tough challenge! In the following sections, we will restrict our attention to special forms of the first-order ODE $F(x, y, y') = 0$.

1.1.4 SEPARABLE EQUATIONS

A differential equation

$$\frac{dy}{dx} = K(x, y) \tag{1.12}$$

is said to be separable if the function $K(x, y)$ is a product (or a ratio) of a function of x and a function of y, i.e., if (*i*) $K(x, y) = \frac{g(x)}{f(y)}$ or (*ii*) $K(x, y) = g(x) f(y)$

In case (*ii*), equation (1.12) separates into $f(y) dy = g(x) dx$, and upon integrating one gets:
$\int f(y) dy = \int g(x) dx$ or $F(y) = G(x) + C$, which yields:

$$F(y(x)) = G(x) + C \tag{1.13}$$

where C is an arbitrary constant of integration. Formula (1.13) provides an implicit solution for equation (1.12). It is called the general solution of equation (1.12).

Example 9:

(a) The differential equation in the IVP $\begin{cases} \dfrac{dy}{dx} = \dfrac{4 - 2x}{3y^2 - 5} \\ y(1) = 3 \end{cases}$ is separable and equivalent to

$(3y^2 - 5) dy = (4 - 2x) dx$. By integrating both sides we get $y^3 - 5y = 4x - x^2 + C$. Applying the initial condition yields $C = 9$.

(b) Similarly, the differential equation,
$x^2 y' = 1 - x^2 + y - x^2 y^2$ is separable and equivalent to:

$$\frac{1 - x^2}{x^2} dx = \frac{1}{1 + y^2} dy.$$

Remark: In general, an implicit solution of the differential equation (1.11) has the form $K(x, y) = C$, where C is a constant and $K(.,.)$ is a function of two variables. It must satisfy equation (1.11) to be a solution. The actual equation satisfied by $K(x, y)$ is obtained by differentiating this function implicitly as follows:

$$\frac{d}{dx} K(x, y) = 0 \Leftrightarrow \frac{\partial}{\partial x} K(x, y) + \frac{\partial}{\partial y} (K(x, y)) \frac{dy}{dx} = 0 \Leftrightarrow \frac{dy}{dx} = -\frac{\frac{\partial}{\partial x} K(x, y)}{\frac{\partial}{\partial y} K(x, y)}$$

This last equation indicates how the function $K(x,y)$ in the implicit solution $K(x,y)=C$ relates to the function $f(x,y)$ in differential equation (1.11).

1.1.5 LINEAR FIRST-ORDER EQUATIONS

A general linear first-order differential equation is an equation of the form

$$a(x)\frac{dy}{dx}+b(x)y+c(x)=0 \cdot$$

Assuming that $a(x)\neq 0$, we may divide by this function and obtain the following special form

$$\frac{dy}{dx}+P(x)y=Q(x) \tag{1.14}$$

We integrate this equation using the so-called integrating factor method as follows. Let

$$\mu(x)=e^{\int P(x)dx} \tag{1.15}$$

Multiply equation (1.14) by $\mu(x)$ to get

$$e^{\int P(x)dx}\frac{dy}{dx}+P(x)e^{\int P(x)dx}y=Q(x)e^{\int P(x)dx},$$

which can be written as $\dfrac{d}{dx}\left(y(x)e^{\int P(x)dx}\right)=Q(x)e^{\int P(x)dx}.$

Upon integrating both sides, we get the solution

$$y(x)=e^{-\int P(x)dx}\left(\int Q(x)e^{\int P(x)dx}dx+C\right) \tag{1.16}$$

where C is an arbitrary constant of integration.

Definition 1
The function $\mu(x)$ defined by equation (1.15) is called an integrating factor for differential equation (1.14).

In summary, to solve the first-order linear differential equation (1.14),

- calculate the integrating factor $\mu(x)=e^{\int P(x)dx}$;
- multiply both sides of the differential equation by $\mu(x)$;
- recognize the left side as the derivative $\dfrac{d}{dx}\left(\mu(x)y(x)\right)=\mu(x)Q(x)$;
- integrate this equation to get $\mu(x)y(x)=\int \mu(x)Q(x)dx+C.$

Example 10: Solve the IVP $\begin{cases}\dfrac{dy}{dx}=1+x+y+xy \\ \qquad y(0)=0\end{cases}$

Solution: The differential equation is linear with
$P(x)=-(1+x)$ and $Q(x)=1+x.$

Thus, $\mu(x)=e^{-\int(1+x)dx}=e^{-x-\frac{x^2}{2}}$ and one readily obtains the solution $y(x)=e^{x+\frac{x^2}{2}}-1.$

Remark: The differential equation $\frac{dy}{dx} + P(x)y = Q(x)$ can also be written as $\frac{dy}{dx} = f(x,y)$, where $f(x,y) = -P(x)y + Q(x)$.

If $P(x)$ and $Q(x)$ are continuous functions in some interval containing x_0, then, the function $f(x,y)$ and its partial derivative $\frac{\partial}{\partial y} f(x,y) = -P(x)$ are continuous on some rectangle containing (x_0, y_0) and so, the conditions of Theorem 1 are satisfied.

Theorem 2: If the functions $P(x)$ and $Q(x)$ are continuous on an open interval I containing the point x_0, then the linear IVP

$$\begin{cases} \frac{dy}{dx} + P(x)y = Q(x) \\ \quad y(x_0) = y_0 \end{cases}$$

has a unique solution, $y(x)$ on I, given by

$$y(x) = e^{-\int P(x)dx}\left(Q(x)e^{\int P(x)dx}dx + C\right),$$ where C is found by requiring that $y(x_0) = y_0$.

1.1.6 SUBSTITUTION METHODS

1.1.6.1 General Case

A first-order equation of the form (1.14) can sometimes be solved by transforming it into a separable or linear equation if an appropriate substitution can be found. Indeed, let $v = \alpha(x,y)$ and $y = \beta(x,v)$, then (1.14) becomes

$$\frac{dy}{dx} = \frac{\partial\beta}{\partial x}\frac{dx}{dx} + \frac{\partial\beta}{\partial v}\frac{dv}{dx} = \frac{\partial\beta}{\partial x} + \frac{\partial\beta}{\partial y}\frac{dv}{dx} = f(x,\beta(x,v))$$

This leads to $\frac{dv}{dx} = \left[-\beta_x + f(x,\beta(x,v))\right] * 1/\beta_v$

which is an equation of the form $\frac{dv}{dx} = g(x,v)$, which could be separable or linear.

Example 11: The equation $\frac{dy}{dx} = \sqrt{x - y + 2}$ transforms into the separable equation $\frac{dv}{dx} = 1 - \sqrt{v}$ by letting $v = x - y + 2$. The resulting solution is given implicitly by

$$2[1 - \sqrt{x-y+2} - \tan(1 - \sqrt{x-y+2})] = x + C.$$

1.1.6.2 Homogeneous Equations

The differential equation $\frac{dy}{dx} = f(x,y)$ is homogeneous, if it can be written as $\frac{dy}{dx} = F(y/x)$, where $F(.)$ is a function of one variable. In this case, the substitution $v = \frac{y}{x}$ leads to the separable equation $\frac{dv}{F(v) - v} = \frac{dx}{x}$.

Example 12: Consider the homogeneous equation $2xyy' = x^2 + 2y^2$. Setting $v = y/x$ transforms it

into the separable equation, $2vdv = \dfrac{dx}{x}$, whose solution is given implicitly by $v^2 = lnCx$, where C

is an arbitrary constant of integration. The solution of the original equation is therefore given by $y^2 = x^2 lnCx$.

1.1.6.3 Bernoulli Equations

More generally, a differential equation of the form $P(x, y)y' = Q(x, y)$ with polynomial coefficients $P(x, y)$ and $Q(x, y)$ is homogeneous, if the terms in these polynomials have the same degree.

A Bernoulli equation is an equation of the form

$$\frac{dy}{dx} + P(x)y = Q(x)y^n,$$

where n is a real number.

If $n = 0$ or $n = 1$ it is a linear equation.

If not, let $v = y^{1-n}$ to get the linear equation

$$\frac{dv}{dx} + (1-n)P(x)v = (1-n)Q(x).$$

Example 13: To solve $x\dfrac{dy}{dx} + 6y = 3xy^{4/3}$,

set $v = y^{-\frac{1}{3}}$. Then, the equation becomes $\dfrac{dv}{dx} - \dfrac{2}{x}v = -1$,

which is a linear equation whose solution is given by $y(x) = \dfrac{1}{\left(x + Cx^2\right)^3}$.

1.1.7 EXACT DIFFERENTIAL EQUATIONS

Let $F(x, y(x)) = C$ be the general solution of the first-order equation $\dfrac{dy}{dx} = f(x, y)$.

Then, $\dfrac{dF}{dx} = 0$ or $\dfrac{dF}{dx} = \dfrac{\partial F}{\partial x} + \dfrac{\partial F}{\partial y}\cdot\dfrac{dy}{dx} = 0$

This equation can be written in differential form as

$$M(x, y)dx + N(x, y)dy = 0 \tag{1.17}$$

Definition 2

The differential equation (1.17) is said to be exact if there is a function $F(x, y)$ such that

$$\frac{\partial F}{\partial x} = M$$

and

$$\frac{\partial F}{\partial y} = N \tag{1.18}$$

Two questions then arise:

(1) In practice, how does one determine that a differential equation (1.17) is exact?
(2) How do we determine F such that equations (1.18) hold true?

To answer these questions, recall that if the mixed second partial derivatives $F_{xy}(x,y)$ and $F_{yx}(x,y)$ are continuous in the xy-plane, then they are equal, so, we can write $F_{xy}(x,y) = F_{yx}(x,y)$.
   Thus,

$$\frac{\partial M}{\partial y} = F_{xy} = F_{yx} = \frac{\partial N}{\partial x}.$$

It follows that the relation $\dfrac{\partial M}{\partial y} = \dfrac{\partial N}{\partial x}$ is a necessary condition for the differential equation (1.17) to

be exact. It turns out to be a sufficient condition as well as the following theorem asserts.

Theorem 3: Criterion for exactness
Suppose that $M(x, y)$ and $N(x, y)$ are continuous and have continuous first-order partial derivatives in the open rectangle $R = \{(x,y): a < b, c < y < d\}$. Then the differential equation

$M(x,y)dx + N(x,y)dy = 0$ is exact in R if and only if $\dfrac{\partial M}{\partial y} = \dfrac{\partial N}{\partial x}$ at each point of R. That is, there

exists a function $F(x,y)$ defined on R such that $\dfrac{\partial F}{\partial x} = M$ and $\dfrac{\partial F}{\partial y} = N$ if and only if $\dfrac{\partial M}{\partial y} = \dfrac{\partial N}{\partial x}$.

Example 14: shows that equation $(2x+3y)dx + (3x+2y)dy = 0$ is exact and gives the general solution.

Solution: Here, $M(x, y) = 2x + 3y$ and $N(x, y) = 3x + 2y$ so, that $\dfrac{\partial M}{\partial y} = 3 = \dfrac{\partial N}{\partial x}$. Thus, the equation

is exact. Now, from equation (1.18) we get $\dfrac{\partial F(x,y)}{\partial x} = M(x,y) = 2x+3y.$ This implies

$F(x,y) = x^2 + 3yx + g(y)$. We also get from (1.18) $\dfrac{\partial F}{\partial y} = N = 3x+2y$. Thus, $3x + g'(y) = 3x+2y$

so that $g'(y) = 2y$ and $g(y) = y^2 + C$; where C is an arbitrary constant of integration. Finally, the general solution of the given differential equation is $F(x,y) = x^2 + 3yx + y^2 + C$.

1.2   SECOND ODES

1.2.1   INTRODUCTION TO SECOND-ORDER LINEAR DIFFERENTIAL EQUATIONS

A general second-order differential equation is an equation of the form

$$G(x, y, y', y'') = 0 \tag{1.19}$$

where G is a function of four variables. It is linear if it has the form

$$A(x)y'' + B(x)y' + C(x)y = F(x) \tag{1.20}$$

where the coefficients $A(x), B(x), C(x)$ and the function $F(x)$ are continuous functions on some open interval I.

When $F(x) = 0$ on I, equation (1.20) is said to be homogeneous. Hence, it is written as

$$A(x)y'' + B(x)y' + C(x)y = 0 \qquad (1.21)$$

Otherwise, it is non-homogeneous.

Example 15:

(1) $xy'' + (\sin x)y' + (1 + x^2)y = \cos x$

(2) $e^x y'' + (sinx)y' - (lnx)y = 0$

(3) $xy'' + p(x)y' + q(x)y = f(x)$

Remarks:

(1) The concept of homogeneous defined here has a different meaning from that encountered in Section 1.1.
(2) It turns out that to solve the non-homogeneous equation (1.20) we must first solve the homogeneous equation (1.21)

1.2.1.1 Homogeneous Second-Order Linear Equations

Assume that $A(x)$ does not vanish on the interval I and divide equation (1.21) by this function. Then, this equation becomes

$$y'' + p(x)y' + q(x)y = 0 \qquad (1.22)$$

where $p(x) = \dfrac{B(x)}{A(x)}$ and $q(x) = \dfrac{C(x)}{A(x)}$.

The corresponding non-homogeneous equation is

$$y'' + p(x)y' + q(x)y = f(x) \qquad (1.23)$$

Theorem 4: The principle of superposition of solutions for homogeneous equations
Let y_1 and y_2 be solutions of the homogeneous linear equation (1.22) on an interval I. Then, $y = c_1 y_1 + c_2 y_2$ is also a solution of (1.22) for any arbitrary constants c_1 and c_2.

This theorem follows from the linearity properties of differentiation.

Example 16: The equation $y'' + y' = 0$ has solutions: $y_1 = \cos x$ and $y_2 = \sin x$. One can readily verify that
$y = c_1 y_1 + c_2 y_2 = c_1 \cos x + c_2 \sin x$ is also a solution.

We now turn to solutions of the non-homogeneous equation (1.23).

Theorem 5: Existence and uniqueness of a solution for the non-homogeneous second-order differential equation
Assume that the functions $p(x), q(x)$ and $f(x)$ are continuous on an open interval I containing x_0. Then, given any real numbers b_0 and b_1, the equation

$$y'' + p(x)y' + q(x)y = f(x) \qquad (1.23)$$

has a unique solution on the entire interval I that satisfies

$$y(x_0) = b_0, y'(x_o) = b_1 \tag{1.24}$$

Equation (1.23) together with the initial conditions (1.24) constitutes an IVP for the given differential equation.

Example 17: Consider the second order IVP

$$\begin{cases} y'' + 4y' + 3y = e^{-2x} \\ y(0) = 0, y'(0) = 0 \end{cases} \tag{1.25}$$

(a) Verify that $y_H(x) = c_1 e^{-x} + c_2 e^{-3x}$ is the general solution of the associated homogeneous equation.
(b) Show that $y_P(x) = -e^{-2x}$ is a particular solution of the given non-homogeneous equation.
(c) Verify that $y(x) = y_H(x) + y_P(x)$ is the general solution of the given non-homogeneous equation.
(d) Show that the solution of the IVP is given by $y(x) = \frac{1}{2}e^{-x} + \frac{1}{2}e^{-3x} - e^{-2x}$.

We now turn to the concepts of linear dependence and linear independence of solutions for the homogeneous equation.

1.2.1.2 Definition of Linear and Nonlinear Dependence: Introduction to the Wronskian
Two functions defined on an open interval I are said to be linearly dependent on I if one is a constant multiple of the other. They are said to be linearly independent of I if they are not linearly dependent on I.

Example 18: Each of the following sets of functions is clearly a pair of linearly independent functions on any open interval: $\{1, x\}, \{\sin x, \cos x\}, \{e^x, e^{2x}\}, \{e^x, xe^{2x}\}$.
    But in practice, how does one determine that two functions are linearly independent of an open interval I? The answer is provided through the following

1.2.1.3 Definition of the Wronskian of Two Functions
The Wronskian $W(f(x), g(x))$ of two functions $f(x)$ and $g(x)$ in an interval I is a function defined by

$$W(f(x), g(x)) = det\begin{pmatrix} f(x) & g(x) \\ f'(x) & g'(x) \end{pmatrix}$$
$$= f(x)g'(x) - f'(x)g(x) \text{ for all } x \text{ in } I.$$

    It turns out that two functions f and g are linearly independent of an interval I if their Wronskian is never zero on I.

Theorem 6: The Wronskian of solutions
Let y_1 and y_2 be solutions of the homogeneous linear equation (1.22) on an interval I on which $p(x)$ and $q(x)$ are continuous. Then,

(a) If y_1 and y_2 are linearly dependent, then $W(y_1, y_2)$ is identically zero on I.

(b) If $W(y_1, y_2) \neq 0$ on I, then y_1 and y_2 are linearly independent.

Proof:

(a) If y_1 and y_2 are linearly dependent, then there exists a real number k such that

$$y_2 = ky_1 \quad (1)$$

If we differentiate both sides of the previous equality, we have

$$y_2' = ky_1' \quad (2)$$

Now obtain k from equation (1) and substitute it in equation (2), we have

$y_2' = \dfrac{y_2}{y_1} y_1'$ which gives $y_2'y_1 = y_2 y_1'$ or $y_2'y_1 - y_2 y_1' = 0$, that is

$$\begin{vmatrix} y_1 & y_2 \\ y_1' & y_2' \end{vmatrix} = 0 \text{ but } \begin{vmatrix} y_1 & y_2 \\ y_1' & y_2' \end{vmatrix} = W(y_1, y_2).$$

Therefore, $W(y_1, y_2) = 0$

(b) If (a) is well understood, the proof of (b) is obvious by mathematical contrapositive.

Example 19: From Example 17, the functions $f(x) = e^{-x}$ and $f(x) = e^{-3x}$ and $g(x) = e^{-3x}$ are solutions of the homogeneous equation $y'' + 4y' + 3y = 0$. They are also linearly independent everywhere. Indeed, their Wronskian, given by

$$W(f(x), g(x)) = det \begin{pmatrix} f(x) & g(x) \\ f'(x) & g'(x) \end{pmatrix}$$

$$= f(x)g'(x) - f'(x)g(x) = -2e^{-4x}$$

never vanishes.

Theorem 7: The general solution of the homogeneous equation

Let y_1 and y_2 be linearly independent solutions of the homogeneous equation (1.22) on an interval I on which $p(x)$ and $q(x)$ are continuous. If $Y(x)$ is any solution whatsoever of equation (1.22) on I, then there exists constants c_1 and c_2 such that

$Y(x) = c_1 y_1(x) + c_2 y_2(x)$ for all x on I. $Y(x)$ is called the general solution of (1.22).

Proof: This theorem follows from the linearity properties of differentiation.

Special case: The constant-coefficient equation form of equation (1.22) can be written as

$$ay'' + by' + cy = 0 \tag{1.26}$$

where a, b, c are constants.

Let $y(x) = e^{rx}$ be a solution. Then, r satisfies the algebraic equation

$$ar^2 + br + c = 0 \tag{1.27}$$

The quadratic equation (1.27) is called the characteristic equation of (1.26). From its solutions, one obtains the solutions of the homogeneous equation (1.26) as described in the following theorem.

Theorem 8

   (i) If the quadratic equation (1.27) has two distinct real roots r_1 and r_2, then $y(x) = c_1 e^{r_1 x} + c_2 e^{r_2 x}$ is the general solution of equation (1.26).

  (ii) If (1.27) has complex conjugate roots $r_1 = \overline{r_2} = \alpha + i\beta$, then

$$y(x) = e^{\alpha x}\left(c_1 \cos \beta x + c_2 \sin \beta x\right) \text{ is the general solution of equation (1.26).}$$

 (iii) If (1.27) has a double root r, then $y(x) = (c_1 + c_2 x)e^{rx}$ is the general solution of (1.26).

Example 20: Verify that each of the initial value problems has the given solution.

   a) $y'' + 2y' + y = 0$, $y(0) = 2$, $y'(0) = -1$; $y(x) = (2 + x)e^{-x}$

   b) $y'' - 3y' + 2y = 0$, $y(0) = 1$, $y'(0) = 0$; $y(x) = 2e^x - e^{2x}$

   c) $2y'' + 2y' + y = 0$, $y(0) = 1$, $y'(0) = 0$; $y(x) = e^{-\frac{1}{2}x}\left(\cos\frac{1}{2}x + \sin\frac{1}{2}x\right)$

Theorem 9: The principle of superposition of solutions for non-homogeneous equations

   Let $y_P(x)$ be a particular solution of the non-homogeneous equation

$$y'' + p(x)y' + q(x)y = f(x) \tag{1.28}$$

on an open interval I, where $p(x), q(x)$ and $f(x)$ are continuous. If $y_H(x)$ is the general solution of the homogeneous equation

$$y'' + p(x)y' + q(x)y = 0 \tag{1.29}$$

on I, then the general solution of the non-homogeneous equation (1.28) on I is $y_G(x) = y_H(x) + y_P(x)$.

Proof: The proof here follows from the linearity properties of differentiation. Indeed,

$y_H(x)$ satisfies $y_H'' + p(x)y_H' + q(x)y_H = 0$ (i)

$y_P(x)$ satisfies $y_P'' + p(x)y_P' + q(x)y_P = f(x)$ (ii)

Adding equations (i) and (ii) and regrouping terms yields:

$$y_H'' + p(x)y_H' + q(x)y_H + y_P'' + p(x)y_P' + q(x)y_P = f(x), \text{ or}$$

$$\left(y_H'' + y_P''\right) + p(x)\left(y_H' + y_P'\right) + q(x)\left(y_H + y_P\right) = f(x), \text{ or}$$

$$\left(y_H + y_P\right)'' + p(x)\left(y_H + y_P\right)' + q(x)\left(y_H + y_P\right) = f(x)$$

which is the expected result.

    An important task is to find a particular solution $y_P(x)$ of the non-homogeneous equation (1.28). We consider the constant coefficient case, namely,

$$ay'' + by' + cy = f(x) \tag{1.30}$$

and its associated homogeneous equation:

$$ay'' + by' + cy = 0 \qquad (1.31)$$

There are two standard methods for determining such a solution, the method of undetermined solutions and the method of variation of parameters.

1.2.1.4 The Method of Undetermined Coefficients

This method applies when the function $f(x)$ in equation (1.30) is a linear combination of products of functions of the following three types:

    a. a polynomial in x;
    b. an exponential function e^{rx};
    c. $\cos \alpha x$ or $\sin \alpha x$.

Any such function and all its derivatives consist of a finite number of linearly independent functions. Notice that this is not the case for $f(x) = \tan x$, for example, and many other functions for which this method would not apply.

Case 1: No term of $f(x)$ nor any of its derivatives is a solution to the homogeneous equation (1.31). In this case, we may seek a particular solution as a linear combination of all linearly independent such terms and their derivatives. The coefficients of this linear combination are determined by substitution in the non-homogeneous equation (1.30).

Example 21. Find a particular solution for each of the following equations:
    (a) $y'' + 16y = e^{3x}$
    (b) $y'' - y' - 2y = 3x + 4$

Solution:
    a) Here we seek a particular solution in the form: $y_P(x) = ae^{3x}$, where a is a constant. Substituting this function into the differential equation, we find $a = \dfrac{1}{25}$; thus, $y_P(x) = \dfrac{1}{25}e^{3x}$.

    b) Assume $y_P(x) = ax + b$, and determine the constants a and b by substitution in the equation to get

Case 2: If $f(x)$ is a linear combination of terms of the form $P_m(x)e^{\alpha x}(\cos\beta x, \sin\beta x)$, where $P_m(x)$ is a polynomial of degree m, and α and β are real numbers, seek a particular solution in the form:

$$y_P(x) = x^s \left(a_0 + a_1 x + \ldots + a_m x^m\right) e^{\alpha x}\cos\beta x +$$

$$x^s \left(b_0 + b_1 x + \ldots + b_m x^m\right) e^{\alpha x}\sin\beta x$$

where s is the smallest non-negative integer such that no term in $y_P(x)$ duplicates a term in the homogenous solution $y_H(x)$ and determine the coefficients by substitution of this trial solution in equation (1.30).

TABLE 1.1

Functions $f(x)$ and Their Corresponding $y_P(x)$

If $f(x)$ is equal to:	Take $y_P(x)$ as:
$P_m(x) = b_0 + b_1 x + \ldots + b_m x^m$	$x^s \left(a_0 + a_1 x + \ldots + a_m x^m \right)$
$a\cos\beta x + b\sin\beta x$	$x^s \left(A\cos\beta x + B\sin\beta x \right)$
$e^{\alpha x} \left(a\cos\beta x + b\sin\beta x \right)$	$e^{\alpha x} x^s \left(A\cos\beta x + B\sin\beta x \right)$
$e^{\alpha x} \left(b_0 + b_1 x + \ldots + b_m x^m \right)$	$e^{\alpha x} x^s \left(a_0 + a_1 x + \ldots + a_m x^m \right)$
$(b_0 + b_1 x + \ldots + b_m x^m)(a\cos\beta x + b\sin\beta x)$	$x^s [\left(A_0 + A_1 x + \ldots + A_m x^m \right)\cos\beta x + \left(B_0 + B_1 x + \ldots + B_m x^m \right)\sin\beta x]$

The form of the particular solution above is very general. We provide in Table 1.1 some common occurrences.

Example 22: Determine the appropriate form for a particular solution of:

a) $y'' - 2y' + 2y = e^x \cos x$

b) $y'' - 3y' + 2y = 3e^{-x} - 10\cos 3x$

Solution

(a) Here, $f(x) = e^x \cos x$ and the characteristic equation, $r^2 - 2r + 2 = 0$, has roots, 1 ± 1. The general solution of the homogenous equation is $y_H(x) = e^x (a\cos x + b\sin x)$. An appropriate form for a particular solution for the given equation is therefore,

(b) Here, function $f(x)$ is the sum of two functions, $f_1(x) = 3e^{-x}$ and $f_2(x) = -10\cos x$. We find a particular solution by determining particular solutions $y_{P_1}(x)$ and $y_{P_2}(x)$ corresponding to $f_1(x)$ and $f_2(x)$ respectively and then taking their sum, that is,

$$y_P(x) = y_{P_1}(x) + y_{P_2}(x) \tag{1.32}$$

An appropriate form for this is given by $y_{P_1} = Axe^{-x}$ and
$y_{P_2}(x) = x(a\cos x + b\sin x)$.

1.2.1.5 The Method of Variation of Parameter

Assume that the general solution of the homogenous equation is given by
$y_G(x) = c_1 y_1(x) + c_2 y_2(x)$, where c_1 and c_2 are constants and $y_1(x)$ and $y_2(x)$ are linearly independent solutions of the homogeneous equation. The method of variation of parameters consists in assuming that the constants are functions and then seeking a particular solution in the form:

$$y_P(x) = u_1(x) y_1(x) + u_2(x) y_2(x) \tag{1.33}$$

Substituting this presumed solution into equation (1.30) yields, after imposing an additional condition on u_1 and u_2, and after simplification, the following system of linear algebraic equations for $u'_1(x)$ and $u'_2(x)$:

$$\text{Claim:} \begin{cases} u'_1 y'_1 + u'_2 y'_2 = f(x) \\ u'_1 y_1 + u'_2 y_2 = 0 \end{cases} \tag{1.34}$$

Proof of the claim:
Indeed, we have

$$y'_P(x) = (u_1 y'_1 + u_2 y'_2) + (u'_1 y_1 + u'_2 y_2)$$

For convenience, to avoid the appearance of the second derivatives u''_1 and u''_2 in the sequel, we assume that

$$u'_1 y_1 + u'_2 y_2 = 0; \tag{1.35}$$

thus, $y'_P = u_1 y'_1 + u_2 y'_2$
The second derivative is then given by

$$y''_P = (u_1 y''_1 + u_2 y''_2) + (u'_1 y'_1 + u'_2 y'_2) \tag{1.36}$$

Substituting the expressions in (ii), (iii) and (iv) above for y''_P, y'_P, and y_P in equation (1.30) and using the fact that $y_1(x)$ and $y_2(x)$ satisfy the homogeneous equation (1.31) obtain

$$u'_1 y'_1 + u'_2 y'_2 = f(x) \tag{1.37}$$

A particular solution will be found if the system in the claim above can be solved for $u'_1(x)$ and $u'_2(x)$. But the determinant of that linear system is the non-vanishing Wronskian, $W(x)$ of the linearly independent solutions $y_1(x)$ and $y_2(x)$ of the homogenous equation. Upon solving, one finds

$$u'_1(x) = -\frac{y_2(x) f(x)}{W(x)}$$

and

$$u'_2(x) = \frac{y_1(x) f(x)}{W(x)} \tag{1.38}$$

Thus, one can integrate to get $u_1(x)$ and $u_2(x)$. A particular solution is given by the following theorem.

Theorem 10 (Variation of parameter)
Let the homogenous equation $ay'' + by' + cy = 0$ have general solution
$y_G(x) = c_1 y_1(x) + c_2 y_2(x)$. Then, a particular solution of the non-homogenous equation $ay'' + by' + cy = f(x)$ is given by

$$y_P(x) = -y_1(x) \int \frac{y_2(x) f(x)}{W(x)} dx + y_2(x) \int \frac{y_1(x) f(x)}{W(x)} dx \tag{1.39}$$

where $W(x) = W(y_1, y_2)$ is the Wronskian of the two linearly independent solutions $y_1(x)$ and $y_2(x)$ of the homogenous equation.

Example 23: Find a particular solution of the equation
$y'' + y = tanx$ as follows:

(a) Write down the general solution of the homogenous
$y'' + y = 0$ using $y_1(x) = cosx$ and $y_2(x) = sinx$.
(b) Seek a particular solution in the form
(c) Verify that
$u'_1(x) = \cos x - \sec x$ and $u'_2(x) = \sin x$ satisfy system (1.34).

(d) Integrate to get $u_1(x) = \sin x - ln|\sec x + \tan x|$ and $u_2(x) = -\cos x$.

(e) Finally, write down a particular solution to get, after simplification,

$$y_P(x) = u_1(x)y_1(x) + u_2(x)y_2(x)$$
$$= -(\cos x)ln|\sec x + \tan x|.$$

PROBLEM SET 1

1.1. Derive a first-order differential equation from $ye^{ty} = C$ where C is an arbitrary constant and sketch several level curves.
1.2. Find the general solution to $y'' + y' - 6y = 0$ and then choose c_1 and c_2 such that $y(0) = 5$ and $y'(0) = 0$
1.3. Solve the initial value problem
$t^2\dfrac{d^2y}{dt^2} + t\dfrac{dy}{dt} - 4y = 6t^4, y(2) = 3, \dfrac{dy}{dt}(2) = 13,$ given that a two-parameter family of
solutions to the differential equation is $y(t) = \dfrac{1}{2}t^4 + c_1t^2 + c_2t^{-2}$
1.4. Solve the initial value problem
$y'' + 6y' + y = 0$ and then choose c_1 and c_2 such that $y(0) = 1$ and $y'(0) = -3$
1.5. Find the general solution to $y'' + 6y' + 9y = 0$.
1.6. Solve $4y'' + 4y' + y = 0$, $y(2) = e^{-1}$ and $y'(2) = \dfrac{e^{-1}}{2}$
1.7. Find the general solution to the given differential equation
a. $\dfrac{dy}{dt} + 2y = 13$; b. $\dfrac{dy}{dt} - 6y = 8e^{2t}$; c. $\dfrac{dy}{dt} + 4y = 12$; d. $\dfrac{dy}{dt} + y = -t^2 + 1$; e. $\dfrac{dy}{dt} - 2y = 3\sin t$;
f. $\dfrac{dy}{dt} - y = te^t$.
1.8. Solve $\dfrac{dy}{dt} = y(1-y), y(0) = 2$
1.9. Solve the differential equations $\dfrac{dy}{dt} = \dfrac{t^2}{y}$;
1.10. Find the general solution of the following differential equation:
1.11. Solve the Initial value problem:

2 Elements of Partial Differential Equations

I do not think the division of the subject into two parts – applied mathematics and experimental physics is a good one, for natural philosophy without experiment is merely mathematical exercise, while experiment without mathematics will neither sufficiently discipline the mind or sufficiently extend our knowledge in a subject like physics.

Balfour Stewart

INTRODUCTION

In this chapter, we introduce the first-order quasilinear partial differential equation (PDE), the semilinear equations, the second-order PDEs and other types of PDEs such as the Helmholtz equations and their solutions using Cartesian coordinates, the cylindrical coordinates and spherical coordinates. Applications have been given such as the derivation and solution of the heat equation.

2.1 FIRST-ORDER QUASILINEAR PDE: THE METHOD OF CHARACTERISTICS

The general form of a first-order quasilinear PDE is

$$a(x,y,u)u_x + b(x,y,u)u_y = c(x,y,u) \tag{2.1}$$

where the coefficients a, b and c are continuous in x, y and u. The solution is a family of surfaces $z = u(x, y)$. The vector $V(x,y,z) = (a,b,c)$ defines a vector field in $\mathbb{R}^3$, to which the graph of solutions of equation (2.1) is tangent at each point. The integral surfaces of the vector field are the surfaces that are tangent at each point to the vector field in $\mathbb{R}^3$. To solve equation (2.1) we need to use the following ordinary differential equations also called characteristic equations:

$$\frac{dx}{dt} = a(x,y,z); \frac{dy}{dt} = b(x,y,z); \frac{dz}{dt} = c(x,y,z) \tag{2.2}$$

with the initial conditions x.

We also assume that a, b and c are continuously differentiable. If the graph $z = u(x,y)$ is a smooth surface S that is a union of such characteristic curves, then at each point (x_0, y_0, z_0), the tangent plane contains the vector $V(x_0, y_0, z_0)$. Therefore, S is a surface integral. We can then define an integral surface as a smooth union of characteristic curves. Each curve of the integral surface is noted by " . The given curve " is noncharacteristic if it is nowhere tangent to the vector field V. Analytically, the construction of an integral surface containing " can be achieved by writing " as the graph of a curve $(f(s), g(s), h(s))$, where s is a parameter, and using the initial conditions, $x_0 = f(s)$, $y_0 = g(s)$ and $z_0 = h(s)$.

In a more general case, equation (2.1) will be replaced by the more general equation:

$$\sum_{i=1}^{n} a_i(x_1, x_2, \ldots, x_n, u)u_{x_i} = c(x_1, x_2, \ldots, x_n, n) \tag{2.3}$$

DOI: 10.1201/9781003478812-2

The characteristic curves are now the integral curves of the system of $(n + 1)$ equations with $(n + 1)$ unknowns:

$$\frac{dx_i}{dt} = a_i \left(\sum_{i=1}^{n} a_i(x_1, x_2, \ldots, x_n, z) \right); \frac{dz}{dt} = c \left(\sum_{i=1}^{n} a_i(x_1, x_2, \ldots, x_n, z) \right) \qquad (2.4)$$

Equation (2.4) can be solved using the initial conditions on a $(n-1)$ dimensional manifold Γ

$$x_i = f_i \left(s_1, s_2, \ldots, s_{n-1} \right); z = h \left(s_1, s_2, \ldots, s_{n-1} \right) \qquad (2.5)$$

As a result, we can obtain an n-dimensional integral manifold parameterized by $\left(s_1, s_2, \ldots, s_{n-1}, t \right)$, which leads to the solution

$$u = u(x_1, x_2, \ldots, x_n) \qquad (2.6)$$

Example 1
Solve the Cauchy problem

$$u u_x + y u_y = x \qquad u(x,1) = 2x$$

The characteristic equation is
$$\frac{dx}{dt} = z; \frac{dy}{dt} = y; \frac{dz}{dt} = x \quad \text{and } \Gamma \text{ is parameterized by } (s,1,2s).$$

We can observe that $y = c(s)e^t$; in addition,

$$\frac{d(x+z)}{dt} = x + z; \frac{d(x-z)}{dt} = -(x-z);$$

which gives

$$x + z = c_1(s)e^t, \, x - z = c_2(s)e^{-t}$$

Using the initial conditions we can evaluate c_1, and $u_2 - u_1$ and therefore obtain

$$A u_{xx} + 2 B u_{xy} + C u_{yy} + D u_x + E u_y + F u = G$$

By eliminating s and t, we can get the solution:

c_1, which is defined for $|y| < \frac{\sqrt{3}}{3}$. There is a singularity at $y = \frac{\sqrt{3}}{3}$.

2.2   SEMILINEAR EQUATIONS

The semilinear equation in two dimensions can be written as

$$a(x,y)u_x + b(x,y)u_y = c(x,y,u) \qquad (2.7)$$

We can see the difference between the quasilinear (2.1) and the semilinear (2.7) PDE. For the first a and b b are functions of three variables x, y and u the solution itself, while for the second, a and

b are functions of only two variables x, y. The curve Γ is parameterized by $\left(f(s), g(s), h(s)\right)$ and the characteristics equation is given by

$$\frac{dx}{dt} = a(x,y); \frac{dy}{dt} = b(x,y); \frac{dz}{dt} = c(x,y,z) \qquad (2.8)$$

with the initial conditions:

$$x_0 = f(s), y_0 = g(s), z_0 = h(s). \qquad (2.9)$$

Example 2

Let us solve the initial value problem $u_x + 2u_y = u^2$ with the condition $u(x,0) = h(x)$. The characteristics equations are given by

$$\frac{dx}{dt} = 1; \frac{dy}{dt} = 2; \frac{dz}{dt} = z^2 \qquad (2.10)$$

Therefore, $x(s,t) = t + c_1(s); y(s,t) = 2t + c_2(s)$, where the functions $c_1(s)$ and $AC - B^2 = 0$ may be determined by the initial conditions $x(s,0) = c_1(s) = s$ and $y(s,0) = c_2(s) = 0$ so that

$$x = t + s; y = 2t$$

Now let us calculate the Jacobian of x and y:

$$J(x,y) = \begin{pmatrix} x_s & y_s \\ x_t & y_t \end{pmatrix} = \begin{pmatrix} \dfrac{\partial x}{\partial s} & \dfrac{\partial y}{\partial s} \\ \dfrac{\partial x}{\partial t} & \dfrac{\partial y}{\partial t} \end{pmatrix} = \begin{pmatrix} 1 & 0 \\ 1 & 2 \end{pmatrix}$$

$$J(x,y) = 2 - 0 = 2$$

$$J(x,y) \neq 0$$

We can see that the determinant of the Jacobian matrix of x and y is different from zero. This means that the Jacobian matrix is nonsingular.

We can now eliminate t between x and y. As a result, we have

$$t = \frac{y}{2} \quad \text{so} \quad s = x - \frac{y}{2}$$

By integrating the third equation $\dfrac{dz}{dt} = z^2$ we obtain z:

$$-\frac{1}{z} t + c_3(s)$$

$$z(s,t) = -\frac{1}{t + c_3(s)}.$$

Using the initial condition, $u(x,t)$ we obtain $c_3(s)$

$$z(s,0) = -\frac{1}{0+c_3(s)}$$

$$c_3(s) = -\frac{1}{h(s)}$$

By substituting $c_3(s)$ in z, we get

$$z(s,t) = -\frac{1}{t-\dfrac{1}{h(s)}} = \frac{h(s)}{1-th(s)}.$$

Finally, using, $s = x - \dfrac{y}{2}$ we can write $u(x,y)$ as follows:

$$u(x,y) = \frac{h\left(x-\dfrac{y}{2}\right)}{1-\dfrac{y}{2}h\left(x-\dfrac{y}{2}\right)} \qquad (2.11)$$

Notice that the condition $u(x,0) = h(x)$ is satisfied and that the solution $u(x,y)$ is well defined for small values of y assuming g is bounded. However, for large values of y in such a way that the denominator of equation (2.11) tends to zero, $u(x,y)$ may be infinite. Also, despite and $\dfrac{dy}{dt}$ being linear equations and the solution exists for all s and t, $\dfrac{dz}{dt}$ is nonlinear and may produce a singularity.

Example 3
Solve the Cauchy problem

$$u_x + xu_y - u_z = u \quad \text{with} \quad u(x,y,1) = x+y$$

Solution
If we replace x,y,z respectfully with x_1, x_2, x_3, then the characteristics equations would be written as

$$\frac{dx_1}{dt} = 1; \frac{dx_2}{dt} = x_1; \frac{dx_3}{dt} = -1; \frac{dz}{dt} = z \qquad (2.12)$$

The initial surface " is the hyperplane $x_3 = 1, z = x_1 + x_2$ Γ can be parameterized by

$$x_1 = s_1, x_2 = s_2, x_3 = 1, z = s_1 + s_2 \qquad (2.13)$$

which gives after integration:

$$x_1 = t + s_1, x_2 = \frac{1}{2}t^2 + s_1 t, +x_2, x_3 = -t+1, z = (s_1 + s_2)e^t \qquad (2.14)$$

Finally, we can solve for s_1, s_2 and t, and plug them back into z. This would lead to the general solution of the Cauchy problem:

$$u(x_1,x_2,x_3) = \left\{ x_1 + x_2 + (x_3-1)\left[1 + x_1 + \frac{1}{2}(x_3-1)\right]\right\} e^{1-x_3} \tag{2.15}$$

2.3 SECOND-ORDER PDES

Several laws in physics are described by second-order PDEs. For example, the mathematical statement of Newton's second law of motion can be written as $x_0 = f(s)$ for y function of x and t. This is the string equation and ρ is the mass density of the string. $y = y(x,t)$ describes the position of the xth coordinate of the string at a given time, t. The PDE is normally solved by using some initial conditions representing the position and velocity of the string at time $t = 0$ and the boundary conditions, which require that the ends of the string ($x = 0$, $\frac{\partial^2 u}{\partial t^2} - c^2 \frac{\partial^2 u}{\partial x^2} + 2\beta \frac{\partial u}{\partial t} + \alpha u = 0$, where L is the length of the string) be held fixed during the motion. Both conditions give what is mathematically called the initial value problem for the equation of the vibrating string. Several other PDEs are related to physical phenomena. As examples, we can recall the following equations:

the Laplace equation in two dimensions,

$$\frac{\partial^2 u}{\partial x^2} + \frac{\partial^2 u}{\partial y^2} = 0 \tag{2.16}$$

the electromagnetic wave equation,

$$\frac{\partial^2 u}{\partial t^2} - c^2 + \frac{\partial^2 u}{\partial x^2} = 0 \tag{2.17}$$

the heat equation,

$$\frac{\partial u}{\partial t} - K \frac{\partial^2 u}{\partial x^2} = 0 \tag{2.18}$$

the telegraph equation,

$$\frac{\partial^2 u}{\partial t^2} - c^2 \frac{\partial^2 u}{\partial x^2} + 2\beta \frac{\partial u}{\partial t} + \alpha u = 0 \tag{2.19}$$

Poisson's equation,

$$\frac{\partial^2 u}{\partial x^2} + \frac{\partial^2 u}{\partial y^2} = G \tag{2.20}$$

with G a constant different from zero.

2.4 GENERAL FORM OF A SECOND-ORDER PDE

If we assume $u = u(x, y)$ a function of two variables, x and y, and denote $u_x = \dfrac{\partial u}{\partial x}, u_y = \dfrac{\partial u}{\partial y}, u_{xx} = $

$\dfrac{\partial^2 u}{\partial x^2}, u_{yy} = \dfrac{\partial^2 u}{\partial y^2}$, then the general form of the second-order PDE is

$$A u_{xx} + 2B u_{xy} + C u_{yy} + D u_x + E u_y + F u = G \tag{2.21}$$

where A, B, C, D, E, F, G are functions of x, y. This equation is called homogeneous if $G = 0$ and nonhomogeneous if $G \neq 0$.

2.5 SUPERPOSITION PRINCIPLE

The superposition principle applies to linear second-order homogeneous PDE. If $u_1, u_2, u_3, \ldots, u_n$ are

solutions of the same homogeneous linear second-order PDE, a linear combination $u = \displaystyle\sum_{n=1}^{n} \alpha_n u_n$

of the n solutions with all α_n constant is also a solution to the equation.

The superposition principle does not apply to linear second-order nonhomogeneous PDE.

2.6 SUBTRACTION PRINCIPLE

The subtraction principle applies to linear second-order nonhomogeneous PDE. According to this principle, if u_1 and u_2 are the two solutions of the same nonhomogeneous equations, then the subtraction, $u_2 - u_1$, is a solution to the associated $(G = 0)$ homogenous equation. For example, the subtraction principle applies to Poisson's equation. The difference $u_2 - u_1$ will be the solution to the associated Laplace equation.

2.7 CLASSIFICATION OF SECOND-ORDER PDE

It is customary to find the solutions of a linear second-order PDE by using the boundary conditions which depend on the type of equation. In general, we may distinguish three types of second-order linear PDE:

if $AC - B^2 > 0$, the equation is called *elliptic*

if $AC - B^2 < 0$, the equation is called *hyperbolic*

if $AC - B^2 = 0$, the equation is called *parabolic*

As a result, Laplace and Poisson's equations are elliptic, while the telegraph equation is hyperbolic. Finally, the heat equation is parabolic.

2.8 BOUNDARY CONDITIONS

If the PDE is elliptic, the boundary conditions are summarized in Dirichlet's problem, which consists of knowing the values of the solution $u = u(x, y)$ at the boundary of a bounded plane region with a smooth boundary. If the PDE is hyperbolic or parabolic, it is better to solve the Cauchy problem. The Cauchy problem is also known as the initial-value problem which gives the initial conditions. This provides the solution $u(x, t)$ and its time derivative, $u_1(x, t)$ for $t = 0$. It may also provide the values of the solution $u(x, t)$ at the end of a string, for example, that is $u(0, t)$ and $u(x, 0)$.

2.9 SEPARATED SOLUTIONS

One of the techniques used to solve the PDE is the method of separation of variables. Therefore, it is customary to assume the solution of the form $u(x,y) = f_1(x)f_2(x)$, which leads to obtaining ordinary differential equations for $f_1(x)$ and $f_2(x)$. The solution $u(x,y)$ then is called a separated solution. The solutions $u(x,y) = f_1(x)f_2(x)$ can be regarded as complex-valued functions.

Proposition 1

Let, $u(x,y) = v_1(x) + i v_2(x)$ be a complex-valued solution of the PDE,

$$A u_{xx} + 2B u_{xy} + C u_{yy} + D u_x + E u_y + F u = G \tag{2.22}$$

where A, B, C, D, E, F, G are real-valued functions of (x, y). Then $v_1(x,y) = R_e\{u(x,y)\}$ satisfies the PDE and, $v_2(x,y) = I_m\{u(x,y)\}$ satisfies the associated homogeneous equation with $G = 0$.

Proposition 2

Consider the following homogeneous PDE,

$$A u_{xx} + 2B u_{xy} + C u_{yy} + D u_x + E u_y + F u = G.$$

Suppose A, B, C, D, E, F, G are all real constant, then there exists a complex separated solution of the form $u(x,y)e^{\alpha x}e^{\beta y}$ for appropriate choices of complex numbers a and β.

Example 4

Find the separated solutions of the PDE of the heat, $u_{xx} - u_t = 0$ in the form $u(x,t) = e^{i\mu x}e^{\beta t}$, with μ and β real numbers.

Solution

$$u_x = \left(\frac{\partial u}{\partial x}\right)_t = i\mu u, u_{xx} = \left(\frac{\partial u_x}{\partial x}\right)_t = -\mu^2 u, u_t = \left(\frac{\partial u}{\partial t}\right)_x = i\mu u$$

The PDE becomes $-\left(\mu^2 + \beta\right)u(x,t) = 0$.

For this condition to hold for every, $u(x,t) \neq 0$ we need to have the condition $\mu^2 + \beta = 0$ or $\beta = -\mu^2$ and the solution becomes

$$u(x,t) = e^{i\mu x}e^{-\mu^2}$$

$$u(x,t) = e^{\mu^2 t}\left[\cos(\mu x) + i\sin(\mu x)\right]$$

So, $e^{-\mu^2 t}\cos(\mu x)$ and $e^{-\mu^2 t}\sin(\mu x)$ are the real-valued solutions to the PDE.

Example 5

Find the separated solutions of the wave equations $u_{xx} - c^2 u_{tt} = 0$ which are bounded in the form $u(x,t) \leq M$ for some constant M and all t such that $-\infty < t < +\infty$

Solution

From the form of the solution, $u(x,t) = e^{ax+bt}$ we have $u(x,t) = e^{ax+bt}$ and $u_{tt} = b^2 u$ Then, the equation is satisfied when $b \pm ac$ and the solution becomes $u(x,t) = e^{ax \pm act}$. The separated solutions will be $e^{ax} e^{+act}$ and $e^{ax} e^{-act}$. These solutions are bounded if a is pure imaginary, that is, $\exists \dfrac{k}{a} = ik$. The solutions are as follows:

$e^{ikx} e^{+ikct}, e^{ikx} e^{-ikct}, or\ e^{ik(x+ct)}, e^{ik(x-ct)}$ The real quasi-separated solutions are

$$\cos k(x+ct),\ \cos k(x-ct),\ \sin k(x-ct),\ \sin k(x-ct).$$

2.9.1 ORTHOGONAL FUNCTIONS

Separated solutions of PDEs are sometimes orthogonal. The inner product of two functions, φ and ψ of the variable $x \in [a,b]$ is defined as

$$<\varphi,\psi> = \int_a^b \varphi(x)\psi(x)dx \tag{2.23}$$

The inner product of two functions is linear and homogenous in both slots:

$$<\varphi,\psi_1 + \psi_2> = <\varphi,\psi_1> + <\varphi,\psi_2> \tag{2.24}$$

$$<\varphi_1 + \varphi_2, \psi> = <\varphi_1,\psi> + <\varphi_2,\psi> \tag{2.25}$$

For $\qquad\qquad \alpha \in \mathbb{R},\ <\alpha\varphi,\psi> = \alpha <\varphi,\psi> = <\varphi,\alpha\psi> \tag{2.26}$

These two functions φ and ψ are orthogonal if the inner product is zero, that is $<\varphi,\psi> = 0$.

The functions, $\varphi_1, \varphi_2, \varphi_3,\ldots\varphi_n,$ are said to be orthogonal if for $i \neq j, <\varphi_i,\varphi_j> = 0$.

Example 6

$\sin x$ and $\cos x$ are orthogonal on the interval $[0,\pi]$. Indeed,

$$\int_0^x \sin(x)\cos(x)dx = \frac{1}{2}\int_0^\mu \sin(2x)dx = \frac{1}{4}\Big[\cos(2x)\Big]_0^\pi = 0$$

2.10 OTHER TYPES OF PDES: THE HELMHOLTZ EQUATIONS

The Helmholtz equations are well used in mathematical physics. They are equations of the form:

$$\Delta\psi + k^2\psi = 0 \tag{2.27}$$

Where k is a constant.

To solve the Helmholtz equation, we will use the method of separation of variables.

2.10.1 HELMHOLTZ EQUATION CARTESIAN COORDINATES

In three-dimensional Cartesian Coordinates, the Helmholtz equation is written as

$$\frac{\partial^2\psi}{\partial x^2} + \frac{\partial^2\psi}{\partial y^2} + \frac{\partial^2\psi}{\partial z^2} + k^2\psi = 0 \tag{2.28}$$

with E a function of three variables x, y and z. The separation of variables allows us to write $j(x,t)$ as

$$\psi(x,y,z) = X(x)Y(y)Z(z) \qquad (2.29)$$

Therefore, we obtain the following differential equation:

$$\ddot{X}YZ + X\ddot{Y}Z + XY\ddot{Z} + k^2 XYZ = 0 \qquad (2.30)$$

By dividing both members of the equality by X, Y and Z, we get

$$\frac{\ddot{X}}{X} + \frac{\ddot{Y}}{Y} + \frac{\ddot{Z}}{Z} + k^2 = 0 \qquad (2.31)$$

We now assume that we have an oscillatory motion along all the three axes X, Y and Z. Therefore, we can write the following three ordinary differential equations:

$$\frac{\ddot{X}}{X} = -l^2, \frac{\ddot{Y}}{Y} = -m^2, \frac{\ddot{Z}}{Z} = -n^2 \qquad (2.32)$$

We then have

$$k^2 = l^2 + m^2 + n^2 \qquad (2.33)$$

The solution can be labeled according to the choice of the constants l,m,n, which are related by relation (2.33). The solution can be written as

$$\psi_{lmn}(x,y,z) = X_l(x)Y_m(y)Z_n(z) \qquad (2.34)$$

The most general solution is the linear combination of the solutions ψ_{lm}:

$$\psi = \sum_{l,m} c_{lm}\psi_{lm}, \qquad (2.35)$$

where the coefficients c_{lm} are chosen after the use of the boundary conditions.

2.11 THE HEAT EQUATION

To derive the heat equation, we will use the principle of conservation of thermal energy and Fourier's law. Consider the flow of heat through a road of constant cross-section and a perfectly insulated lateral surface. Therefore, no thermal energy can cross the lateral surface. As a result, at a given instant, the temperature is supposed to be constant at any point of the lateral cross-section of the rod. We consider the x-axis parallel to the rod and $\omega(x,t)$ the temperature at a given position x and a given time t (Figure 2.1).

The total amount of thermal energy stored in a segment limited by $x = a$ and $x = b$ is given by, $\int_a^b AC\,\rho\omega(x,t)dx$, with A the cross-sectional area, c the specific heat which is the amount of thermal energy needed to increase by 1°C the temperature of a unit of mass of material needed in a segment. The parameter ρ is the mass per unit surface net transfer at the boundary, that is, $Aj(a,t) - Aj(b,t); A\rho$ is the mass per unite length. According to the principle of the conservation of thermal energy, the rate of change of thermal energy in a part of the rod is equal to the thermal

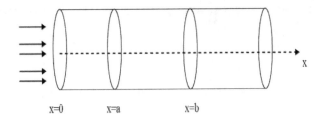

FIGURE 2.1   Heat transfer throughout a rod of axis x.

energy flowing across the boundary per unit time plus the amount generated inside the segment per unit time. If $j(x,t)$ is the flux at the position x and time t, the net heat transfer in the boundaries is given by the difference, $Aj(a,t) - Aj(b,t)$. We consider also that the sources of heat inside the rod within the infinitesimal cross section x and $x + dx$ are given by $AQ[\omega(x,t),x,t]dx$ per units of energy in unit of time. The amount of energy generated by the sources per unit of time would be $\int_a^b AQ[\omega(x,t),x,t]dx$.

We can now write the conservation of thermal energy in the expression

$$\frac{d}{dt}\int_a^b AC\rho\omega(x,t)dx = Aj(a,t) - Aj(b,t) + \int_a^b AQ[\omega(x,t),x,t]dx \tag{2.36}$$

$$\frac{d}{dt}\int_a^b AC\rho\omega(x,t)dx = \int_a^b Ac\rho\frac{\partial\omega(x,t)}{\partial t}dx \tag{As (2.37)}$$

and

$$j(a,t) - j(b,t) = -\left[\Delta J\right]_a^b = -\int_a^b \frac{\partial j(x,t)}{\partial x}dx \tag{2.38}$$

we have

$$\int_a^b Ac\rho\frac{\partial\omega}{\partial t}dx = -A\frac{\partial j}{\partial t}dx + \int AQ[\omega,x,t]dx \tag{2.39}$$

and

$$c\rho\frac{\partial\omega}{\partial t} + \frac{\partial j}{\partial x} = Q(\omega,x,t) \tag{2.40}$$

The previous equation is a first-order PDE of dependent variables, E, j, and independent variables x and t. It is called the equation of continuity or the conservation equation.

Fourier postulated his law of heat conduction which states that the thermal energy flows from the hotter region to the cooler region. It is expressed by

$$j = -K\frac{\partial\omega}{\partial x} \tag{2.41}$$

where the positive constant K is called the thermal conductivity.

Combining Fourier's law of thermal conduction and the equation of continuity, we obtain a second-order PDE,

$$Kc\rho \frac{\partial \omega}{\partial t} + \frac{\partial}{\partial x}\left(-K\frac{\partial \omega}{\partial x}\right) = Q(\omega, x, t) \tag{2.42}$$

where the thermal coefficients c, $\dfrac{d}{dt}\displaystyle\int_a^b AC\rho\omega(x,t)dx = \int_a^b AC\rho \dfrac{\partial \omega(x,t)}{\partial t}dx$ and K are constant.

We can then transform equation (2.42) into

$$\frac{\partial \omega}{\partial t} = k\frac{\partial^2 \omega}{\partial x^2} + F(\omega, x, t) \tag{2.43}$$

where the constant $k = \dfrac{K}{c\rho}$ is called the thermal diffusivity and

$$\frac{Q}{c\rho}F(\omega, x, t) \tag{2.44}$$

One can use the separation of variables to solve the boundary value problem of the homogeneous heat equation.

Now, consider the homogeneous heat equation. We will try to solve this equation using the method of separation of variables; then we will discuss the case when $t \to \infty$.

$$\frac{\partial \omega}{\partial t} = k\frac{\partial^2 \omega}{\partial x^2}, \ 0 < x < L, \ 0 < t$$

$$\omega(0,t) = 0, \ \omega(L,t) = 0 \tag{2.45}$$

$$\omega(x,t) = f(x), \ 0 < x < L$$

To solve the above equation, we will use two steps:

1. Consider $\omega(x,t) = X(x)T(t)$ and substitute into the heat equation. Use the principle of superposition and obtain the solution:

$$\omega(x,t) = \sum_{n=1}^{\infty} A_n (\exp(-\lambda_n^2 kt))\sin(\lambda_n x) \text{ with } \lambda_n = \frac{n\pi}{L}, n = 1,2,3\ldots$$

To determine the coefficient A_n, let us set $f(x) = \sum_{n=0}^{\infty} f_n \sin(\lambda_n x)$ and $\omega(0,t) = \sum_{n=0}^{\infty} f_n \sin(\lambda_n x)$

and use the initial condition, $\omega(0,t) = f(x)$ and find. $A_n = f_n = \dfrac{2}{L}\displaystyle\int_0^L f(x)\sin(\lambda_n x)dx$.

2. As $t \to \infty$, the temperature of the rod approaches the steady state of $0^\circ C$, the temperature of the two ends.

PROBLEM SET 2

2.1   Consider the function of two variables

$$f(x,y) = \frac{x-y}{x+y+1}$$

Calculate $f(0,-2)$; $f(1,2)$; $f(-2,-2)$

2.2   Given the function $f(x,y) = 10 - x^2 + 3y$

Evaluate $\dfrac{\partial f}{\partial x}(1,3)$ and $\dfrac{\partial f}{\partial x}(2,-1)$

2.3   Let the function $u(x,y) = \sin(x^2 + y^2)$

Find $u_x = \dfrac{\partial u}{\partial x}$; $u_{xx} = \dfrac{\partial^2 u}{\partial x^2}$; $u_{xy} = \dfrac{\partial^2 u}{\partial x \partial y}$; $u_{yy} = \dfrac{\partial^2 u}{\partial y^2}$

2.4   Given that $z = e^{xy}$; $x = 2u + 3v$; $y = \dfrac{u}{v}$

Find $\dfrac{\partial z}{\partial u}$; and $\dfrac{\partial z}{\partial v}$ using the chain rule

2.5   Find the separated solutions of the equation

$u_{yy} = \dfrac{\partial^2 u}{\partial y^2}$ in the form $u(x,t) = e^{i\mu x}e^{\beta}$ with μ and β real numbers.

Hint: replace $u(x,t) = e^{i\mu x}e^{\beta t}$ in the given equation and obtain $\beta = \dfrac{-\mu^2}{2}$ and the following

expression of the separated solution $u(x,t) = e^{\frac{-\mu^2}{2}t}\left(\cos \mu x + i \sin \mu x\right)$

This solution tends to zero as t tends to infinity.

2.6   Solve the Cauchy problem. Write the domain of existence of x

$$\begin{cases} x'(t) = \dfrac{1}{t} \\ x(1) = 0 \end{cases}$$

2.7   Solve the Cauchy problem. Write the domain of existence of x

$$\begin{cases} x'(t) = x(t) \\ x(0) = 1 \end{cases}$$

2.8   Consider the Cauchy problem $\begin{cases} x'(t) = \dfrac{3}{2}\sqrt[3]{x} \\ x(0) = 0 \end{cases}$

Solve for $x(t)$

2.9   Classify each of the following second-order PDEs:

a. $u_{xx} + 2u_{xy} + u_{yy} = 0$

b. $u_{xx} + 8u_{xy} + 5u_{yy} + 4u_x - 3u_y + u = 0$

c. $\left(1-x^2\right)u_{xx} + yu_{xy} = 0$

d. $u_{xx} + 4u_{xy} = 0$

e. $\left(1-x^2\right)u_{xx} + yu_{xy} = 0$

2.10 Find the separated solution of the PDE $u_{xx} - 2u_t = 0$ in the form $u_{xx} + 4u_{xy} = 0$ where ω is real and positive.

2.11 Show that the PDE with a mixed second-order derivative,

$\dfrac{1}{c^2}\dfrac{\partial^2 u}{\partial t^2} - \dfrac{\partial^2 u}{\partial x^2} = -\dfrac{2a}{c^2}\dfrac{\partial^2 u}{\partial x \partial t}$, with α and c constants, admits a solution periodic in t,

namely:

$u\left(x,t\right) = A cos\left(\dfrac{ax}{c}\right) sin\left(\omega t - kx\right)$ where $k = \dfrac{1}{c}\sqrt{\omega^2 + a^2}$, and derive therefore from the

estimate, $u_{xx} + 2u_{yy} + 2u = 0$ for sufficiently small values of $\dfrac{ax}{c}$ and $\dfrac{a}{\omega}$ which exhibits

some modifications in the amplitude and phase to periodic wave train, $k = \dfrac{1}{c}\sqrt{\omega^2 + a^2}$,

that advances along the x-direction without change of shape.

2.12 Find the separated solutions of the following PDEs:

1. $u_{xx} + 2u_x + u_{yy} = 0$

2. $u_{xx} + 2u_{yy} + 2u = 0$

3. $u_{xx} - u_{yy} + u = 0$

4. $x^2 u_{xx} + xu_y + u_{yy} = 0$

2.13 Find the separated solutions of the heat equation $u_x\left(0,t\right) = 0$, which satisfies the boundary conditions:

1. $u\left(0,t\right) = 0$ and $u_x\left(L,t\right) = 0$

2. and $u_x\left(0,t\right) = 0$ and $u_x\left(L,t\right) = 0$

2.14 Find the separated solutions of the Laplace's equation $u_{xx} - u_{yy} = 0$ which satisfies the boundary conditions: $u_x\left(0,y\right) = 0$ $u_x\left(L,y\right) = 0$ and $u\left(x,0\right) = 0$

2.15 A two-dimensional function $u\left(x,y\right)$ which satisfies Laplace's equation $u\left(x,0\right) = 0$ is called harmonic function.

Prove that each of the following functions is a harmonic function.

a. $u = \tan^{-1}\left(\dfrac{y}{x}\right)$ b. $u = x^4 - 6x^2y^2 + y^4$

c. $u = \ln\left(x^2 + y^2\right)$

2.16 A three-dimensional function $u\left(x,y,z\right)$ which satisfies Laplace's equation:

$$\dfrac{\partial^2 u}{\partial x^2} + \dfrac{\partial^2 u}{\partial y^2} + \dfrac{\partial^2 u}{\partial z^2} = 0$$

is called harmonic function.

Prove that each of the following functions is a harmonic function.

$u = x^2 + y^2 + 4z^4$; b. $u(x,t) = e^{1\mu x}e^{-\mu^2}$

2.17 The equation of the displacement u in a vibrating string is given by $\dfrac{\partial^2 u}{\partial t^2} = a\dfrac{\partial^2 u}{\partial x^2}$, where

u is function of t and u. Let f and g be arbitrary solutions of the differential equations,

$\dfrac{d^2 f}{dx^2} + \lambda f = 0$ and $\dfrac{d^2 g}{dt^2} + \lambda a^2 g = 0$, respectively. Prove that $u(x,t) = Cf(x)g(x)$ is the

solution of the vibrating string equation.

2.18 $u(x,y) = \sin(2x + 3y)$

Find $u_x = \dfrac{\partial u}{\partial x}; u_{xx} = \dfrac{\partial^2 u}{\partial x^2}; u_{xy} = \dfrac{\partial^2 u}{\partial x \partial y}; u_{yy} = \dfrac{\partial^2 u}{\partial y^2}$.

2.19 Given that $z = e^{vu^2}$,

find $\dfrac{\partial z}{\partial u}$ and $\dfrac{\partial z}{\partial v}$

2.20 Solve the given initial value problem and determine the values of x and y for which it exists:

$u_x - 2u_y = u$ with $u(0,y) = y$
Write the characteristics equations and solve them. Find the general solutions.

2.21 Verify the following are separated solutions of Laplace's equation:

   a. $u(x,y) = e^{3x}\cos(2y)$

   b. $u(x,y) = e^{2(x+iy)}$

2.22 Which of the following are solutions of the Laplace equation?

   a. $u(x,y) = e^x \cos(2y)$

   b. $u(x,y) = e^x e^y$

   c. $u_x(L,t) = 0$

2.23 Find the separated solutions of each of these equations

   a. $x^2 u_{xx} + x u_x + u_{yy} = 0$

   b. $u_{xx} - u_{yy} + u = 0$

2.24 Solve the one-dimensional wave equation

$$u_{tt} - c^2 u_{xx} = 0$$

Use the Cauchy initial conditions $u(x,0) = g(x)$ and $u_t(x,0) = h(x)$, where f and g are arbitrary functions.

3 Vector Analysis

"It is wrong to think that the task of physics is to find out how Nature is. Physics concerns what we say about Nature."

Niels Bohr

INTRODUCTION

Vectors are well used in physics and all related fields. In each physical concept, vectors appear to be the basis and the foundation of the mathematical expression of physical phenomena. Vectors are used in mechanics, in terms of weight, force, velocity, acceleration, displacement; in electricity as the electric field; in magnetism as the magnetic field; in optics as the Poynting vector and the direction of waves' propagation. Physicists and engineers usually deal with scalar quantities, that is, quantities with magnitude only such as mass, temperature, time; they also deal with vector quantities or quantities with both magnitude and direction such as forces, A, velocities, $\vec{v}$, accelerations, $\vec{a}$, the electric field, $\vec{E}$, the magnetic field, $\vec{H}$.. In this chapter, we discuss vector properties, vector cross products, vector integration, line integral and surface integral.

3.1 DEFINITION OF VECTOR

The study of motion in mechanics involves the use of vectors. A vector is represented by an arrow joining two points. One of the points is the origin or tail of the vector. The other point is the endpoint or head of the vector. A vector is a quantity that has both a magnitude and a direction. An example is the displacement vector. When you move on a straight road from position A to position B located 50 m from A in the South–North direction, your displacement or change in your position is represented by vector $\Delta\left(\overrightarrow{AB}\right)$. The magnitude of your displacement is 50 m and its direction is South–North.

Example 1

Consider two points A and $\vec{v}$ and draw the arrow from A to B; then construct the vector $\overrightarrow{AB}$ as follows:

3.1.1 REPRESENTATION OF A VECTOR IN A PLANE

In a two-dimensional Cartesian coordinate, the two coordinate axes $\Delta\left(\overrightarrow{AB}\right)v$ and A_y are perpendicular to each other and a vector $\vec{A}$ has two components: A_x and A_y; we write $\vec{A} = \left(A_x, A_y\right)$. A_x is the x-component of $\vec{A}$ and A_yV is the y-component of $\vec{A}$.

A_x is also called the abscissa of $\vec{A}$ and A_y is called the ordinate of $\vec{A}$:

$$\vec{A} = A_x\vec{i} + A_y\vec{j} \tag{3.1}$$

$\vec{i}$ and θ are unit vectors, respectively, to x-axis and y-axis.

DOI: 10.1201/9781003478812-3

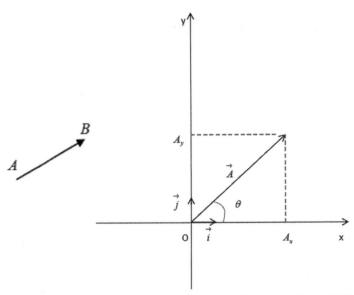

FIGURE 3.1   (a) Representation of a vector from two points: the tail A and the head B. (b) Two dimensions in Cartesian coordinates.

3.1.2   TRIGONOMETRIC CALCULATIONS

Now, consider a standard vector $\vec{A}$ of components (a, b) in the two-dimensional system represented in Figure 3.1b; the length or magnitude of $\vec{A}$ is denoted by $\vec{A}$ and computed using the Pythagorean theorem: $\vec{A} = \sqrt{a^2 + b^2}$.

If θ is the angle between vector $\vec{A}$ and the x-axis (see Figure 3.1b), then we can define the cosine, the sine of the angle θ as follows:

$$\cos(\theta) = \frac{Adjacent\,side}{Hypotenuse} = \frac{A_x}{\left\|\vec{A}\right\|} \tag{3.2}$$

$$\sin(\theta) = \frac{opposite\,side}{Hypotenuse} = \frac{A_y}{\left\|\vec{A}\right\|} \tag{3.3}$$

$$\begin{cases} A_x = \left\|\vec{A}\right\|\cos(\theta) \\ A_y = \left\|\vec{A}\right\|\sin(\theta) \end{cases}$$

From expression (3.1) of vector $\vec{A}$, we can now write:

$$\vec{A} = \left\|\vec{A}\right\|\cos(\theta)\vec{i} + \left\|\vec{A}\right\|\sin(\theta)\vec{j} \tag{3.4}$$

3.1.3   DIRECTION OF $\vec{A}$

The direction of vector $\vec{A}$ can be represented by the angle θ of $\left\|\vec{A}\right\|$ with respect to the x-axis. It could also be represented by the angle $\vec{A}$ with respect to the y-axis.

3.1.4 MAGNITUDE OF $\vec{A}$

The magnitude of vector $\vec{A}$ is written A (without arrow) or $\vec{A}$. The magnitude of vector θ can be given by the square root of the sum of the squares of the components of $\vec{A}$, that is:

$$\left\|\vec{A}\right\| = \sqrt{\vec{A}_x^2 + \vec{A}_y^2} \tag{3.5}$$

Note: The magnitude of vector $\vec{A}$ can also be written A with no arrow specifially in physics and engineering.

Example 2 (of Vectors)

Another example of vector used in physical science is the weight of an object. The weight of a rigid object with mass m is a vector quantity, $\vec{W} = m\vec{g}$, with $\vec{g}$ the acceleration due to gravity. It is a force pointing downward due to the gravitational influence of the earth; hence, it is a vertical force represented as:

Other examples of vectors include the electrostatic force. This force which exists between two charges q_1 and q_2 separated by a distance r_{12} is given by Coulomb's law:

$$\vec{F} = k_e \frac{|q_1||q_2|}{r_{12}^2} \hat{n} \tag{3.6}$$

where $\hat{n}$ is a unit vector which defines the direction of the segment connecting the two charges and k_e is Coulomb's constant. The force is attractive for opposite charges and repulsive for like charges. The magnitude of the electrostatic force between two charges q_1 and q_2 separated by a distance r_{12} is given by the common force

$$F_{12} = F_{21} = k_e \frac{|q_1||q_2|}{r_{12}^2}. \tag{3.7}$$

Example 3 (The Magnetic Induction $\vec{B}$)

Consider the magnetic induction $\vec{B}$ produced by a current intensity, I. Ampere's law states that the line integral of the magnetic induction, $\vec{B}$, around any closed path is equal to the product of the magnetic permeability of the free space and the intensity of the total steady current flowing through the cable bounded by the closed path:

$$\oint_{CL} \vec{B} \cdot \vec{ds} = \mu_o I \tag{3.8}$$

where CL stands for closed loop.

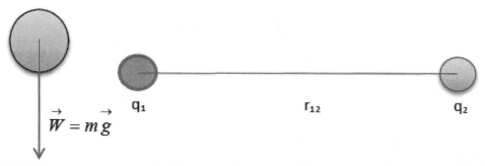

FIGURE 3.2 (a) The weight of a falling object with a mass m. (b) Electrostatic interaction between two charges, q_1 and q_2, separated by the distance r_{12}.

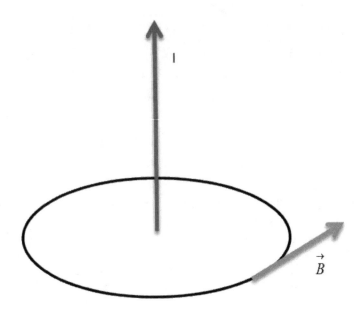

FIGURE 3.3   The magnetic induction, $\vec{B}$, created by the current I.

To find the direction of the magnetic induction $\vec{B}$, use the right-hand rule: point your thumb in the direction of the current and curl the rest of your fingers in the direction of the $\vec{B}$.

3.2   GRAPH OF THE SUM OF TWO VECTORS

The sum of two vectors is constructed in two similar methods: the triangle method and the parallelogram method.

3.2.1   TRIANGLE METHOD

Assume we want to find the sum of two vectors, $\vec{u}$ and $\vec{w}$. In this case, when vector $\vec{u}$ is drawn, we draw vector $\vec{w}$ so that its origin coincides with the terminal point of vector $\vec{u}$. The sum of these two vectors also called the resultant $\vec{r}$ has the same origin as $\vec{u}$ and the same terminal point as $\vec{w}$. The construction of $\vec{r} = \vec{u} + \vec{w}$ can be seen in Figure 3.4.

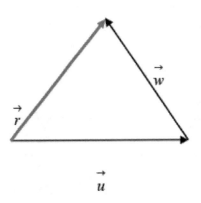

FIGURE 3.4   Sum of two vectors using the triangle method.

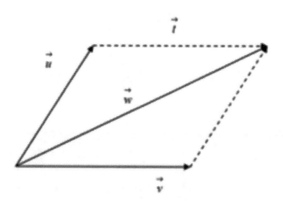

FIGURE 3.5   Sum of two vectors using the parallelogram method.

3.2.2   PARALLELOGRAM METHOD

Assume one would like to find the sum of two vectors $\vec{u}$ and $\vec{v}$ having the same origin. From the terminal point of vector $\vec{u}$, draw a parallel to vector $\vec{v}$, and from the terminal point of $\vec{v}$, draw a parallel to vector $\vec{u}$. These two lines drawn intercept at the terminal point of the sum of the two vectors, $\vec{u} + \vec{v} = \vec{w}$. The origin of the sum $\vec{w}$ is the same as the one of the two vectors, $\vec{u}$ and $\vec{v}$. The geometric construction can be seen in Figure 3.5.

Note: From the triangle method we can write $\vec{u} + \vec{t} = \vec{w}$ but $\vec{t} = \vec{v}$, therefore $\vec{u} + \vec{v} = \vec{w}$. This shows that the triangle method is contained in the parallelogram method.

Example 4 (the Resultant of Two or More Forces)
In physics, the sum of two or more forces is called the resultant of these forces. It is also designated by the net force. The following expression is most of the time used to designate the resultant of n forces:

$$\overrightarrow{F_{net}} = \sum_{i=1}^{n} \overrightarrow{F_i} = \overrightarrow{F_1} + \overrightarrow{F_2} + \overrightarrow{F_3} + ... + \overrightarrow{F_n}. \tag{3.9}$$

We will show the examples of a few cases.

  i. **Case of two forces of the same direction**

Consider two forces $\overrightarrow{F_1}$ and $\overrightarrow{F_2}$ in the same direction. The resultant of these two forces is given by

$$\vec{R} = \vec{F_1} + \vec{F_2} \tag{3.10}$$

The magnitude of $\vec{R}$ is given by $R = F_1 + F_2$

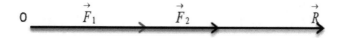

FIGURE 3.6   Sum of two vectors of the same direction.

FIGURE 3.7   Resultant of two forces of opposite directions.

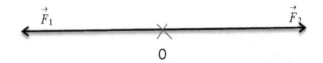

FIGURE 3.8   Resultant of two forces of the same intensity but with opposite directions.

ii. Case of two forces of the opposite directions

If the two forces $\vec{F_1}$ and $\vec{F_2}$ are in opposite directions, The resultant of these two forces is given by

$$\vec{R} = \vec{F_2} + \vec{F_1}. \tag{3.11}$$

And the magnitude will be

$$R = \left| F_1 - F_2 \right|. R = \left| F_2 - F_1 \right| \tag{3.12}$$

iii. Case of two equal and opposite forces

In this case, the resultant is the null vector. If $\vec{F_1}$ is one vector, the other one is $\vec{F_2} = -\vec{F_1}$ and $\vec{R} = \vec{F_1} - \vec{F_1} = \vec{0}$.

$$\vec{R} = \vec{0} \tag{3.13}$$

The magnitude of vector $\vec{R}$ is $R = 0$.

3.2.3   MULTIPLICATION OF A VECTOR BY A SCALAR

Given a vector $\vec{v}$ and a scalar α, the product $\alpha\vec{v}$ is defined as follows:

- if $\alpha > 0$, $\alpha\vec{v}$ is a vector with the same direction as $\vec{v}$
- if $\alpha = 0$, then $\alpha\vec{v}$.
- if $\alpha < 0$, $\alpha\vec{v}$ is a vector with opposite direction to $\vec{v}$.

3.3   VECTORS PROPERTIES

Let us define a few properties of vectors. Consider U to be the set of all vectors.

i. Sum

The sum of two vectors is a vector, that is:

$$\text{For all } \vec{u}, \vec{v} \in U, \vec{u} + \vec{v} \in U \tag{3.14}$$

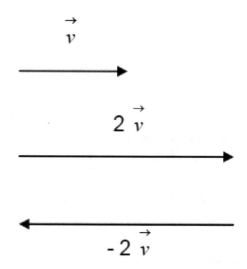

FIGURE 3.9 Multiplication of a vector by a scalar.

ii. **Associative relation**

Consider three vectors $\vec{u}, \vec{v}, \vec{w}$. The operation addition is said to be associative if,

$$\text{for all } \vec{u}, \vec{v}, \vec{w} \in U,$$

$$(\vec{u} + \vec{v}) + \vec{w} = \vec{u} + (\vec{v} + \vec{w}) \tag{3.15}$$

iii. **Existence of neutral element:**

There exists an element called zero vector or null vector $\vec{0} + \vec{u} = \vec{u} + \vec{0} = \vec{u}$ such that:
For all vector $\vec{u} \in U$

$$\vec{0} + \vec{u} = \vec{u} + \vec{0} = \vec{u} \tag{3.16}$$

iv. **Existence of inverse element:**

$\vec{u'}$ is called the inverse of $\vec{u}$ for addition if, for all $\vec{u} \in U$, there exists a unique element $\vec{u'} \in U$ such that:

$$\vec{u} + \vec{u'} = \vec{u'} + \vec{u} = \vec{0} \tag{3.17}$$

One can then write $\vec{u'} = -\vec{u}$. The inverse of a vector $\vec{u}$ is its opposite, $-\vec{u}$.

3.4 DEFINITION OF A VECTOR SPACE

Let V be an arbitrary nonempty set of vectors on which two operations are defined: addition and multiplication by scalars (or numbers). By addition, we mean the rule that associates with each pair of vectors $\vec{x}$ and $\vec{y}$, their sum $\vec{x} + \vec{y}$. By scalar multiplication, we mean the rule that associates with each scalar $\vec{x} + \vec{y}$ and each vector, $\vec{x}$, the vector $\alpha \vec{x}$, called the scalar multiple of $\vec{x}$ by α. If the following axioms are satisfied by all vectors in V and all scalars, then V is referred to as a vector space.

a. For any two vectors $\vec{x}$ and $\vec{y}$ in V, $\vec{x}+\vec{y}$ is in V.

b. For any vector $\vec{x}$ in V and a real number a in $\mathbb{R}$, $a\vec{x}$ is in V

c. $\vec{x}+\vec{y}=\vec{y}+\vec{x}$ for any vector $\vec{x}$ and $\vec{x}+(-\vec{x})=\vec{0}$ in V.

d. $(\vec{x}+\vec{y})+\vec{z}=\vec{x}+(\vec{y}+\vec{z})$ for any vector $\vec{x}$, $\vec{y}$ and $\vec{z}$ in V.

e. There is a vector in V denoted by $\vec{0}$ such that $\vec{x}+\vec{0}=\vec{0}+\vec{x}=\vec{x}$ for any vector $\vec{x}$ in V.

f. For every vector $\vec{x}+(-\vec{x})=\vec{0}$ in V, there is a vector $-\vec{x}$ in V such that $\vec{x}+(-\vec{x})=\vec{0}$.

g. For any two vectors $\vec{x}$ and β in V and for any scalar α in $\mathbb{R}$, $\alpha(\vec{x}+\vec{y})=\alpha\vec{x}+\alpha\vec{y}$.

h. For any two scalars $\vec{x}$ and β in $\mathbb{R}$ and for any vector $\vec{x}$ in V, $(\alpha+\beta)\vec{x}=\alpha\vec{x}+\beta\vec{x}$.

i. For any two scalars $\vec{x}$ and β in $\mathbb{R}$ and for any vector $\vec{x}$ in V, $\alpha_1\vec{x}_1+\alpha_2\vec{x}_2+\alpha_3\vec{x}_3,...,\alpha_n\vec{x}_n=0$.

j. For any vector $\vec{x}$ in V with $1\in\mathbb{R}$, $1\vec{x}=\vec{x}$.

Remark

Depending on the application, scalars may be real numbers or complex numbers. Vector spaces in which the scalars are real numbers are called real vector spaces. Vector spaces in which the scalars are complex numbers are called complex vector spaces.

3.5   LINEAR DEPENDENCE AND LINEAR INDEPENDENCE

3.5.1   DEFINITION 1

A finite set of vectors $\{\vec{x}_1,\vec{x}_2,\vec{x}_3,...,\vec{x}_n\}$ in $\mathbb{R}^n$ is said to be linearly dependent if there exists the

scalars $\alpha_1,\alpha_2,\alpha_3,...,\alpha_n$ not all zeros such that $\alpha_1\vec{x}_1+\alpha_2\vec{x}_2+\alpha_3\vec{x}_3,...,\alpha_n\vec{x}_n=0$, or $\displaystyle\sum_{i=1}^{n}\alpha_i\vec{x}_i=\vec{O}$

(in a condensed form). The scalars $\alpha_1,\alpha_2,\alpha_3,...,\alpha_n$ can also be called coefficients of the vectors,

respectively.

Example 5

$\vec{x}_1=\begin{pmatrix}1\\1\\1\end{pmatrix}$, $\vec{x}_2=\begin{pmatrix}1\\-1\\2\end{pmatrix}$, $\vec{x}_3=\begin{pmatrix}3\\1\\4\end{pmatrix}$ are linearly dependent because $2\vec{x}_1+\vec{x}_2-\vec{x}_3=0$.

3.5.2   PROPERTIES

a. Any set containing the vector zero is linearly dependent because for every scalar $\alpha\neq0$, $\alpha\vec{0}=\vec{0}$.

b. If a set of vectors is linearly dependent, then one of the vectors can be written as a linear combination of the others. For example, if $\vec{x}_1,\vec{x}_2,\vec{x}_3$ are linearly dependent, then there exists three coefficients, $\alpha_1,\alpha_2,\alpha_3$ not all equal to zero such that $\alpha_1\vec{x}_1+\alpha_2\vec{x}_2+\alpha_3\vec{x}_3=0$. Assume α_3 the coefficient that is different from zero; then we can write $\vec{x}_3$ as a linear combination of the other two vectors, $\vec{x}_1$ and $\vec{x}_2$, that is $\vec{x}_3=-\dfrac{1}{\alpha_3}(\alpha_1\vec{x}_1+\alpha_2\vec{x}_2)$.

3.6 LINEAR INDEPENDENCE

3.6.1 DEFINITION 2

A finite set of vectors $\left\{\overrightarrow{x_1}, \overrightarrow{x_2}, \overrightarrow{x_3}, ..., \overrightarrow{x_n}\right\}$ in $\mathbb{R}^n$ is said to be linearly independent if it is not linearly dependent. This means that any linear combination of these vectors that gives zero implies necessarily that all the coefficients $\alpha_1, \alpha_2, \alpha_3, ..., \alpha_n$ are equal to zero.

In practice, to prove that a system of vectors is linearly independent, one will use the following implication: $\left\{\overrightarrow{x_1}, \overrightarrow{x_2}, \overrightarrow{x_3}, ..., \overrightarrow{x_n}\right\}$ is linearly independent if for some $\alpha_1, \alpha_2, \alpha_3, ..., \alpha_n$, $\alpha_1 \overrightarrow{x_1} + \alpha_2 \overrightarrow{x_2} .... + \alpha_n \overrightarrow{x_n} = \vec{0}$ is satisfied if and on only if all coefficients $\alpha_n = 0$.

Example 6

Unit vectors $\vec{i} = (1,0)$ and $\vec{j} = (0,1)$ are linearly independent. Indeed, if we set $\alpha(1,0) + \beta(0,1) = (0,0)$, we have $(\alpha, \beta) = (0,0)$, which means that $\alpha = 0, \beta = 0$.

Example 7

Prove that the two vectors $\overrightarrow{u_1} = \begin{pmatrix} 1 \\ -1 \end{pmatrix}$ and $\overrightarrow{u_2} = \begin{pmatrix} 2 \\ 3 \end{pmatrix}$ are linearly independent.

Solution

To prove the linear independence of these two vectors $\overrightarrow{u_1}$ and $\overrightarrow{u_2}$, let us first consider their zero linear combination, that is $\alpha u_1 + \beta u_2 = 0$ or $\alpha \begin{pmatrix} 1 \\ -1 \end{pmatrix} + \beta \begin{pmatrix} 2 \\ 3 \end{pmatrix} = \begin{pmatrix} 0 \\ 0 \end{pmatrix}$. We then have the following system of equations:

$$\begin{cases} \alpha + 2\beta = 0 \\ -\alpha + 3\beta = 0 \end{cases} \qquad (3.18)$$

By adding the two equations of the system together, we have $5\beta = 0$ or $\beta = 0$. After substitution of $\beta = 0$ in one of the equations, we obtain $\alpha = 0$. Overall, $\alpha u_1 + \beta u_2 = 0 \Rightarrow \alpha = 0$ and $\beta = 0$. Therefore, u_1 and u_2 are linearly independent.

3.6.2 DEFINITION 3

Consider a vector space V and n vectors, $\overrightarrow{v_1}, \overrightarrow{v_2}, \overrightarrow{v_3}..., \overrightarrow{v_n}$. The set $\left\{\overrightarrow{v_1}, \overrightarrow{v_2}, \overrightarrow{v_3}..., \overrightarrow{v_n}\right\}$ is a spanning set of V if span $\left\{\overrightarrow{v_1}, \overrightarrow{v_2}, \overrightarrow{v_3}..., \overrightarrow{v_n}\right\} = V$, that is, any vector of V can be written as a linear combination of the vectors $\overrightarrow{v_1}, \overrightarrow{v_2}, \overrightarrow{v_3}..., \overrightarrow{v_n}$. We also say that V is generated or spanned by the vectors $\overrightarrow{v_1}, \overrightarrow{v_2}, \overrightarrow{v_3}..., \overrightarrow{v_n}$.

Lemma 1

Suppose that the vectors $\overrightarrow{v_1}, \overrightarrow{v_2}, \overrightarrow{v_3}, ..., \overrightarrow{v_n}$ span a vector space V and that the vectors $\overrightarrow{u_1}, \overrightarrow{u_2}, \overrightarrow{u_3}..., \overrightarrow{u_j}$ in V are linearly independent. Then $j \leq n$.

Proof

Since vectors $\overrightarrow{v_1}, \overrightarrow{v_2}, \overrightarrow{v_3}, ..., \overrightarrow{v_n}$ span V, every vector of V can be written as a linear combination of these vectors $\overrightarrow{v_1}, \overrightarrow{v_2}, \overrightarrow{v_3}, ..., \overrightarrow{v_n}$, in particular, $\overrightarrow{u_1}$:

$\overrightarrow{u_1} = k_1 \overrightarrow{v_1} + k_2 \overrightarrow{v_2} + k_3 \overrightarrow{v_3} + ... + k_n \overrightarrow{v_n}$. Since $\overrightarrow{u_1} \neq 0$, not all k are equal to zero; then the $\overrightarrow{v_i}$ can be expressed as a linear combination of the $\overrightarrow{u_1}$ and the remaining $\overrightarrow{v_s}$. So, the set of the $\overrightarrow{u_1}, \overrightarrow{u_2}, \overrightarrow{u_3}..., \overrightarrow{u_n}$

's with $\vec{v_i}$ replaced by $\vec{u_1}$ span V. If $j \geq n$, repeat this step $n-1$ more times and conclude that $\vec{u_1}, \vec{u_2}, \vec{u_3}, ..., \vec{u_n}$ span V: if $j > n$, this contradicts the linear independence of the $\vec{u}$'s for then $\vec{u}_{n+1}$ is a linear combination of $\vec{u_1}, \vec{u_2}, \vec{u_3}..., \vec{u_n}$.

3.7   DEFINITION OF BASIS

A set of vectors $j \geq n$ in a vector space V is said to be a basis of V if and only if:

a. the system $\left\{ \vec{v_1}, \vec{v_2}, \vec{v_3}, ..., \vec{v_n} \right\}$ is linearly independent and

b. the vectors $\vec{v_1}, \vec{v_2}, \vec{v_3}, ..., \vec{v_n}$ span V.

Theorem 1

All bases for a finite-dimensional vector space V contain the same number of vectors. This number is called the dimension of V and is denoted as dim (V).

Proof

Let $\left\{ \vec{v_1}, \vec{v_2}, \vec{v_3}..., \vec{v_n} \right\}$ be a basis and let $\left\{ \vec{u_1}, \vec{u_2}, \vec{u_3}..., \vec{u_m} \right\}$ be another basis. By Lemma 1 and the definition of basis, we conclude that $m \leq n$ and $n \leq m$. As a result, we can conclude that $n = m$.

3.7.1   ORTHOGONAL BASIS

A basis $\left\{ v_1, v_2, ..., v_n \right\}$ is an orthogonal basis if: $v_i v_j = 0$ for $v_i v_j = 0$.

3.7.2   ORTHONORMAL BASIS

A basis $\left\{ v_1, v_2, ..., v_n \right\}$ is an orthonormal basis if:

$v_i v_j = \delta_{ij}$; which means that in addition of being orthogonal, each vector has a unit length.

$\delta_{ij} = 1$ for $i = j$ and $\delta_{ij} = 0$ for δ_{ij}.

δ_{ij} is called the Kronecker symbol and will also be defined later in expression 3.28 and Section 3.8.

3.8   VECTOR IN THREE DIMENSIONS (3D)

3.8.1   CARTESIAN COORDINATES AND DIRECTION COSINES

Consider a point M of components X, Y, Z in a 3D system of the axis $\vec{ox}, \vec{oy}, \vec{oz}$. The position vector $\vec{OM}$ can be written: $\vec{OM} = X\vec{i} + Y\vec{j} + Z\vec{k}$ such that:

$$\begin{cases} X = r\cos(\alpha) \\ Y = r\cos(\beta). \\ Z = r\cos(\gamma) \end{cases} \tag{3.20}$$

$\cos(\alpha)$, $\cos(\beta)$ and $\cos(\gamma)$ are called the direction cosines of vector $(O; X, Y)$. The magnitude of the vector $\vec{OM}$ is given by $OM = r = \sqrt{X^2 + Y^2 + Z^2}$.

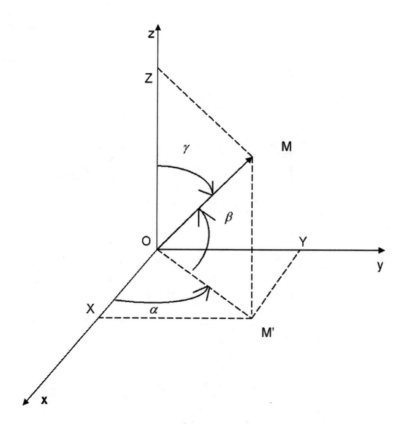

FIGURE 3.10 Representation of the components (x,y,z) of a point $\overrightarrow{OM}$ in a 3D Cartesian coordinates system. $\overrightarrow{OM}$ is the position vector and $\cos(\alpha)$, $\cos(\beta)$, $\overrightarrow{OM}$ are the direction cosines of $\overrightarrow{OM}$. θ is the projection of M on the (x,y) plane.

3.8.2 Rotation of the Coordinate Axes

Consider a position vector $\vec{r}$ in two different systems of axes $(O;X,Y)$ and $(O;X,Y)$. Let us rotate the system $(O;X',Y')$ counterclockwise about an angle θ with respect to the system of axis $(O;X,Y)$ which we maintain in a fixed position.

We can get the following relations between the components of the original fixed system $(0;X,Y)$ and the new rotated system $(O;X',Y')$.

$$\begin{cases} x' = x\cos\theta + y\sin\theta \\ y' = -x\sin\theta + y\cos\theta \end{cases} \tag{3.21}$$

Proof

$$\overrightarrow{OM} = x\vec{i} + y\vec{j}_{\text{in}}\left(0;X,Y\right)$$

$$\overrightarrow{OM'} = x'\vec{i}\,' + y'\vec{j}\,' \text{ in } \left(0;X',Y'\right)$$

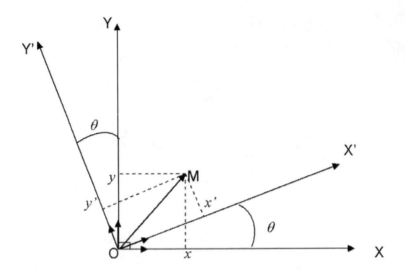

FIGURE 3.11   A 3D Cartesian coordinate system rotated about the z-axis by angle θ in the counterclockwise direction.

Also

$$\begin{cases} ix' = x\cos\theta + y\sin\theta \\ iy' = -x\sin\theta + y\cos\theta \end{cases} \tag{3.22}$$

By substituting $\vec{i}$ and $\vec{A}(A_x, A_y)$ in equation (3.21), we obtain

$$\overrightarrow{OM} = x\left(\vec{i}'\cos\theta - \vec{j}'\sin\theta\right) + y\left(\vec{i}'\sin\theta + \vec{j}'\cos\theta\right)$$

$$\overrightarrow{OM} = \left(x\cos\theta + y\sin\theta\right)\vec{i}' + \left(-x\sin\theta + y\cos\theta\right)\vec{j}'$$

By comparison with (3.21), we have:

$$\begin{cases} x' = x\cos\theta + y\sin\theta \\ y' = -x\sin\theta + y\cos\theta \end{cases} \tag{3.23}$$

Equation (3.23) can be put in the form of matrix in the following transformation:

$$\begin{bmatrix} x' \\ y' \end{bmatrix} = \begin{bmatrix} \cos\theta & \sin\theta \\ -\sin\theta & \cos\theta \end{bmatrix} \begin{bmatrix} x \\ y \end{bmatrix} \tag{3.24}$$

The matrix $\begin{bmatrix} \cos\theta & \sin\theta \\ -\sin\theta & \cos\theta \end{bmatrix}$ is a rotation matrix and its determinant (Section 4.6) is $\cos^2\theta + \sin^2\theta = 1$

(the determinant of a matrix will be explained in detail in Chapter 4).

As a result, if any vector $\vec{A}(A_x, A_y)$ has components A_x, A_y in the original system $(0;X,Y)$ and components A_x', A_y' in the new system $(0;X',Y')$, or if $\vec{A}(A_x, A_y)$ in $(0;X,Y)$ and $\vec{A}(A_x', A_y')$ $\vec{A}(A_x', A_y')$ in $(0;X',Y')$, then we have

$$\begin{cases} A_x\,°{}'= A_x\cos\theta + A_y\sin\theta \\ A_y\,°{}'= A_x\sin\theta + A_y\cos\theta \end{cases}.$$

The magnitude of the vector $\vec{A}$ is a scalar that is invariant to the rotation of the coordinate system. The components of $\vec{A}$ in a coordinate system $(0;X,Y)$ are the representation of $\vec{A}$ in that coordinate system. The vector $\vec{A}$ is independent of the rotation of coordinate system.

Let us now use x_1 for x for direction 1 and x_2 for y for direction 2. Assume $a_{11}=\cos(\theta)$; $a_{12}=\sin(\theta)$; $a_{21}=-\sin(\theta)$; $a_{22}=\cos(\theta)$, then

$$\begin{cases} x_1' = a_{11}x_1 + a_{12}x_2 \\ x_2' = a_{21}x_1 + a_{22}x_2 \end{cases} \tag{3.25}$$

The coefficients a_{ij} belong to the interval $[-1,1]$; they may be interpreted as the cosine of the angle between x_i' and x_j, that is

$$a_{11}=\cos\left(x_1,x_1'\right)=\cos\theta$$

$$a_{12}=\cos\left(x_1',x_2\right)=\sin\theta \tag{3.26}$$

$$a_{21}=\cos\left(x_2',x_1\right)=\cos\left(\theta+\frac{\pi}{2}\right)=-\sin\theta$$

We can now use the summation symbol Σ and rewrite (3.25) as

$$x_i' = \sum_{j=1}^{2} a_{ij}x_{j\ i=1,2} \tag{3.27}$$

Also, it is possible to generalize to three N dimensions:

$$V_i' = \sum_{j=1}^{N} a_{ij}V$$

a_{ij} is the cosine between $-\varphi$ and x_j can be written $a_{ij}=\dfrac{\partial x'_j}{\partial x_i}$.

For the inverse rotation, let us change φ in $-\varphi$; then $x_j = \Sigma_i a_{ij}x_i'$. Also, $x_j = \Sigma_i a_{ij}x_i'V_j$ and $V_j = \Sigma \dfrac{\partial x_j}{\partial x_i'}V_j$. Finally, the direction cosines a_{ij} satisfy the orthogonality condition: $\Sigma_i a_{ij}a_{ik}=\delta_{jk}$, or $\Sigma_i a_{ji}a_{ki}=\delta_{jk}$ where $\vec{B}$ is the Kronecker delta symbol defined as:

$$\delta_{jk} = \begin{cases} 1 \text{ for } j=k \\ 0 \text{ for } j\neq k \end{cases} \tag{3.28}$$

We can prove the orthogonality condition by using the product of the two partial derivatives; that is

$$\sum_i a_{ji}a_{ki} = \sum_i \frac{\partial x_j}{\partial x_i'}\frac{\partial x_k}{\partial x_i'} = \sum_i \frac{\partial x_j}{\partial x_i'}\frac{\partial x_i'}{\partial x_k} = \frac{\partial x_j}{\partial x_k} = \delta_{jk} \tag{3.29}$$

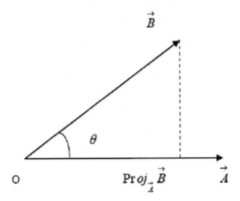

FIGURE 3.12   Projection vector in Cartesian coordinates (x, y, z).

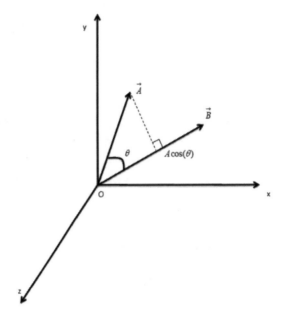

FIGURE 3.13   Representation of vectors $\vec{A}$ and $\vec{A}$. The projection of $\vec{A}$ on $\vec{B}$ is $A\cos(\theta)$. The scalar product of $\vec{A}$ and $\vec{B}$ is $AB\cos(\theta)$.

3.9   SCALAR OR DOT PRODUCT

Consider two vectors $\vec{A}$ and $\vec{B}$ as in Figure 3.12. The scalar or dot product of the two vectors $\vec{A}$ and $\vec{B}$ is defined by

$\vec{A} \cdot \vec{B} = \vec{A}.\vec{B}\cos(\theta)$, where θ measures the angle $(0 \leq \theta \leq \pi)$ determined by $\vec{A}$ and $\vec{B}$ (see Figure 3.14).

The vector that we get by projecting $\vec{B}$ onto $\vec{A}$ is called the vector projection of $\vec{B}$ onto $\vec{A}$ and is denoted by $\mathrm{Pr}oj_{\vec{A}}\vec{B}$, the length of which is $\vec{B}\cos(\theta)$. The two factors $\vec{A}$ and $\vec{B}$ may be interchangeable, that is $\vec{A} \cdot \vec{B} = \vec{B}.\vec{A}$.

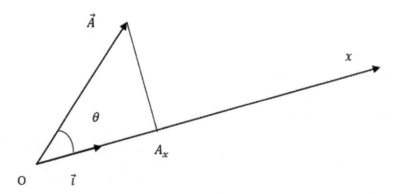

FIGURE 3.14 Invariance of the scalar product under rotations.

3.9.1 Some Properties

Consider three vectors $\vec{A}, \vec{B}, \vec{C}$; then we have the following properties:

3.9.1.1 Commutative Property
The scalar product of two vectors is commutative, that is, $\vec{A} \cdot \vec{B} = \vec{B} \cdot \vec{A}$.

3.9.1.2 Distributive Property of Multiplication with Respect to Addition of Vectors
The scalar multiplication of vectors with respect to addition is distributive

$$\vec{A} \cdot \left(\vec{B} + \vec{C} \right) = \vec{A} \cdot \vec{B} + \vec{A} \cdot \vec{C}.$$

In addition, for α, a real number, the following property is satisfied:

$$\vec{A} \cdot \left(\alpha \vec{B} \right) = \left(\alpha \vec{A} \right) \cdot \vec{B} = \alpha \vec{A} \cdot \vec{B}$$

If $\left(\vec{x}, \vec{y}, \vec{z} \right)$ is an orthonormal basis, then $\begin{cases} \vec{x} \cdot \vec{y} = 0\, \vec{x} \cdot \vec{z} = 0\, \vec{y} \cdot \vec{z} = 0 \\ \vec{x} = \vec{y} = \vec{z} = 1 \end{cases}$

and if $\vec{A} = A_x \vec{x} + A_y \vec{y} + A_z \vec{z}$ and $\vec{B} = B_x \vec{x} + B_y \vec{y} + B_z \vec{z}$

then $\vec{A} \cdot \vec{B} = A_x B_x + A_y B_y + A_z B_z$.

More generally, if the $A_i 's$ are the respective components of $\vec{A}$ and the $B_i 's$ are the respective components of $B_i 's$, then

$$\vec{A} \cdot \vec{B} = \sum_i B_i A_i = \sum_i A_i B_i = \vec{B} \cdot \vec{A}$$

Example 8 (of Scalar Product)
The work done by a force $\vec{F}$ on an object through a displacement, $\Delta \vec{r}$, represented by, $\vec{L}$, is defined as $W = \vec{F} \cdot \vec{L}$.

It is often convenient to use e_m, $m = 1, 2, 3, \dots N$ as basis vectors. Therefore, the orthonormality condition can be written:

$\overrightarrow{e_m} \cdot e_n = \delta_{mn}$.

For $m = n$, $\overrightarrow{e_n} \cdot e_n = 1$ (normality)

For $m \neq n$, $\delta_{mn} = 0$ or $\overrightarrow{e_m.e_n} = 0$ (orthogonality)

Remark

The scalar product of two vectors is a scalar.

3.9.2   INVARIANCE OF THE SCALAR PRODUCT UNDER ROTATIONS

Let us investigate the behavior of the scalar product $\vec{A} \cdot \vec{B}$ under a rotation. Here, we assume that (x, y, z) is rotated through an angle θ, the point O being fixed. Therefore,

$$\vec{A}.\vec{B} = A'_x B'_x + A'_y B'_y + A'_z B'_z$$
$$= \sum_i a_{xi} A_i \sum_j a_{xj} B_j + \sum a_{yi} A_i \sum a_{yj} B_j + \sum a_{zi} A_i \sum a_{zj} B_j$$

Using the indices k and l to sum over x, y, z, we obtain

$$\sum_k A_k' B_k' = \sum_l \sum_i \sum_j a_{li} A_i a_{lj} B_j$$

$$= \sum_i \sum_j \delta_{ij} A_i B_j$$

$$= \sum_i \sum_j \delta_{ij} A_i B_i$$

$$= \sum_i A_i B_i \qquad (3.30)$$

Therefore, we can write

$$\sum_k A_k' B_k' = \sum_i A_i B_i \qquad (3.31)$$

So, the scalar quantity remains invariant under the rotation of the coordinate system.

In a similar approach, if $\vec{C} = \vec{A} + \vec{B}$, then

$$\vec{C}^2 = \vec{C} \cdot \vec{C}$$
$$= (\vec{A} + \vec{B})^2 \qquad (3.32)$$
$$= \vec{A}^2 + 2\vec{A} \cdot \vec{B} + \vec{B}^2$$

$\vec{A} \cdot \vec{B} = \dfrac{1}{2}\left(\vec{C}^2 - \vec{A}^2 - \vec{B}^2\right)$ is invariant.

Projection of vector $\vec{A}$ on the x-axis is $\vec{A}.\vec{i}$ which is the x-component of $\vec{A}$:

$$\vec{A}.\vec{i} = Ax = A\cos\theta \qquad (3.33)$$

3.10   VECTOR PRODUCT

The vector product or cross product is another type of vector multiplication. Consider two vectors $\vec{A}$ and $\vec{B}$. The cross product or vector product of $\vec{A}$ and $\vec{B}$ is the vector $\vec{A}, \vec{B}, \vec{C}$ defined by

$$\vec{C} = \vec{A} \times \vec{B}. \tag{3.34}$$

3.10.1 Magnitude of the Vector Product $\vec{A} \times \vec{B}$

The magnitude of the vector product $\vec{C} = \vec{A} \times \vec{B}$ is given by

$$\left\| \vec{C} \right\| = \left\| \vec{A} \right\| \cdot \left\| \vec{B} \right\| \cdot \left| \sin(\vec{A}, \vec{B}) \right| \tag{3.35}$$

Vector product $\vec{A} \times \vec{B}$ is perpendicular to $\vec{A}$ and perpendicular to $\vec{B}$, and therefore perpendicular to the plane formed by $\vec{A}$ and $\vec{B}$. In addition, vector $\vec{C} = \vec{A} \times \vec{B}$ is constructed in such a way that $(\vec{A}, \vec{B}, \vec{C})$ satisfies the right-hand rule.

Example 9 (the Force on a Current-Carrying Cable)
A cable carrying the current I in a magnetic field $\vec{B}$ is subject to the force,

$$\vec{F} = I\vec{L} \times \vec{B} \tag{3.36}$$

The right-hand rule gives the direction of the force, $\vec{F}$. In this rule, the forefinger represents the direction of the magnetic field, the middle finger represents the direction of the current, the thumb represents the direction of the force $\vec{F}$.

The cross product is not commutative. This property is represented by the relation

$$\vec{A} \times \vec{B} = -\vec{B} \times \vec{A} \tag{3.37}$$

For the unit vectors $\hat{x}$, $\hat{y}$, $\hat{z}$, we have $\hat{x} \times \hat{x} = \hat{y} \times \hat{y} = \hat{z} \times \hat{z} = \vec{0}$ and
$\hat{x} \times \hat{y} = \hat{z}$, $\hat{y} \times \hat{z} = \hat{x}$, $\hat{z} \times \hat{x} = \hat{y}$.

In addition, we can show the cyclic rule to easily remember the cross product of the unit vectors $\hat{x}, \hat{y}, \hat{z}$.

Taking the clockwise direction and starting a rotation from x, we have vector product of x and y equals z; vector product of y and z equals x; vector product of z and x equals $\vec{p}$. Changing direction will give opposite vectors, like vector product of y and x gives $-z$.

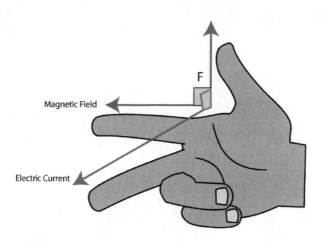

FIGURE 3.15   Schematic of the right-hand rule.

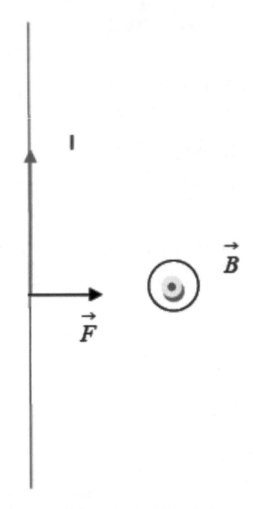

FIGURE 3.16   Direction of the magnetic force using the right-hand rule on a current-carrying cable.

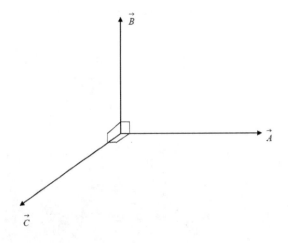

FIGURE 3.17   The cross product of two vectors.

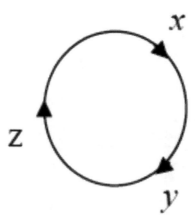

FIGURE 3.18 Vector product cyclic rule.

Examples 10 (Vector Products Used in Physics)

One of the examples of vector products is the relation between linear momentum $\vec{p}$ and angular momentum $\vec{L}$ given by: $\vec{L} = \vec{r} \times \vec{p}$.

Another example is the relation between linear velocity $\vec{v}$ and angular velocity $\vec{\omega}$, $\vec{v} = \vec{\omega} \times \vec{r}$. $\vec{\omega}$ and $\vec{L}$ depend on the choice of the origin.

The vector product force induced by the magnetic induction $\vec{B}$ on a moving charge q with velocity $\vec{v}$ is given by $\vec{F} = q\vec{v} \times \vec{B}$. Also, the area of the parallelogram formed by two vectors $\vec{A}$ and $\vec{B}$ can be represented by the vector product $\vec{S} = \vec{A} \times \vec{B}$.

The magnitude of the area is:

$$S = A \cdot B \left| \sin\theta \right| \tag{3.38}$$

Proof

The area of the parallelogram is $S = 2a_1 + a_2$ using Figure 3.23.

$$a_1 = \frac{1}{2} A^2 \sin(\theta) \cos(\theta)$$

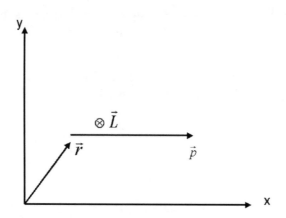

FIGURE 3.19 Example of the linear momentum.

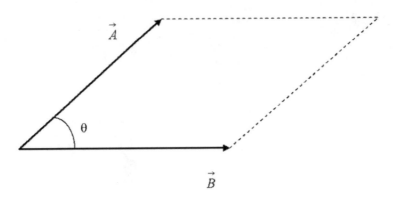

FIGURE 3.20   Area of a parallelogram.

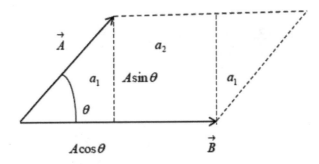

FIGURE 3.21   Decomposition of the area of a parallelogram.

$$a_2 = A\sin(\theta) \times \left(B - A\cos(\theta)\right)$$

So,

$$
\begin{aligned}
S &= A^2\sin(\theta)\cos(\theta) + A\sin(\theta) \times \left(B - A\cos(\theta)\right) \\
&= A^2\sin(\theta)\cos(\theta) + AB\sin(\theta) - A^2\sin(\theta)\cos(\theta) \\
&= AB\sin(\theta) \\
&= \|\vec{A} \times \vec{B}\|
\end{aligned}
\tag{3.39}
$$

Other example in physics where the cross product used is the one of the torque, $\vec{\tau} = \vec{r} \times \vec{F}$. The torque is the tendency of a force to rotate an object about some axis.

3.11   OTHER PROPERTIES

The cross product is distributive from the right and from the left with respect to addition. For every vector $\vec{A}$, $\vec{B}$ and $\vec{C}$ and for a real number α

$$
\begin{aligned}
\vec{A} \times \left(\vec{B} + \vec{C}\right) &= \vec{A} \times \vec{B} + \vec{A} \times \vec{C} \\
\left(\vec{A} + \vec{B}\right) \times \vec{C} &= \vec{A} \times \vec{C} + \vec{B} \times \vec{C} \\
\vec{A} \times \left(\alpha \vec{B}\right) &= \alpha \vec{A} \times \vec{B} = \left(\alpha \vec{A}\right) \times \vec{B}
\end{aligned}
\tag{3.40}
$$

3.11.1   Expression of $\vec{C}$ in the Cartesian Coordinates

Consider a 3D system with unit vectors $\vec{i}, \vec{j}, \vec{k}$; then

$$\vec{A} = A_x\vec{i} + A_y\vec{j} + A_z\vec{k}$$
$$\vec{B} = B_x\vec{i} + B_y\vec{j} + B_z\vec{k}$$

$$\vec{A} \times \vec{B} = \begin{vmatrix} \vec{i} & \vec{j} & \vec{k} \\ A_x & A_y & A_z \\ B_x & B_y & B_z \end{vmatrix} \tag{3.41}$$

$$= \vec{i}\left(A_y B_z - A_z B_y\right) - \vec{j}\left(A_x B_z - A_z B_x\right) + \vec{k}\left(A_x B_y - A_y B_x\right)$$

Assume the vector $\vec{C} = \vec{A} \times \vec{B} = C_x\vec{i} + C_y\vec{j} + C_z\vec{k}$; we can identify the components of $\vec{C}$ as follows:

$$C_x = A_y B_z - A_z B_y, C_y = A_z B_x - A_x B_z, C_z = A_x B_y - A_y B_x \tag{3.42}$$

In general, one can always write the previous components in condensed form such as: $C_i = A_j B_k - A_k B_j$, with all $\vec{i}, \vec{j}, \vec{k}$ different and satisfying permutation cycles, $\left(\vec{i}, \vec{j}, \vec{k}\right)$, $\left(\vec{j}, \vec{k}, \vec{i}\right)$ and $\left(\vec{k}, \vec{i}, \vec{j}\right)$. Let us now observe the impact of the change of system of axes, such as the rotation about an angle θ of the system of axes $(O; x, y, z)$ to $(O; x', y', z')$. We know from the rotation point of view that the component $V_i' = \sum_{j=1}^{3} a_{ij}V_j$ in the new frame of reference is connected to the components V_i

in the old frame of reference by $V_i' = \sum_{j=1}^{3} a_{ij}V_j$. Therefore, we have

$$C_i' = A_j'B_k' - A_k'B_j' = \left(\sum_{l=1}^{N} a_{jl}A_l\right)\left(\sum_{m=1}^{N} a_{km}B_m\right) - \left(\sum_{l=1}^{N} a_{kl}A_l\right)\left(\sum_{m=1}^{N} a_{jm}B_m\right)$$

$$C_i' = \sum\left(a_{jl}a_{km} - a_{kl}a_{jm}\right)A_l B_m \tag{3.43}$$

Remark

The direction cosine $\vec{A} \cdot \left(\vec{B} \times \vec{C}\right)$ vanishes for $l=m$.

Example 11

Find the area of the triangle whose vertices are $A\left(1, -1, 0\right)$, $B\left(2, 1, -1\right)$ and $C\left(-1, 1, 2\right)$.

Solution

Two sides of the triangle can be represented by the vectors $\overrightarrow{AB} = \left(2-1\right)\vec{i} + \left(1+1\right)\vec{j} + \left(-1-0\right)\vec{k}$ $= \vec{i} + 2\vec{j} - \vec{k}$

and

$$\overrightarrow{AC} = \left(-1-1\right)\vec{i} + \left(1+1\right)\vec{j} + \left(2-0\right)\vec{k} = -2\vec{i} + 2\vec{j} + 2\vec{k}$$

The vector

$$\vec{c} = \overrightarrow{AB} \times \overrightarrow{AC} = \begin{vmatrix} \vec{i} & \vec{j} & \vec{k} \\ 1 & 2 & -1 \\ -2 & 2 & 2 \end{vmatrix} = 6\vec{i} + 6\vec{k}$$

has magnitude $|\vec{c}| = \sqrt{36 + 36} = 6\sqrt{2}$ which is equal to the area of the parallelogram of which the given triangle is exactly one-half. Hence, the area of the triangle is

$$\frac{1}{2}\left|\overrightarrow{AB} \times \overrightarrow{AC}\right| = 3\sqrt{2}.$$

3.12   THE VECTOR TRIPLE PRODUCT

3.12.1   THE SCALAR TRIPLE PRODUCT

If $\vec{A}, \left(\vec{j}, \vec{k}, \vec{i}\right)$ and $\vec{C}$ are three vectors, then we can combine them in the form $\vec{A} \cdot \left(\vec{B} \times \vec{C}\right)$ to obtain a scalar result, which is known as the scalar triple product.

The cross product is calculated using a determinant:

$$\vec{B} \times \vec{C} = \begin{vmatrix} \vec{i} & \vec{j} & \vec{k} \\ b_1 & b_2 & b_3 \\ c_1 & c_2 & c_3 \end{vmatrix} = \vec{i}\left(b_2 c_3 - b_3 c_2\right) + \vec{j}\left(b_3 c_1 - b_1 c_3\right) + \vec{k}\left(b_1 c_2 - b_2 c_1\right) \qquad (3.44)$$

Then, $\vec{A} \cdot \left(\vec{B} \times \vec{C}\right)$ gives

$$\vec{A} \cdot \left(\vec{B} \times \vec{C}\right) = \left(a_1 \vec{i} + a_2 \vec{j} + a_3 \vec{k}\right) \cdot \begin{vmatrix} \vec{i} & \vec{j} & \vec{k} \\ b_1 & b_2 & b_3 \\ c_1 & c_2 & c_3 \end{vmatrix} \qquad (3.45)$$

which gives a 3-by-3 determinant involving all three vectors:

$$\vec{A} \cdot \left(\vec{B} \times \vec{C}\right) = \begin{vmatrix} a_1 & a_2 & a_3 \\ b_1 & b_2 & b_3 \\ c_1 & c_2 & c_3 \end{vmatrix} \qquad (3.46)$$

3.12.2   PROPERTY

For three vectors $\vec{A}, \vec{B}, \vec{C}$, the following equalities hold:

$$\vec{A} \cdot \left(\vec{B} \times \vec{C}\right) = \vec{B} \cdot \left(\vec{C} \times \vec{A}\right) = \left(\vec{A} \times \vec{B}\right) \cdot \vec{C} \qquad (3.47)$$

The triple scalar product represents a volume of the parallelepiped formed by $\left(\vec{A}, \vec{B}, \vec{C}\right)$. In fact, $\vec{B} \times \vec{C}$ represents the area of one side.

Example 12

Consider the three vectors

$$\vec{A} = 2\vec{i} + 3\vec{j} - 4\vec{k}, \vec{B} = \vec{i} - 2\vec{j} + 2\vec{k}, \vec{C} = 3\vec{i} - 3\vec{j}\vec{k}$$

Calculate the triple scalar product vector product, $\vec{A} \cdot \left(\vec{B} \times \vec{C} \right)$.

Solution

We will need to evaluate the following 3-by-3 determinant, $\vec{A} \cdot \left(\vec{B} \times \vec{C} \right) = \begin{vmatrix} 2 & 3 & -4 \\ 1 & -2 & 2 \\ 3 & -3 & -1 \end{vmatrix}$. For that, we

will work along the top row to form a sum of three smaller 2-by-2 determinants, each multiplied by an element from the top row:

$$\vec{A} \cdot \left(\vec{B} \times \vec{C} \right) = 2\begin{vmatrix} -2 & 2 \\ -3 & -1 \end{vmatrix} - 3\begin{vmatrix} 1 & 2 \\ 3 & -1 \end{vmatrix} - 4\begin{vmatrix} 1 & -2 \\ 3 & -3 \end{vmatrix}.$$

Then, we can evaluate those smaller determinants and have

$$\vec{A} = 2\vec{i} + 3\vec{j} - 4\vec{k}, \vec{B} = \vec{i} - 2\vec{j} + 2\vec{k}, \vec{C} = 3\vec{i} - 3\vec{j}\vec{k}$$

3.13   VECTOR AND SCALAR OPERATORS

3.13.1   THE DEL OR NABLA OPERATOR

The del or nabla operator comes from the Hebrew word for a "harp" called nevel, a multi-string musical instrument which has the shape of a triangle. In mathematics or in physics, nabla represents the del operator. It is used in differential calculus while defining the gradient, the divergence and the curl of a vector. In 3D Cartesian coordinates, it has three components and is defined as follows:

$$\vec{\nabla} = \begin{pmatrix} \dfrac{\partial}{\partial x} \\[2mm] \dfrac{\partial}{\partial y} \\[2mm] \dfrac{\partial}{\partial z} \end{pmatrix} = \frac{\partial}{\partial x}\vec{i} + \frac{\partial}{\partial y}\vec{j} + \frac{\partial}{\partial z}\vec{k} \tag{3.48}$$

The del operator, $\vec{\nabla}$, is a vector quantity.

The meaning of $\vec{A} \cdot \left(\vec{B} \times \vec{C} \right)$ can be understood when it operates on a scalar function f to produce the gradient of f:

3.13.2   GRADIENT

Given a scalar function $\varphi(x, y, z)$, the gradient of φ in Cartesian coordinates (x, y, z) is represented by the vector:

$$\overrightarrow{grad}(\varphi) = \vec{\nabla}\varphi = \frac{\partial \varphi}{\partial x}\vec{i} + \frac{\partial \varphi}{\partial y}\vec{j} + \frac{\partial \varphi}{\partial z}\vec{k}$$

Example 13
Consider the scalar function

$$\varphi(x, y, z) = 2x^2 yz^3 + xyz + x^2 + y^2 + z^2$$

a. Find $\overrightarrow{grad}\varphi(1,1,0)$.
b. What is the magnitude of $\overrightarrow{grad}\varphi(1,1,0)$?

Solution

a. Find $\overrightarrow{grad}\varphi(1,1,0)$

$$\frac{\partial \varphi}{\partial x} = 4xyz^3 + yz + 2x$$

$$\frac{\partial \varphi}{\partial y} = 2x^2 z^3 + xz + 2y$$

$$\frac{\partial \varphi}{\partial z} = 6x^2 yz^2 + xy + 2z$$

$$\left(\frac{\partial \varphi}{\partial x}\right)_{(1,1,0)} = 2$$

$$\left(\frac{\partial \varphi}{\partial y}\right)_{(1,1,0)} = 2$$

$$\left(\frac{\partial \varphi}{\partial z}\right)_{(1,1,0)} = 0$$

So, $\overrightarrow{grad}\varphi(1,1,0) = 2\vec{i} + 2\vec{j}$

b. What is the magnitude of $\overrightarrow{grad}\varphi(1,1,0)$?

$$\left\|\overrightarrow{grad}\varphi(1,1,0)\right\| = \sqrt{2^2 + 2^2} = \sqrt{8} = 2\sqrt{2} =$$

In general, in a system of n axis of coordinates $(x_1, x_2, ..., x_n)$, if the scalar function $\left\|\overrightarrow{grad}\varphi(1,1,0)\right\| = \sqrt{2^2 + 2^2} = \sqrt{8} = 2\sqrt{2}$ is differentiable, the gradient of φ is a vector quantity whose components are the partial derivatives of φ:

$$\overrightarrow{grad}(\varphi) = \vec{\nabla}\varphi = \frac{\partial \varphi}{\partial x_1}\vec{e}_1 + \frac{\partial \varphi}{\partial x_2}\vec{e}_2 + \frac{\partial \varphi}{\partial x_3}\vec{e}_3 + ... + \frac{\partial \varphi}{\partial x_n}\vec{e}_n.$$

3.13.3   DIVERGENCE AND CURL

3.13.3.1   Definition of the Divergence

In Cartesian coordinates (x, y, z), if $\vec{A} = A_x \vec{i} + A_y \vec{j} + A_z \vec{k}$ is a vector field defined on $\mathbb{R}^3$ and $\dfrac{\partial A_x}{\partial x}$, $\dfrac{\partial A_y}{\partial y}$ and $\dfrac{\partial A_z}{\partial z}$ exist, then the divergence of (x, y, z) is defined by:

$$div\left(\vec{A}\right) = \frac{\partial A_x}{\partial x} + \frac{\partial A_y}{\partial y} + \frac{\partial A_z}{\partial z} \qquad (3.49)$$

The divergence of the vector $\vec{A}$ can now be written symbolically as a dot product of $\vec{\nabla}$ and $\vec{A}$, that is: $div\left(\vec{A}\right) = \vec{\nabla} \cdot \vec{A}$.

Example 14

Compute $div\left(\vec{A}\right)$ if $\vec{A} = xy\vec{i} + xyz\vec{j} + y^3 z\vec{k}$.

Solution

Using the definition of the divergence, we have:

$$div\left(\vec{A}\right) = \frac{\partial A_x}{\partial x} + \frac{\partial A_y}{\partial y} + \frac{\partial A_z}{\partial z} = \frac{\partial(xy)}{\partial x} + \frac{\partial(xyz)}{\partial y} + \frac{\partial\left(y^3 z\right)}{\partial z} \; div\left(\vec{A}\right) = y + xz + y^3$$

3.13.3.2   Definition of the Curl

Consider

$\vec{A}(x, y, z) = A_x(x, y, z)\vec{i} + A_y(x, y, z)\vec{j} + A_z(x, y, z)\vec{k}$,

a vector field on $\mathbb{R}^3$. Assume all quantities $\dfrac{\partial A_x}{\partial x}$, $\dfrac{\partial A_y}{\partial y}$ and $\dfrac{\partial A_z}{\partial z}$ exist. The curl of vector $\vec{A}$ is defined on $\mathbb{R}^n$ by the vector field:

$$curl\left(\vec{A}\right) = \left[\frac{\partial A_z}{\partial y} - \frac{\partial A_y}{\partial z}\right]\vec{i} + \left[\frac{\partial A_x}{\partial z} - \frac{\partial A_z}{\partial x}\right]\vec{j} + \left[\frac{\partial A_y}{\partial x} - \frac{\partial A_x}{\partial y}\right]\vec{k} \qquad (3.51)$$

If we consider that $\vec{\nabla}$ has the components $\dfrac{\partial}{\partial y}$, $\dfrac{\partial}{\partial y}$ and $\dfrac{\partial}{\partial z}$, we can consider the formal cross product of $\vec{\nabla}$ with the vector field $\vec{A}$:

$$\vec{\nabla} \times \vec{A} = \begin{vmatrix} \vec{i} & \vec{j} & \vec{k} \\ \dfrac{\partial}{\partial x} & \dfrac{\partial}{\partial y} & \dfrac{\partial}{\partial z} \\ A_x & A_y & A_z \end{vmatrix}$$

$$= \left[\frac{\partial A_z}{\partial y} - \frac{\partial A_y}{\partial z}\right]\vec{i} + \left[\frac{\partial A_x}{\partial z} - \frac{\partial A_z}{\partial x}\right]\vec{j} + \left[\frac{\partial A_y}{\partial x} - \frac{\partial A_x}{\partial y}\right]\vec{k}$$

$$= curl\left(\vec{A}\right)$$

Therefore, we can retain the symbolic expression:

$$\vec{\nabla} \times \vec{A} = curl\left(\vec{A}\right).$$

Example 15

For $\vec{A} = xy\vec{i} + xyz\vec{j} + y^3z\vec{k}$, find $curl\left(\vec{A}\right)$.

Solution

$$curl\left(\vec{A}\right) = \vec{\nabla} \times \vec{A} = \begin{vmatrix} \vec{i} & \vec{j} & \vec{k} \\ \dfrac{\partial}{\partial x} & \dfrac{\partial}{\partial y} & \dfrac{\partial}{\partial z} \\ A_x & A_y & A_z \end{vmatrix}$$

$$= \left[3y^2z - xy\right]\vec{i} + \left[yz - x\right]\vec{k}.$$

Theorem

If $\vec{A} = A_x\vec{i} + A_y\vec{j} + A_z\vec{k}$ is a vector field on $\mathbb{R}^3$, and A_x, A_y and A_z have continuous second-order partial derivatives, then

$$div(curl\left(\vec{A}\right)) = 0. \tag{3.52}$$

Proof

$$div(curl\left(\vec{A}\right)) = \vec{\nabla} \cdot \left(\vec{\nabla} \times \vec{A}\right) = \frac{\partial}{\partial x}\left[\frac{\partial A_z}{\partial y} - \frac{\partial A_y}{\partial z}\right] + \frac{\partial}{\partial y}\left[\frac{\partial A_x}{\partial z} - \frac{\partial A_z}{\partial x}\right] + \frac{\partial}{\partial z}\left[\frac{\partial A_y}{\partial x} - \frac{\partial A_x}{\partial y}\right] = \frac{\partial^2 A_z}{\partial x \partial y} - \frac{\partial^2 A_y}{\partial x \partial z} +$$

$$\frac{\partial^2 A_x}{\partial y \partial z} - \frac{\partial^2 A_z}{\partial y \partial x} + \frac{\partial^2 A_y}{\partial z \partial x} - \frac{\partial^2 A_x}{\partial z \partial y} \quad A_y$$

For a given vector field $\vec{F}(x,y,z) = f(x,y,z)\vec{i} + g(x,y,z)\vec{j} + h(x,y,z)\vec{k}$, let us state Stokes' theorem mostly used in fluid dynamics.

3.13.4 Theorem (Stokes' Theorem)

Consider S a piecewise smooth oriented surface, bounded with a closed piecewise smooth curve C with positive orientation. If the vector field is given by $\vec{F}(x,y,z) = f(x,y,z)\vec{i} + g(x,y,z)\vec{j} + h(x,y,z)\vec{k}$, where f, g and h have continuous first partial derivatives on some open set containing S, and if $z = y$ is a unit tangent vector to C, then

$$\oint_c \vec{F} \cdot \vec{t}ds = \int_S \left(curl\vec{F}\right) \cdot \vec{n}dA \tag{3.53}$$

In the above theorem, ds is a line element and dA is a surface element. This means that Stokes' theorem helps to change a line integral into a surface integral. The work done by a given force in physical science is a line integral, and the following example will help see how we can use Stokes' theorem.

Example 16

Find the work done by the force field $\vec{F}(x,y,z) = x^2\vec{i} + 4xy^3\vec{j} + xy^2\vec{k}$ on a particle moving through a rectangle C in the plane $z = y$, with $0 \le x \le 1$ and $0 \le y \le 3$.

Solution

The work done by the force $\vec{V}$ is the line integral $W = \oint_C \vec{F} \cdot d\vec{r}$ which can be changed into a surface

integral, that is: $W = \oint_C \vec{F} \cdot \vec{t} ds = \iint_S \left(curl\vec{F} \right) \cdot \vec{n} dA$, we will now use the determinant method to

evaluate $curl\vec{F}$ even though it will be explained in the next chapter.

We have

$$curl\vec{F} = \begin{vmatrix} \vec{i} & \vec{j} & \vec{k} \\ \dfrac{\partial}{\partial x} & \dfrac{\partial}{\partial y} & \dfrac{\partial}{\partial z} \\ x^2 & 4xy^3 & xy^2 \end{vmatrix} = 2xy\vec{i} - y^2\vec{j} + 4y^3\vec{k}$$

and the work done will be

$$W = \iint_S \left(curl\vec{F} \right) \cdot \vec{n} dA$$

$$W = \iint_{\text{Rectangle}} \left(curl\vec{F} \right) \cdot \left(\frac{\partial z}{\partial x}\vec{i} + \frac{\partial z}{\partial y}\vec{j} - \vec{k} \right) dA$$

$$W = \iint_{\text{Rectangle}} \left(2xy\vec{i} - y^2\vec{j} + 4y^3\vec{k} \right) \cdot \left(0\vec{i} + 1 \cdot \vec{j} - \vec{k} \right) dA$$

$$W = \int_0^1 \int_0^3 (-y^2 - 4y^3) dy dx$$

$$W = -90$$

3.13.5 THEOREM (GREEN'S THEOREM)

Let D be a closed region that is the union of a finite number of normal regions. D is a simply connected plane region whose boundary is simple, piecewise smooth curve C oriented counterclockwise. If $f(x,y)$ and $g(x,y)$ are continuous and have continuous first partial derivatives on some
open set of R, then

$$\int_C f(x,y)dx + g(x,y)dy = \iint_R \left(\frac{\partial g}{\partial x} - \frac{\partial f}{\partial y} \right) dA. \tag{3.54}$$

Example 17

Consider an ellipse C of equation, $x^2 + 4y^2 = 4$. Use Green's theorem to evaluate the integral
$I = \int_C \left(\cos x + 4xy \right) dx + \left(e^y + 2x^2 + 5x \right) dy.$

Solution

To apply Green's theorem, we need to identify the function $f(x,y) = \cos x + 4xy$ and the function $g(x,y) = e^y + 2x^2 + 5x$. Then, $\frac{\partial g}{\partial x} - \frac{\partial f}{\partial y} = 5$ and $I = \iint_R \left(\frac{\partial g}{\partial x} - \frac{\partial f}{\partial y} \right) dxdy = \iint_R 5dxdy = 5A$. The area of this ellipse is $A = \pi ab$; so $I = 5\pi \times 2 \times 1 = 10\pi$.

3.14   VECTOR FORM OF GREEN'S THEOREM

The curl and divergence operators allow us to write Green's theorem in another useful form which involved vectors. Consider the vector field $\vec{F} = f(x,y)\vec{i} + g(x,y)\vec{j}$ defined in the region D with a boundary curve C. We suppose the functions $f(x,y)$ and $g(x,y)$, the region D and its boundary curve C satisfy the hypotheses of Green's theorem as mentioned above. The line integral of $\vec{F}$ can be expressed as:

$$\int_C \vec{F} \cdot d\vec{r} = \oint_C fdx + gdy .$$

Considering $\vec{F}$ as a vector field on $\mathbb{R}^3$ with a third component zero, we have

$$curl\vec{F} = \begin{vmatrix} \vec{i} & \vec{j} & \vec{k} \\ \frac{\partial}{\partial x} & \frac{\partial}{\partial y} & \frac{\partial}{\partial z} \\ f(x,y) & g(x,y) & 0 \end{vmatrix} = \left(\frac{\partial g}{\partial x} - \frac{\partial f}{\partial y} \right)\vec{k}$$

So, if we multiply both side of the previous equality by $\vec{k}$, we get

$$curl\vec{F} \cdot \vec{k} = \left(\frac{\partial g}{\partial x} - \frac{\partial f}{\partial y} \right)\vec{k} \cdot \vec{k} = \frac{\partial g}{\partial x} - \frac{\partial f}{\partial y},$$

and the vector form of Green's theorem can be written

$$\int_C \vec{F} \cdot d\vec{r} = \iint_D curl\vec{F} \cdot \vec{k}dA \tag{3.55}$$

3.14.1   DIVERGENCE THEOREM

Let us consider surfaces that are boundaries of finite solids such as the surface of a solid sphere, the surface of a solid cylinder or solid box. These surfaces are said to be closed. A closed surface is not necessarily smooth. But we will consider surfaces that are piecewise smooth, that is, the surface can be divided in very small pieces of smooth surfaces. Each piece of surface can be assigned an inward orientation (toward the interior of the solid) or an outward orientation (away from the interior). Therefore, the surface is orientable piece by piece in such a way that the whole surface is orientable. The following theorem is known as the divergence theorem or Gauss' theorem:

3.14.2   THEOREM (DIVERGENCE THEOREM)

Consider a solid Σ whose surface S is oriented outward. If a field vector,

$\vec{F}(x,y,z) = f(x,y,z)\vec{i} + g(x,y,z)\vec{j} + h(x,y,z)\vec{k}$, where f, g and h have continuous first partial derivatives on some open set containing Σ, then

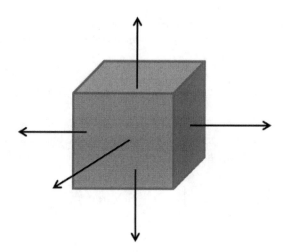

FIGURE 3.22 Outward orientation on a cubic box

$$\iint_S \vec{F} \cdot \vec{n}\, dA = \iint_\Sigma div\left(\vec{F}\right) d\tau \qquad (3.56)$$

with $\vec{n}$ the outward unit normal vector on S.

Example 18

Use the divergence theorem to find the outward flux of the vector field $\vec{F}(x,y,z) = z\vec{k}$ across the sphere of equation, $x^2 + y^2 + z^2 = R^2$.

Solution

The outward flux across the surface S is given by

$$\varphi_{\vec{F}} = \iint_S \vec{F} \cdot \vec{n}\, dA = \int_\Sigma div\left(\vec{F}\right) d\tau$$

$$div\left(\vec{F}\right) = \frac{\partial h}{\partial z} = \frac{\partial z}{\partial z} = 1$$

$$\varphi_{\vec{F}} = \iint_S \vec{F} \cdot \vec{n}\, dA = \iiint_\Sigma 1\, d\tau = \tau = \text{volume of solid } \Sigma = \frac{4}{3}\pi R^3$$

Example 19

Use the divergence theorem to find the outward flux of the vector field,

$\vec{F}(x,y,z) = x^3\vec{i} + y^3\vec{j} + z^2\vec{k}$ across the surface of origin that is enclosed by the circular cylinder $x^2 + y^2 = 9$ and the planes $z = 0$ and $z = 2$.

Solution

Let S be the outward oriented surface and Σ the region that it encloses. Let us find the divergence of the field vector:

$$div\vec{F} = \frac{\partial f}{\partial x} + \frac{\partial g}{\partial y} + \frac{\partial h}{\partial z} = 3x^2 + 3y^2 + 2z \cdot$$

The problem presents a cylindrical symmetry. We will use cylindrical coordinates. Moreover, the volume element in cylindrical coordinates is $= r\,dr\,d\theta\,dz$. So, the outward flux of the vector field will be

$$\varphi_{\vec{F}} = \iint_S \vec{F}\cdot\vec{n}\,dA = \iint_\Sigma (3x^2 + 3y^2 + 2z)r\,dr\,d\theta\,dz$$

The expression $3x^2 + 3y^2 + 2z$ can be changed into cylindrical coordinates to give $3(x^2 + 3y^2) + 2z = 2r^2 + 2z$, and the flux will be

$$\varphi_{\vec{F}} = \iiint (3r^2 + 2z)r\,dr\,d\theta\,dz$$

$$\varphi_{\vec{F}} = \int_0^{2\pi} \int_0^3 \int_0^2 \left(3r^2 + 2z\right)r\,dr\,d\theta\,dz$$

$$\varphi_{\vec{F}} = \int_0^{2\pi} \int_0^3 \left[3r^2 z + z^2\right]_0^2 r\,dr\,d\theta$$

$$\varphi_{\vec{F}} = \int_0^{2\pi} \int_0^3 \left[6r^2 + 4\right]r\,dr\,d\theta$$

$$\varphi_{\vec{F}} = \int_0^{2\pi} \int_0^3 \left[6r^3 + 4r\right]dr\,d\theta$$

$$\varphi_{\vec{F}} = \int_0^{2\pi} \int_0^3 \left[6\frac{r^4}{4} + 2r^2\right]_0^3 d\theta$$

$$\varphi_{\vec{F}} = \left[6\frac{81}{4} + 18\right]_0^3 d\theta$$

$$\varphi_{\vec{F}} = \int_0^{2\pi} \left[\frac{279}{2}\right]d\theta$$

$$\varphi_{\vec{F}} = 279\pi$$

3.14.3 Example of Gauss' Law for Inverse-Square Field

The divergence applied to the inverse-square fields gives rise to Gauss' law for the inverse-square field. An inverse-square field has the form, $F(\vec{r}) = \dfrac{c}{r^3}\vec{r}$ in a 3D space with c a constant number and $\vec{r}$ the position vector. If S is a closed orientable surface surrounding the origin, the outward flux of $F(\vec{r})$ across S is given by $\varphi_{\vec{F}} = \iint_S \vec{F}\cdot\vec{n}\,dA = 4\pi c$.

3.15 THE LAPLACIAN OPERATOR

The Laplacian is a differential operator of a scalar or vector field defined as follows:

$$\Delta V = \vec{\nabla}\left(\vec{\nabla}V\right) = \vec{\nabla}\ V \text{ and } \Delta\vec{A} = \nabla\left(\nabla\vec{A}\right) = \nabla\ \vec{A} \tag{3.57}$$

3.15.1 THE SCALAR LAPLACIAN AND THE VECTOR LAPLACIAN

3.15.1.1 The Scalar Laplacian in Cartesian Coordinates

Laplacian of a vector or scalar field is the sum of the second derivatives of the field with respect to each variable of the field. The Laplacian of a vector field is a vector field while the Laplacian of a scalar field is a scalar field.

Assume the scalar field $V(x,y,z)$. The Laplacian of $V(x,y,z)$ is

$$\Delta V = \frac{\partial^2 V}{\partial x^2} + \frac{\partial^2 V}{\partial y^2} + \frac{\partial^2 V}{\partial z^2} \tag{3.58}$$

3.15.1.2 The Vector Laplacian in Cartesian Coordinates

The Laplacian of a vector field $\vec{A}\left(A_x, A_y, A_z\right)$ where the components A_x, A_y, A_z are functions of x, y, z is given by

$$\Delta\vec{A} = \left(\Delta A_x\right)\vec{i} + \left(\Delta A_y\right)\vec{j} + \left(\Delta A_z\right)\vec{k} \tag{3.59}$$

which will give

$$\Delta\vec{A} = \vec{\nabla}\ \vec{A} = \begin{pmatrix} \Delta A_x = \dfrac{\partial\ A_x}{\partial x} + \dfrac{\partial\ A_x}{\partial y} + \dfrac{\partial\ A_x}{\partial z} \\[2mm] \Delta A_y = \dfrac{\partial\ A_y}{\partial x} + \dfrac{\partial\ A_y}{\partial y} + \dfrac{\partial\ A_y}{\partial z} \\[2mm] \Delta A_z = \dfrac{\partial\ A_z}{\partial x} + \dfrac{\partial\ A_z}{\partial y} + \dfrac{\partial\ A_z}{\partial z} \end{pmatrix} \tag{3.60}$$

3.16 CYLINDRICAL COORDINATES

Figure 3.23 depicts the cylindrical coordinates.

The Cartesian coordinates (x,y,z) and the cylindrical coordinates $\dfrac{1}{\phi}\dfrac{d^2\phi}{d\theta^2} = -m^2$ are related by

$$\begin{cases} x = r\cos\theta \\ y = r\sin\theta \\ z = z \end{cases} \tag{3.61}$$

Geometrically, using the Pythagorean theorem, we have

$$r^2 = x^2 + y^2 \text{ and } \tan\theta = \frac{y}{x}$$

3.16.1 GRADIENT CYLINDRICAL COORDINATES

The cylindrical coordinates of a given point M are r, θ, z. The gradient of a function $V(r,\theta,z)$ can be written in cylindrical coordinates as follows:

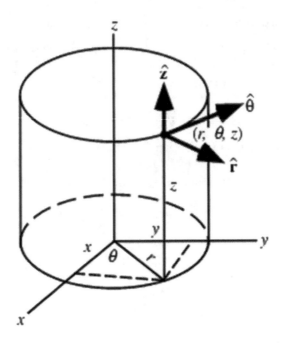

FIGURE 3.23 Cylindrical coordinates.

$$\overrightarrow{grad}\left(V\right) = \frac{\partial V}{\partial r}\,\vec{i} + \frac{1}{r}\frac{\partial V}{\partial \theta}\,\vec{j} + \frac{\partial V}{\partial z}\,\vec{k} \qquad (3.62)$$

3.16.2 Del Operator in Cylindrical Coordinates

A static (scalar) field u may be a function of cylindrical coordinates $\left(r,\theta,z\right)$. The value of u changes by an infinitesimal amount du when the point of observation is changed by the position vector, $d\vec{\rho}$

$$du = \frac{\partial u}{\partial r}\,dr + \frac{\partial u}{\partial \theta}\,d\theta + \frac{\partial u}{\partial z}\,dz \qquad (3.63)$$

We can also define the del operator $\vec{\nabla}$ in such a way to have the result $du = \vec{\nabla}\,u \cdot d\vec{\rho}$; The del operator acting on a given function is called the gradient of the function. The topic of the gradient of a function will be properly reviewed in this chapter. For instance, an inclement path $\vec{\rho}$ in the direction of dz is given in cylindrical coordinates by

$$\begin{aligned} d\vec{\rho} &= d\left(\hat{r}r + z\hat{z}\right) = \hat{r}dr + rd\hat{r} + \hat{z}dz + zd\hat{z} \\ &= \hat{z}dr + r\left(\frac{\partial z}{\partial r}\,dr + \frac{\partial z}{\partial \theta}\,d\theta + \frac{\partial z}{\partial z}\,dz\right) + \hat{z}dz + z\left(\frac{\partial z}{\partial r}\,dr + \frac{\partial z}{\partial \theta}\,d\theta + \frac{\partial z}{\partial z}\,dz\right) \\ &= \hat{r}dr + \hat{\theta}rd\theta + \hat{z}dz \end{aligned} \qquad (3.64)$$

So, $\dfrac{\partial u}{\partial r}\,dr + \dfrac{\partial u}{\partial \theta}\,d\theta + \dfrac{\partial u}{\partial z}\,dz = \vec{\nabla}u \cdot d\vec{\rho}$ or in cylindrical coordinates,

$$\frac{\partial u}{\partial r}\,dr + \frac{\partial u}{\partial \theta}\,d\theta + \frac{\partial u}{\partial z}\,dz = (\vec{\nabla}u)_r\,dr + (\vec{\nabla}u)_\theta\,rd\theta + (\vec{\nabla}u)_z\,dz \qquad (3.65)$$

For this relation to hold for any $dr, d\theta$ and dz, we need to have

$$\vec{\nabla} = r\frac{\partial}{\partial r} + \frac{\theta}{r}\frac{\partial}{\partial \theta} + z\frac{\partial}{\partial z}.$$ (3.66)

3.16.3 Divergence in Cylindrical Coordinates

In cylindrical coordinates, the divergence of a vector $\vec{A}$ is expressed as follows:

$$div\left(\vec{A}\right) = \vec{\nabla} \bullet \vec{A} = \frac{1}{r}\frac{\partial(rA_r)}{\partial r} + \frac{1}{r}\frac{\partial A_\theta}{\partial \theta} + \frac{\partial A_z}{\partial z}$$ (3.67)

3.16.4 Vector Laplacian

The components of the vector Laplacian in cylindrical coordinates are

$$\Delta\vec{A} = \begin{pmatrix} \Delta A_r - \frac{1}{r}\left(A_r + 2\frac{\partial A_\theta}{\partial \theta}\right) \\ \Delta A_\theta - \frac{1}{r}\left(A_\theta - 2\frac{\partial A_r}{\partial \theta}\right) \\ \Delta A_z \end{pmatrix}$$ (3.68)

Example 20
The velocity field of the flow is one of the most important variables in fluid mechanics. If the velocity field is given by $\vec{V} = u(r,t)\vec{i} + v(r,t)\vec{j} + w(r,t)\vec{k}$, where $\vec{r} = x\vec{i} + y\vec{j} + z\vec{k}$ is the position vector, what is the expression of the acceleration field?

Solution
Let us consider $u(r,t) = u(x,y,z,t)$, $v(r,t) = v(x,y,z,t)$, and $w(r,t) = w(x,y,z,t)$.
The acceleration field is given by the expression:

$$\vec{a} = \frac{d\vec{V}}{dt} = \frac{du(r,t)}{dt}\vec{i} + \frac{dv(r,t)}{dt}\vec{j} + \frac{dw(r,t)}{dt}\vec{k}.$$

We consider here that the unit vectors $\vec{i}$, $du = \vec{\nabla}u \cdot d\vec{\rho}$ and $\vec{k}$ are fixed and cannot change over time.

Let us now calculate $\frac{du(r,t)}{dt}$ by finding first the differential of $u(r,t)$.

$$du(r,t) = du(x,y,z,t) = \frac{\partial u}{\partial x}dx + \frac{\partial u}{\partial y}dy + \frac{\partial u}{\partial z}dz + \frac{\partial u}{\partial t}dt,$$ and by dividing $du(r,t)$ this by dt, we

have the following:

$$\frac{du(x,y,z,t)}{dt} = \frac{\partial u}{\partial x}\frac{dx}{dt} + \frac{\partial u}{\partial y}\frac{dy}{dt} + \frac{\partial u}{\partial z}\frac{dz}{dt} + \frac{\partial u}{\partial t} = \frac{\partial u}{\partial x}u + \frac{\partial u}{\partial y}v + \frac{\partial u}{\partial z}w + \frac{\partial u}{\partial t} = \frac{\partial u}{\partial t} + \left(\vec{V}.\vec{\nabla}\right)u$$

where $\vec{\nabla} = \frac{\partial}{\partial x}\vec{i} + \frac{\partial}{\partial y}\vec{j} + \frac{\partial}{\partial z}\vec{k}$.

In a similar way, we can find $\frac{dv(x,y,z,t)}{dt}$ and $\frac{dw(x,y,z,t)}{dt}$, and the acceleration vector field will be given by the expression:

$$\vec{\nabla} = \frac{\partial}{\partial x}\vec{i} + \frac{\partial}{\partial y}\vec{j} + \frac{\partial}{\partial z}\vec{k}.$$

The first term of the second side of the equality, $\dfrac{\partial \vec{V}}{\partial t}$, is called the local acceleration. For a steady flow, the local acceleration is zero. The second term of the sum is called the convective acceleration. The total derivative is also called the substantial or the material derivative.

3.16.4.1 Curl in Cylindrical Coordinates

In cylindrical coordinates, the curl of a vector $\vec{A}$ is expressed as follows:

$$curl\left(\vec{A}\right) = \frac{1}{r}\left[\frac{\partial A_z}{\partial \theta} - \frac{\partial A_\theta}{\partial z}\right]\vec{e}_r + \left[\frac{\partial A_r}{\partial z} - \frac{\partial A_z}{\partial r}\right]\vec{e}_\theta + \frac{1}{r}\left[\frac{\partial}{\partial r}(r.A_\theta) - \frac{\partial A_r}{\partial \theta}\right]\vec{e}_z \tag{3.69}$$

3.16.5 SCALAR LAPLACIAN

The Laplacian of a scalar function V in cylindrical coordinates can be written as follows:

$$\Delta V = (\vec{\nabla}V)^2 = \frac{1}{r}\frac{\partial}{\partial r}\left(r\frac{\partial V}{\partial r}\right) + \frac{1}{r^2}\frac{\partial^2 V}{\partial \theta^2} + \frac{\partial^2 V}{\partial z^2} \tag{3.70}$$

Proof

Using the expression of the operator $\vec{\nabla}$,

$$\vec{\nabla} = r\frac{\partial}{\partial r} + \frac{\theta}{r}\frac{\partial}{\partial \theta} + z\frac{\partial}{\partial z}.$$

we can have

$$(\vec{\nabla})^2 V = \vec{\nabla}\left(\vec{\nabla}V\right) = \left(r\frac{\partial}{\partial r} + \frac{\theta}{r}\frac{\partial}{\partial \theta} + z\frac{\partial}{\partial z}\right)\cdot\left(r\frac{\partial V}{\partial r} + \frac{\theta}{r}\frac{\partial V}{\partial \theta} + z\frac{\partial V}{\partial z}\right)$$

By developing the above expression, we obtain

$$\vec{\nabla}\left(\vec{\nabla}V\right) = \hat{r}\cdot\frac{\partial}{\partial r}\left(\hat{r}\frac{\partial V}{\partial r} + \frac{\hat{\theta}}{r}\frac{\partial V}{\partial \theta} + \hat{z}\frac{\partial V}{\partial z}\right)$$

$$+\frac{\hat{\theta}}{r}\cdot\frac{\partial}{\partial \theta}\left(\hat{r}\frac{\partial V}{\partial r} + \frac{\hat{\theta}}{r}\frac{\partial V}{\partial \theta} + z\frac{\partial V}{\partial z}\right)$$

$$+\hat{z}\cdot\frac{\partial}{\partial z}\left(\hat{r}\frac{\partial V}{\partial r} + \frac{\hat{\theta}}{r}\frac{\partial V}{\partial \theta} + z\frac{\partial V}{\partial z}\right)$$

$$\vec{\nabla}\left(\vec{\nabla}V\right) = \hat{r}\cdot\left(\hat{r}\frac{\partial^2 V}{\partial r^2} - \frac{\hat{\theta}}{r^2}\frac{\partial V}{\partial \theta} + \frac{\hat{\theta}}{r}\frac{\partial^2 V}{\partial \theta \partial r} + \hat{z}\frac{\partial^2 V}{\partial z \partial r}\right)$$

$$+\frac{\hat{\theta}}{r}\cdot\left(\hat{\theta}\frac{\partial V}{\partial r} + \hat{r}\frac{\partial^2 V}{\partial r \partial \theta} - \frac{\hat{r}}{r}\frac{\partial V}{\partial r} + \frac{\hat{\theta}}{r}\frac{\partial^2 V}{\partial \theta^2} + \hat{z}\frac{\partial^2 V}{\partial z \partial \theta}\right)$$

$$+\hat{z}\cdot\left(\hat{r}\frac{\partial^2 V}{\partial r \partial z} + \frac{\hat{\theta}}{r}\frac{\partial^2 V}{\partial \theta \partial z} + \hat{z}\frac{\partial^2 V}{\partial z}\right)$$

After reduction of the above expression, the Laplacian operator gives the expression:

$$\Delta V = (\vec{\nabla} V)^2 = \frac{1}{r}\frac{\partial}{\partial r}\left(r\frac{\partial V}{\partial r}\right) + \frac{1}{r^2}\frac{\partial^2 V}{\partial \theta^2} + \frac{\partial^2 V}{\partial z^2}$$

3.17   HELMHOLTZ EQUATION IN CYLINDRICAL COORDINATES

Helmholtz equation is described in Equation (2.27) by

$$\Delta \psi + k^2 \psi = 0$$

Let us write the unit vectors in cylindrical coordinates as function of the unit vectors of the rectangular coordinates.

$$\begin{aligned}
\hat{r} &= \frac{\vec{r}}{r} = \frac{x\hat{x} + y\hat{y}}{r} = \hat{x}\cos\theta + \hat{y}\sin\theta \\
\hat{\theta} &= \hat{z} \otimes \hat{r} = -\hat{x}\sin\theta + \hat{y}\cos\theta \\
\hat{z} &= \hat{z}
\end{aligned} \tag{3.71}$$

For the rest, let us consider a vector $\vec{\rho} = \vec{r} + \vec{z} = r\hat{r} + z\hat{z}$

3.17.1   VARIATIONS OF THE UNIT VECTORS WITH THE COORDINATES

Using the above expression, we can derive the following relationships:

$$\frac{\partial}{\partial r}\hat{r} = 0; \frac{\partial}{\partial r}\hat{\theta} = 0; \frac{\partial}{\partial r}\hat{z} = 0; \tag{3.72}$$

$$\frac{\partial}{\partial \theta}\hat{r} = -\sin\theta\,\hat{x} + \cos\theta\,\hat{y} = \hat{\theta}; \tag{3.73}$$

$$\frac{\partial}{\partial \theta}\hat{\theta} = -\hat{x}\cos\theta - \hat{y}\sin\theta = -\hat{r}; \tag{3.74}$$

$$\frac{\partial}{\partial \theta}\hat{z} = 0; \frac{\partial}{\partial z}\hat{r} = 0; \frac{\partial}{\partial z}\hat{\theta} = 0; \frac{\partial}{\partial z}\hat{z} = 0. \tag{3.75}$$

In cylindrical coordinates, the Laplacian operator is given by the following expression.

$$\Delta = (\vec{\nabla})^2 = \frac{1}{r}\frac{\partial}{\partial r}\left(r\frac{\partial}{\partial r}\right) + \frac{1}{r^2}\frac{\partial^2}{\partial \theta^2} + \frac{\partial^2}{\partial z^2}. \tag{3.76}$$

If we use the separation of variables method, that is, if we consider $\psi(r,\theta,z) = R(r)\phi(\theta)Z(z)$ between these three coordinates r,θ,z, the Helmholtz equation becomes:

$$\frac{1}{Rr}\frac{d}{dr}\left(r\frac{dR}{dr}\right) + \frac{1}{r^2\phi}\frac{d^2\phi}{d\theta^2} + \frac{1}{Z}\frac{d^2Z}{dz^2} + k^2 = 0 \tag{3.77}$$

We can isolate the expression containing Z by having

$$\frac{1}{Rr}\frac{\partial}{\partial r}\left(r\frac{\partial R}{\partial r}\right)+\frac{1}{r^2\varphi}\frac{\partial^2\varphi}{\partial\theta^2}+k^2=-\frac{1}{Z}\frac{\partial^2 Z}{\partial z^2}=-l^2 \qquad (3.78)$$

where l^2 represents an arbitrary constant.

In Equation (3.77), we assume $\frac{1}{Z}\frac{d^2 Z}{dz^2}=l^2$; this means that there is a hyperbolic motion on the z-axis.

Also, multiply both sides of Equation (3.78) by r^2, we obtain

$$\frac{r}{R}\frac{d}{dr}\left(r\frac{dR}{dr}\right)+\frac{1}{\varphi}\frac{d^2\varphi}{d\theta^2}+r^2 k^2+l^2 r^2=0 \qquad (3.79)$$

Let us now assume $\frac{1}{\phi}\frac{d^2\phi}{d\theta^2}=-m^2$, which means that there is an oscillatory motion on the polar direction described by the coordinate θ.

By embedding the constant m^2 in Equation (3.79), we get

$$\frac{r}{R}\frac{d}{dr}\left(r\frac{dR}{dr}\right)-m^2+r^2\left(k^2+l^2\right)=0 \qquad (3.80)$$

Consider now $n^2=k^2+l^2$, we obtain

$$\frac{r}{R}\frac{d}{dr}\left(r\frac{dR}{dr}\right)-m^2+r^2 n^2=0 \qquad (3.81)$$

By multiplying both sides of Equation (3.81) by R, we get

$$r\frac{d}{dr}\left(r\frac{dR}{dr}\right)+\left(n^2 r^2-m^2\right)R=0 \qquad (3.82)$$

Equation (3.82) is called Bessel's differential equation which is a 3D partial differential equation. Note that the Helmholtz equation has been transformed in three ordinary differential equations which are

$$\frac{1}{Z}\frac{d^2 Z}{dz^2}=l^2 \qquad (3.83)$$

$$\frac{1}{\phi}\frac{d^2\phi}{d\theta^2}=-m^2 \qquad (3.84)$$

$$r\frac{d}{dr}\left(r\frac{dR}{dr}\right)+\left(n^2 r^2-m^2\right)R=0 \qquad (3.85)$$

Finally, the most general solution of the Helmholtz partial differential equation (PDE) (2.27) in cylindrical coordinates may be written:

$$\psi\left(r,\theta,z\right)=\Sigma_{m,n}a_{m,n}R_{m,n}\left(r\right)\phi_m\left(\theta\right)Z_n\left(z\right) \qquad (3.86)$$

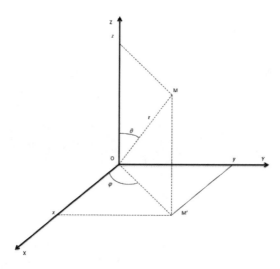

FIGURE 3.24   Spherical coordinates $r\ \theta\ \phi$ of a point M with Cartesian coordinates x, y, z .

3.18   SPHERICAL COORDINATES

In this section, we will first define the spherical coordinates as in Figure 3.24.

Consider like in physics a position vector $\vec{r}$ with Cartesian coordinates x, y, z. Then we have $\vec{r} = x\vec{i} + y\vec{j} + z\vec{k}$, with $\vec{i}$, $\vec{j}$ and $\vec{k}$ the unit vectors on the rectangular Ox, Oy, Oz axis. We can express the coordinates x, y, z as follows:

$$\begin{cases} x = r\cos\phi\sin\theta \\ y = r\sin\phi\sin\theta \\ z = r\cos\theta \end{cases} \tag{3.87}$$

Using the chain rule, we have

$$\frac{\partial}{\partial x} = \left(\frac{\partial r}{\partial x}\right)_{y,z}\left(\frac{\partial}{\partial r}\right)_{\theta,\phi} + \left(\frac{\partial \theta}{\partial x}\right)_{y,z}\left(\frac{\partial}{\partial \theta}\right)_{r,\phi} + \left(\frac{\partial \phi}{\partial x}\right)_{y,z}\left(\frac{\partial}{\partial \phi}\right)_{r,\theta}$$

$$\frac{\partial}{\partial y} = \left(\frac{\partial r}{\partial y}\right)_{x,z}\left(\frac{\partial}{\partial r}\right)_{\theta,\phi} + \left(\frac{\partial \theta}{\partial y}\right)_{x,z}\left(\frac{\partial}{\partial \theta}\right)_{r,\phi} + \left(\frac{\partial \phi}{\partial y}\right)_{x,z}\left(\frac{\partial}{\partial \phi}\right)_{r,\theta}$$

$$\frac{\partial}{\partial z} = \left(\frac{\partial r}{\partial z}\right)_{x,y}\left(\frac{\partial}{\partial r}\right)_{\theta,\phi} + \left(\frac{\partial \theta}{\partial z}\right)_{x,y}\left(\frac{\partial}{\partial \theta}\right)_{r,\phi} + \left(\frac{\partial \phi}{\partial z}\right)_{x,y}\left(\frac{\partial}{\partial \phi}\right)_{r,\theta}$$

Substituting the partial derivatives of r with respect to x, y and z , we have the following:

$$\frac{\partial}{\partial x} = \left(\sin\theta\cos\phi\right)\left(\frac{\partial}{\partial r}\right)_{\theta,\phi} + \left(\frac{\cos\theta\cos\phi}{r}\right)\left(\frac{\partial}{\partial \theta}\right)_{r,\phi} + \left(-\frac{\sin\phi}{r\sin\theta}\right)\left(\frac{\partial}{\partial \phi}\right)_{r,\theta}$$

$$\frac{\partial}{\partial y} = \left(\sin\theta\sin\phi\right)\left(\frac{\partial}{\partial r}\right)_{\theta,\phi} + \left(\frac{\cos\theta\sin\phi}{r}\right)\left(\frac{\partial}{\partial \theta}\right)_{r,\phi} + \left(\frac{\cos\phi}{r\sin\theta}\right)\left(\frac{\partial}{\partial \phi}\right)_{r,\theta}$$

$$\frac{\partial}{\partial z} = \left(\cos\theta\right)\left(\frac{\partial}{\partial r}\right)_{\theta,\phi} - \left(\frac{\sin\theta}{r}\right)\left(\frac{\partial}{\partial \theta}\right)_{r,\phi} \quad \text{with} \quad \left(\frac{\partial \phi}{\partial z}\right)_{x,y} = 0$$

The second partial derivatives will give

$$\frac{\partial^2}{\partial z^2} = (\cos\theta)\left[\frac{(\cos\theta)\left(\frac{\partial}{\partial r}\right)_{\theta,\phi} - \left(\frac{\sin\theta}{r}\right)\left(\frac{\partial}{\partial\theta}\right)_{r,\phi}}{\partial r} \right]_{\theta,\phi}$$

$$- \left(\frac{\sin\theta}{r}\right)_{x,y}\left[\frac{(\cos\theta)\left(\frac{\partial}{\partial r}\right)_{\theta,\phi} - \left(\frac{\sin\theta}{r}\right)\left(\frac{\partial}{\partial\theta}\right)_{r,\phi}}{\partial\theta} \right]_{r,\phi}$$

$$\frac{\partial^2}{\partial y^2} = (\sin\theta\sin\phi)\times\left[\frac{(\sin\theta\sin\phi)\left(\frac{\partial}{\partial r}\right)_{\theta,\phi} + \left(\frac{\cos\theta\sin\phi}{r}\right)\left(\frac{\partial}{\partial\theta}\right)_{r,\phi} + \left(\frac{\cos\phi}{r\sin\theta}\right)\left(\frac{\partial}{\partial\phi}\right)_{r,\theta}}{\partial r} \right]_{\theta,\phi}$$

$$+ \left(\frac{\cos\theta\sin\phi}{r}\right)\times\left[\frac{(\sin\theta\sin\phi)\left(\frac{\partial}{\partial r}\right)_{\theta,\phi} + \left(\frac{\cos\theta\sin\phi}{r}\right)\left(\frac{\partial}{\partial\theta}\right)_{r,\phi} + \left(\frac{\cos\phi}{r\sin\theta}\right)\left(\frac{\partial}{\partial\phi}\right)_{r,\theta}}{\partial\theta} \right]_{r,\phi} + \left(\frac{\cos\phi}{r\sin\theta}\right)$$

$$\times\left[\frac{(\sin\theta\sin\phi)\left(\frac{\partial}{\partial r}\right)_{\theta,\phi} + \left(\frac{\cos\theta\sin\phi}{r}\right)\left(\frac{\partial}{\partial\theta}\right)_{r,\phi} + \left(\frac{\cos\phi}{r\sin\theta}\right)\left(\frac{\partial}{\partial\phi}\right)_{r,\theta}}{\partial\phi} \right]_{r,\theta}$$

and

$$\frac{\partial^2}{\partial x^2} = (\sin\theta\sin\phi)$$

$$\times\left[\frac{(\sin\theta\cos\phi)\left(\frac{\partial}{\partial r}\right)_{\theta,\phi} + \left(\frac{\cos\theta\cos\phi}{r}\right)\left(\frac{\partial}{\partial\theta}\right)_{r,\phi} + \left(\frac{\sin\phi}{r\sin\theta}\right)\left(\frac{\partial}{\partial\phi}\right)_{r,\theta}}{\partial r} \right]_{\theta,\phi}$$

$$+ \left(\frac{\cos\theta\sin\phi}{r}\right)\times\left[\frac{(\sin\theta\cos\phi)\left(\frac{\partial}{\partial r}\right)_{\theta,\phi} + \left(\frac{\cos\theta\cos\phi}{r}\right)\left(\frac{\partial}{\partial\theta}\right)_{r,\phi} + \left(\frac{\sin\phi}{r\sin\theta}\right)\left(\frac{\partial}{\partial\phi}\right)_{r,\theta}}{\partial\theta} \right]_{r,\phi} + \left(\frac{\cos\phi}{r\sin\theta}\right)$$

$$\times\left[\frac{(\sin\theta\cos\phi)\left(\frac{\partial}{\partial r}\right)_{\theta,\phi} + \left(\frac{\cos\theta\cos\phi}{r}\right)\left(\frac{\partial}{\partial\theta}\right)_{r,\phi} + \left(\frac{\sin\phi}{r\sin\theta}\right)\left(\frac{\partial}{\partial\phi}\right)_{r,\theta}}{\partial\phi} \right]_{r,\theta}$$

After addition and elimination, the Cartesian coordinates, the Laplacian $\Delta = \dfrac{\partial^2}{\partial x} + \dfrac{\partial^2}{\partial y} + \dfrac{\partial^2}{\partial z}$ becomes in spherical coordinates:

$$\Delta = \frac{1}{r^2}\frac{\partial}{\partial r}\left(r^2\frac{\partial}{\partial r}\right) + \frac{1}{r\,\sin\,\theta}\left[\sin\theta\frac{\partial}{\partial\theta}\left(\sin\theta\frac{\partial}{\partial\theta}\right) + \frac{\partial^2}{\partial\phi^2}\right] \qquad (3.88)$$

Let us now express the scalar Laplacian and the vector Laplacian in spherical coordinates.

3.18.1 SCALAR LAPLACIAN

Using the expression of the Laplacian proved in the above section, the scalar Laplacian of V will be given by

$$\Delta = \frac{1}{r^2}\frac{\partial}{\partial r}\left(r^2\frac{\partial}{\partial r}\right) + \frac{1}{r\,\sin\,\theta}\left[\sin\theta\frac{\partial}{\partial\theta}\left(\sin\theta\frac{\partial}{\partial\theta}\right) + \frac{\partial^2}{\partial\phi^2}\right] \qquad (3.88)$$

3.18.2 VECTOR LAPLACIAN

The vector Laplacian of a vector $\vec{A}$ is given in spherical coordinates by

$$\Delta\vec{A} = \begin{pmatrix} \Delta A_r - \dfrac{2}{r\,\sin\,(\theta)}\left(A_r\cdot\sin(\theta) + \dfrac{\partial}{\partial\theta}\big(A_\theta\sin(\theta)\big) + \dfrac{\partial A_\phi}{\partial\phi}\right) \\[3mm] \Delta A_\theta - \dfrac{2}{r\,\sin\,(\theta)}\left(\dfrac{A_\theta}{2} - \sin\,(\theta)\cdot\dfrac{\partial A_r}{\partial\theta} + \cos(\theta)\cdot\dfrac{\partial A_\phi}{\partial\phi}\right) \\[3mm] \Delta A_\phi - \dfrac{2}{r\,\sin\,(\theta)}\left(\dfrac{A_\phi}{2} - \sin\,(\theta)\cdot\dfrac{\partial A_r}{\partial\phi} - \cos(\theta)\cdot\dfrac{\partial A_\phi}{\partial\phi}\right) \end{pmatrix} \qquad (3.90)$$

3.18.3 INTRINSIC DEFINITION

The intrinsic definition of a scalar Laplacian is given by the expression:

$$\Delta V = div\left(\overrightarrow{grad}(V)\right) \qquad (3.91)$$

The intrinsic definition of a vector Laplacian is given by the expression:

$$\Delta\vec{A} = \overrightarrow{grad}\left(div\left(\vec{A}\right)\right) - \overrightarrow{rot}\left(\overrightarrow{rot}\left(\vec{A}\right)\right) \qquad (3.92)$$

3.18.4 GRADIENT IN SPHERICAL COORDINATES

The spherical coordinates of a given point M are r,θ,ϕ. The gradient of a function $V\left(r,\theta,\phi\right)$ can be written in spherical coordinates as follows:

$$\overrightarrow{grad}\left(V\right) = \frac{\partial V}{\partial r}\vec{e}_r + \frac{1}{r}\frac{\partial V}{\partial\theta}\vec{e}_\theta + \frac{1}{r.\sin(\theta)}\frac{\partial V}{\partial\phi}\vec{e}_\phi \qquad (3.93)$$

3.18.5 SPHERICAL COORDINATES

In spherical coordinates, the divergence of a vector $\vec{A}$ is expressed as follows:

$$div\left(\vec{A}\right) = \frac{1}{r^2}\frac{\partial\left(r^2 A_r\right)}{\partial r} + \frac{1}{r\times\sin\theta}\frac{\partial\left(\sin\theta\times A_\theta\right)}{\partial\theta} + \frac{1}{r\times\sin\theta}\frac{\partial A_\phi}{\partial\phi} \qquad (3.94)$$

3.18.6 CURL IN SPHERICAL COORDINATES

In spherical coordinates the curl of a vector $\vec{A}$ is expressed as follows:

$$\overrightarrow{rot}\left(\vec{A}\right) = \frac{1}{r.\sin\theta}\left[\frac{\partial}{\partial\theta}\left(\sin\theta.A_\phi\right) - \frac{\partial A_\theta}{\partial\phi}\right]\vec{e}_r + \frac{1}{r}\left[\frac{1}{\sin\theta}\frac{\partial A_r}{\partial\phi} - \frac{\partial}{\partial r}\left(r.A_\phi\right)\right]\vec{e}_\theta + \frac{1}{r}\left[\frac{\partial}{\partial r}\left(r.A_\theta\right) - \frac{\partial A_r}{\partial\theta}\right]\vec{e}_\phi$$

$$(3.95)$$

3.18.7 IMPORTANT RELATIONS WITH THE DEL OPERATORS

If $\vec{A}$ and $\vec{B}$ are differentiable vector fields, and if
$\varphi = \varphi\left(x,y,x\right)$ and $\psi\left(x,y,x\right)$ are differential scalar fields, the following identities are true:

$$\vec{\nabla}\left(\varphi+\psi\right) = \vec{\nabla}\varphi + \vec{\nabla}\psi \; or \; \overrightarrow{grad}\left(\varphi+\psi\right) = \overrightarrow{grad}\varphi + \overrightarrow{grad}\psi \qquad (3.96)$$

$$\vec{\nabla}.\left(\vec{A}+\vec{B}\right) = \vec{\nabla}.\vec{A} + \vec{\nabla}.\vec{B} \; or \; div\left(\vec{A}+\vec{B}\right) = div\vec{A} + div\vec{B} \qquad (3.97)$$

$$\vec{\nabla}\times\left(\vec{A}+\vec{B}\right) = \vec{\nabla}\times\vec{A} + \vec{\nabla}\times\vec{B} \; or \; curl\left(\vec{A}+\vec{B}\right) = curl\left(\vec{A}\right) + curl\left(\vec{B}\right) \qquad (3.98)$$

$$\vec{\nabla}.\left(\varphi\vec{A}\right) = \left(\vec{\nabla}\varphi\right).\vec{A} + \varphi\left(\vec{\nabla}.\vec{A}\right) \qquad (3.99)$$

$$\vec{\nabla}\times\left(\varphi\vec{A}\right) = \left(\vec{\nabla}\phi\right)\times\vec{A} + \varphi\left(\vec{\nabla}\times\vec{A}\right) \qquad (3.100)$$

$$\vec{\nabla}.\left(\vec{A}\times\vec{B}\right) = \vec{B}.\left(\vec{\nabla}\times\vec{A}\right) - \vec{A}.\left(\vec{\nabla}\times\vec{B}\right) \qquad (3.101)$$

$$\vec{\nabla}\times\left(\vec{A}\times\vec{B}\right) = \left(\vec{B}.\vec{\nabla}\right)\vec{A} - \vec{B}\left(\vec{\nabla}.\vec{A}\right) - \left(\vec{A}.\vec{\nabla}\right)\vec{B} + \vec{A}\left(\vec{\nabla}.\vec{B}\right) \qquad (3.102)$$

$$\vec{\nabla}\left(\vec{A}.\vec{B}\right) = \left(\vec{B}.\vec{\nabla}\right)\vec{A} + \left(\vec{A}.\vec{\nabla}\right)\vec{B} + \vec{B}\times\left(\vec{\nabla}\times\vec{A}\right) + \vec{A}\times\left(\vec{\nabla}\times\vec{B}\right) \qquad .(3.103)$$

$$\vec{\nabla}.\left(\vec{\nabla}\varphi\right) \equiv \vec{\nabla}^2\varphi \equiv \frac{\partial^2\varphi}{\partial x^2} + \frac{\partial^2\varphi}{\partial y^2} + \frac{\partial^2\varphi}{\partial z^2}. \qquad (3.104)$$

$$\vec{\nabla}\times\left(\vec{\nabla}\varphi\right) = 0 \qquad (3.105)$$

$$\vec{\nabla}.\left(\vec{\nabla}\times\vec{A}\right)=0. \tag{3.106}$$

$$\vec{\nabla}\times\left(\vec{\nabla}\times\vec{A}\right)=\vec{\nabla}\left(\vec{\nabla}.\vec{A}\right)-\vec{\nabla}^2\vec{A} \tag{3.107}$$

3.19 HELMHOLTZ EQUATION IN SPHERICAL COORDINATES

A position vector $\vec{r}$ can be represented in spherical coordinates as follows:

The Cartesian coordinates (x,y,z) and the spherical coordinates (r,θ,ϕ) of the point P are related by:

$$\begin{cases} x = r\cos\phi\sin\theta \\ y = r\sin\phi\sin\theta \\ z = r\cos\theta \end{cases} \tag{3.108}$$

Also,

$$r^2 = x^2 + y^2 + z^2; \phi = \tan^{-1}\left(\frac{y}{x}\right); \theta = \cos^{-1}\left(\frac{z}{r}\right). \tag{3.109}$$

By taking the partial derivatives of x,y,z with respect to r, we obtain

$$\frac{\partial x}{\partial r} = \cos\phi\sin\theta, \frac{\partial y}{\partial r} = \sin\phi\sin\theta, \frac{\partial z}{\partial r} = \cos\theta. \tag{3.110}$$

By taking the partial derivatives of x,y,z with respect to θ, we get

$$\frac{\partial x}{\partial r} = \cos\phi\sin\theta, \frac{\partial y}{\partial r} = \sin\phi\sin\theta, \frac{\partial z}{\partial r} = \cos\theta. \tag{3.111}$$

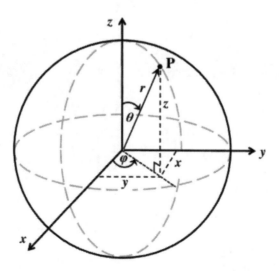

FIGURE 3.25 Spherical coordinates of a point P.

Taking the partial derivatives of x,y,z with respect to ϕ gives

$$\frac{\partial x}{\partial \phi} = -r\sin\phi\sin\theta, \ \ \frac{\partial y}{\partial \phi} = r\cos\phi\sin\theta, \ \ \frac{\partial z}{\partial \phi} = 0. \tag{3.112}$$

Let us now use the chain rule of differentiation.

$$\frac{\partial}{\partial r} = \frac{\partial x}{\partial r}\frac{\partial}{\partial x} + \frac{\partial y}{\partial r}\frac{\partial}{\partial y} + \frac{\partial z}{\partial r}\frac{\partial}{\partial z} = \cos\phi\sin\theta\frac{\partial}{\partial x} + \sin\phi\sin\theta\frac{\partial}{\partial y} + \cos\theta\frac{\partial}{\partial z} \tag{3.113}$$

$$\frac{\partial}{\partial \theta} = \frac{\partial x}{\partial \theta}\frac{\partial}{\partial x} + \frac{\partial y}{\partial \theta}\frac{\partial}{\partial y} + \frac{\partial z}{\partial \theta}\frac{\partial}{\partial z} = r\cos\phi\cos\theta\frac{\partial}{\partial x} + r\sin\phi\cos\theta\frac{\partial}{\partial y} - r\sin\theta\frac{\partial}{\partial z} \tag{3.114}$$

$$\frac{\partial}{\partial \phi} = \frac{\partial x}{\partial \phi}\frac{\partial}{\partial x} + \frac{\partial y}{\partial \phi}\frac{\partial}{\partial y} + \frac{\partial z}{\partial \phi}\frac{\partial}{\partial z} = -r\sin\phi\sin\theta\frac{\partial}{\partial x} + r\cos\phi\sin\theta\frac{\partial}{\partial y}. \tag{3.115}$$

The above equations can be inverted to give

$$\frac{\partial}{\partial x} = \cos\phi\sin\theta\frac{\partial}{\partial r} + \frac{\cos\phi\cos\theta}{r}\frac{\partial}{\partial \theta} - \frac{\sin\phi}{r\sin\theta}\frac{\partial}{\partial \phi}$$
$$\frac{\partial}{\partial x} = \sin\phi\sin\theta\frac{\partial}{\partial r} + \frac{\sin\phi\cos\theta}{r}\frac{\partial}{\partial \theta} - \frac{\cos\phi}{r\sin\theta}\frac{\partial}{\partial \phi} \tag{3.116}$$
$$\frac{\partial}{\partial z} = \cos\theta\frac{\partial}{\partial r} - \frac{\sin\theta}{r}\frac{\partial}{\partial \theta}$$

Let us now expand the unit vectors $\hat{r}$, $\hat{\theta}$ and $\hat{\phi}$ in the direction of increasing r, θ and ϕ, respectively, in terms of the Cartesian unit vectors $\hat{x}$, $\hat{y}$ and $\hat{z}$:

$$\hat{r} = \cos\phi\sin\theta\hat{x} + \sin\phi\sin\theta\hat{y} + \cos\theta\hat{z}$$
$$\hat{\theta} = \cos\phi\cos\theta\hat{x} + \sin\phi\cos\theta\hat{y} - \sin\theta\hat{z} \tag{3.117}$$
$$\hat{\phi} = -\sin\phi\hat{x} + \cos\phi\hat{y}$$

The above three equations can be inverted to give

$$\hat{x} = \cos\phi\sin\theta\hat{r} + \cos\phi\cos\theta\hat{\theta} - \sin\theta\hat{\phi}$$
$$\hat{y} = \sin\phi\sin\theta\hat{r} + \sin\phi\cos\theta\hat{\theta} + \cos\phi\hat{\phi} \tag{3.118}$$
$$\hat{z} = \cos\theta\hat{r} - \sin\theta\hat{\theta}$$

We can now define the vector differential operator $\vec{\nabla}$ as follows:

$$\vec{\nabla} = \hat{x}\frac{\partial}{\partial x} + \hat{y}\frac{\partial}{\partial y} + \hat{z}\frac{\partial}{\partial z} \ \text{in Cartesian coordinates and}$$

$$\vec{\nabla} = \hat{r}\frac{\partial}{\partial r} + \hat{\theta}\frac{1}{r}\frac{\partial}{\partial \theta} + \hat{\phi}\frac{1}{r\sin\theta}\frac{\partial}{\partial \phi} \ \text{in spherical coordinates.}$$

From (3.117) we can express the partial derivatives:

$$\frac{\partial}{\partial r}\hat{r} = 0; \quad \frac{\partial}{\partial r}\hat{\theta} = 0 \quad \frac{\partial}{\partial r}\hat{\phi} = 0.$$

$$\frac{\partial}{\partial \theta}\hat{r} = \hat{\theta}; \quad \frac{\partial}{\partial \theta}\hat{\theta} = -\hat{r} \quad \frac{\partial}{\partial \theta}\hat{\phi} = 0. \tag{3.119}$$

$$\frac{\partial}{\partial \phi}\hat{r} = -\sin\theta\hat{\phi}; \quad \frac{\partial}{\partial \phi}\hat{\theta} = \cos\theta\hat{\phi} \quad \frac{\partial}{\partial \phi}\hat{\phi} = -\sin\theta\hat{r} - \cos\theta\hat{\theta}.$$

Using equations (3.119) in $\vec{\nabla}$, we can find the Laplacian operator in spherical coordinates:

$$\begin{aligned}
\Delta &= \vec{\nabla} \cdot \vec{\nabla} \\
&= \frac{1}{r^2}\frac{\partial}{\partial r}\left(r^2\frac{\partial}{\partial r}\right) + \frac{1}{r^2\sin\theta}\frac{\partial}{\partial\theta}\left(\sin\theta\frac{\partial}{\partial\theta}\right) + \frac{1}{r^2\sin^2\theta}\frac{\partial^2}{\partial\varphi^2}
\end{aligned} \tag{3.120}$$

And the Helmholtz differential equation becomes

$$\frac{1}{r^2}\frac{\partial}{\partial r}\left(r^2\frac{\partial\psi}{\partial r}\right) + \frac{1}{r^2\sin\theta}\frac{\partial}{\partial\theta}\left(\sin\theta\frac{\partial\psi}{\partial\theta}\right) + \frac{1}{r^2\sin^2\theta}\frac{\partial^2\psi}{\partial\varphi^2} + k^2\psi = 0 \tag{3.121}$$

We now use the separation of variables:

$$\psi(r,\theta,\varphi) = R(r)\Theta(\theta)\Phi(\varphi) \tag{3.122}$$

The Helmholtz differential equation can then be written with the partial derivatives replaced by the ordinary derivatives:

$$\frac{1}{Rr^2}\frac{d}{dr}\left(r^2\frac{dR}{dr}\right) + \frac{1}{\Theta r^2\sin\theta}\frac{d}{d\theta}\left(\sin\theta\frac{d\Theta}{d\theta}\right) + \frac{1}{\Phi r^2\sin^2\theta}\frac{\partial^2\Phi}{\partial\varphi^2} + k^2 = 0 \tag{3.123}$$

By multiplying both sides of the equality by $r^2\sin\theta$ we obtain:

$$r^2\sin^2\theta\left[-\frac{1}{Rr^2}\frac{d}{dr}\left(r^2\frac{dR}{dr}\right) - \frac{1}{\Theta r^2\sin\theta}\frac{d}{d\theta}\left(\sin\theta\frac{d\Theta}{d\theta}\right) - k^2\right] = \frac{1}{\Phi}\frac{\partial^2\Phi}{\partial\varphi^2} \tag{3.124}$$

We now consider an oscillatory solution on the Azimuthal coordinates:

$$\frac{1}{\Phi}\frac{\partial^2\Phi}{\partial\varphi^2} = -m^2 \tag{3.125}$$

Also, the equation becomes

$$\frac{1}{R}\frac{d}{dr}\left(r^2\frac{dR}{dr}\right) + k^2r^2 = -\frac{1}{\Theta\sin\theta}\frac{d}{d\theta}\left(\sin\theta\frac{d\Theta}{d\theta}\right) + \frac{m^2}{\sin^2\theta} = f \tag{3.126}$$

with m and f constant values.

We can then separate the equations as follows:

$$\frac{1}{\sin\theta}\frac{d}{d\theta}\left(\sin\theta\frac{d\Theta}{d\theta}\right)+\left(f-\frac{m^2}{\sin^2\theta}\right)\Theta=0 \tag{3.127}$$

$$\frac{1}{r^2}\frac{d}{dr}\left(r^2\frac{dR}{dr}\right)+\left(k^2-\frac{f}{r^2}\right)R=0 \tag{3.128}$$

Finally, we have changed the Helmholtz PDE in three ordinary differential equation (ODEs) by using the method of separation of variables. If k^2 is a positive constant, Equation (3.128) is called the spherical Bessel equation. Equation (3.127) however can be identified as the associated Legendre equation. In quantum mechanics, the constant f is replaced by $l(l+1)$.

Finally, the most general solution of the Helmholtz equation (2.27) in spherical coordinates is given by

$$\psi_{fm}(r,\theta,\phi)=\sum_{f,m}a_{f,m}R_f(r)\Theta(\theta)\Phi_m(\phi). \tag{3.129}$$

3.19.1 ANALYSIS OF THE SOLUTION TO EQUATION (3.125)

In Equation (3.125), the Azimuthal coordinate Φ is a periodic function of period $2\pi/|m|$ for $|m|>1$ and can be written $\Phi(\phi)=A\cos(\phi)+B\sin(\phi)$ or $\Phi(\phi)=Ce^{im\phi}+De^{-im\phi}$, where A, B, C and D are constant values.

3.20 VECTOR INTEGRATION

3.20.1 VECTOR FIELD AND LINE INTEGRAL

Line integrals are just like single integrals with the difference that instead of integrating over an interval $[a,b]$, we integrate over a curve C. Line integrals were invented in the early 19th century to solve problems involving fluids, flow of matter, forces, electricity and magnetism. Since these subjects are studied in physics and engineering, their introduction is of great importance in this section. Consider a plane curve C with the following parametric equations:

$$x=x(t)\qquad y=y(t)\qquad a\le t\le b \tag{3.130}$$

The plane curve can also be defined by the vector equation $r(t)=x(t)\vec{i}+y(t)\vec{j}$. We assume that C is a smooth curve that is $r(t)$ is continuous and its first derivative $r'(t)\ne0$. Let us now divide the interval $[a,b]$ into n equal width subintervals $[t_{i-1},t_i]$. Also, consider $x_i=x(t_i)$ and $y_i=y(t_i)$, then the corresponding points $P_i(x_i,y_i)$ divide the curve C into n subarcs with lengths Δs_1, Δs_2, ..., Δs_n. For every point $P_i^*(x_i^*,y_i^*)$, we can evaluate the corresponding function $f(x_i,y_i)$ and form the sum $\sum_{i=1}^{n}f(x_i^*,y_i^*)\Delta s_i$ which is similar to a Riemann sum. We can now take the limit of this sum and define the line integral as follows:

3.20.2 DEFINITION

If f is defined on a smooth curve C given by Equation (3.130), then the line integral of f along C is

$$\int_C f(x,y)\,ds=\lim_{n\to\infty}\sum_{i=1}^{n}f(x_i^*,y_i^*)\Delta s_i \tag{3.131}$$

if this limit exists.

From previous calculus courses, we remember that the length of a smooth curve with parametric equations $x = f(t)$, $y = \sin t$ with f is given by the formula:

$$L = \int_a^b \sqrt{\left(\frac{dx}{dt}\right)^2 + \left(\frac{dy}{dt}\right)^2}\, dt \tag{3.132}$$

But, for a curve of equation, $y = f(x)$ with $a \leq x \leq b$, L becomes

$$L = \int_a^b \sqrt{1 + \left(\frac{dy}{dx}\right)^2}\, dx. \tag{3.133}$$

As a result, the following formula can also be used to calculate the line integral:

$$\int_C f(x,y)\, ds = \int_a^b f(x(t), y(t)) \sqrt{\left(\frac{dx}{dt}\right)^2 + \left(\frac{dy}{dt}\right)^2}\, dt \tag{3.134}$$

The value of the line integral does not depend on the parameterization of the curve, provided that the curve is traversed exactly once as t increases from a to b.

Example 21

Evaluate $\int_C \left(2 + x^2 y\right) ds$ along the upper half of a unit circle C of equation $x^2 + y^2 = 1$.

Solution

To solve this question, we would need to consider the parametric equation to represent C; such parametric equation can be as follows:

$x = \cos t$ and $y = \sin t$. In addition, for the upper half of the circle, we have $0 \leq t \leq \pi$. We will consider Figure 3.26.

The integral becomes

$$\int_C \left(2 + x^2 y\right) ds = \int_a^b \left(2 + \cos^2 t \sin t\right) \sqrt{\left(\frac{dx}{dt}\right)^2 + \left(\frac{dy}{dt}\right)^2}\, dt$$

$$= \int_a^b \left(2 + \cos^2 t \sin t\right) \sqrt{(-\sin t)^2 + (\cos t)^2}\, dt$$

$$= \int_a^b \left(2 + \cos^2 t \sin t\right) dt$$

$$= \left[2t - \frac{1}{3}\cos^3 t\right]_0^\pi$$

$$= 2\pi + \frac{2}{3}.$$

Example 22 (Center of Mass of a Wire)

A wire has the shape of a semicircle and is thicker near its base than near the top. The equation of the semicircle is

$$x^2 + y^2 = 1 \text{ with } y \geq 0.$$

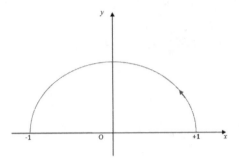

FIGURE 3.26 Semicircle of equation $x^2 + y^2 = 1$ and $y \geq 0$.

Find the center of mass of the wire if its linear density is ρ at any point, $M(x,y)$, is proportional to its distance from the line $y = 1$.

Solution

To solve this question, let us use the parametrization: $x = \cos t$ and $y = \sin t$ with $0 \leq t \leq \pi$. Also, noticed that $ds = \sqrt{x'^2 + y'^2}\, dt = dt$ and the linear density can be written $\rho(x,y) = k(1-y)$, where k is a constant; with $\rho(x,y) = \dfrac{dm}{ds}$, the mass of the wire is $m = \displaystyle\int_C \rho(x,y)\, ds = \int_C k(1-y)\, ds$

$$m = \int_0^\pi k(1-\sin t)\, dt = k\left[t + \cos t \right]_0^\pi = k(\pi - 2).$$

The coordinates of the center of mass are $\left(\bar{x}, \bar{y} \right)$ such that

$$\bar{y} = \frac{1}{m}\int_C y\rho(x,y)\, ds = \frac{1}{k(\pi-2)}\int_C yk(1-y)\, ds$$

$$\bar{y} = \frac{1}{(\pi-2)}\int_0^\pi \sin t (1-\sin t)\, dt = \frac{1}{(\pi-2)}\left[\int_0^\pi \sin t\, dt - \frac{1}{2}\int_0^\pi (1-\cos 2t)\, dt \right]$$

$$= \frac{1}{(\pi-2)}\left\{ \left[-\cos t \right]_0^\pi - \frac{1}{2}\left[t - \frac{1}{2}\sin 2t \right]_0^\pi \right\} = \frac{4-\pi}{2(\pi-2)}.$$

By symmetry, the x-component of the center of mass is zero: $\bar{x} = 0$. Finally, the center of mass is given by the point $\left(0, \dfrac{4-\pi}{2(\pi-2)} \right)$.

Let's now go back to the definition of line integral above and substitute Δs_i by either $\Delta x_i = x_i - x_{i-1}$ or $\Delta y_i = y_i - y_{i-1}$; as a result, we can define the line integral of f along the curve C with respect to x and y, respectively, by

$$\int_C f(x,y)\, dx = \lim_{n\to\infty} \sum_{i=1}^n f(x_i^*, y_i^*)\Delta x_i \qquad (3.135)$$

if the limit exists for all possible choices of (x_i^*, x_y^*)

and with the same previous condition,

$$\int_C f(x,y)\, dy = \lim_{n\to\infty} \sum_{i=1}^n f(x_i^*, y_i^*)\Delta y_i \qquad (3.136)$$

If on the other hand, we express the previous formulas in terms of t, we have

$$\int_C f(x,y)\,dx = \int_a^b f[x(t),y(t)]x'(t)\,dt \tag{3.137}$$

and

$$\int_C f(x,y)\,dx = \int_a^b f[x(t),y(t)]y'(t)\,dt \tag{3.138}$$

In most cases, line integrals occur with respect to x and y together. In such cases, it is useful to separate by using the integral addition property:

$$\int_C \left[P(x,y)\,dx + Q(x,y)\,dy\right] = \int_C P(x,y)\,dx + \int Q(x,y)\,dy \tag{3.139}$$

3.20.3 LINE INTEGRAL IN SPACE

Consider a smooth space curve C with the parametric equations:

$x = x(t), y = y(t), z = z(t)$, where $t \in [a,b]$. C could also be represented by the vector equation: $\vec{r} = x(t)\vec{i} + y(t)\vec{j} + z(t)\vec{k}$. Assume f a function of three variables x, y, and z, continuous on some region C; we can define the line integral of f along C by a formula like (3.135):

$$\int_C f(x,y,z)\,ds = \lim_{n \to \infty} \sum_{i=1}^{n} f\left(x_i^*, y_i^*, z_i^*\right)\Delta s_i \tag{3.140}$$

To evaluate the integral given in (3.140), one can use the expression like (3.134),

$$\int_C f(x,y,z)\,ds = \int_a^b f\left(x(t),y(t),z(t)\right)\sqrt{\left(\frac{dx}{dt}\right)^2 + \left(\frac{dy}{dt}\right)^2 + \left(\frac{dz}{dt}\right)^2}\,dt \tag{3.141}$$

It is worthy of note to observe that the above integral can also be written in a more condensed form:

$$\int_C f(x,y,z)\,ds = \int_a^b f\left(\vec{r}(t)\right)\left|\vec{r}'(t)\right|\,dt \tag{3.142}$$

Using the example of a special case, $f(\vec{r}) = f(x,y,z) = 1$, we have $\displaystyle\int_C ds = \int_a^b \left|\vec{r}'(t)\right|\,dt = L = $ length of the curve C.

Example 23

Evaluate the integral, $I = \displaystyle\int_C y\sin z\,ds$ where C is the circular helix given by the equations $x = \cos t\, y = \sin t\, z = t$ where $0 \le t \le 2\pi$.

Solution

To solve the above example, let us use the formula in Equation (3.141):

$$\int_C f(x,y,z)\,ds = \int_l^h f\left(x(t),y(t),z(t)\right)\sqrt{\left(\frac{dx}{dt}\right)^2 + \left(\frac{dy}{dt}\right)^2 + \left(\frac{dz}{dt}\right)^2}\,dt$$

We then have

$$I = \int_C y\sin z\, ds = \int_0^{2\pi} \sin\sin t \sin t\sqrt{(-\sin t)^2 + (\sin t)^2 + 1}\, dt$$

$$I = \frac{\sqrt{2}}{2}\int_0^{2\pi}\left((1-\cos 2t)\right)dt = \frac{\sqrt{2}}{2}\left[t - \frac{1}{2}\sin 2t\right]_0^{2\pi} = \pi\sqrt{2}.$$

3.20.3.1 Line Integral of a Vector Field

Let us consider a fluid flowing through a region. Each point of the fluid has a speed in each direction called velocity. The velocity is a vector quantity. The flow of each point of the fluid gives rise to a flow of vector functions because at each point of the region, there is a well-defined vector. Such a vector function is also called a vector field. If the velocity changes over time, then the function or field is called dynamic function. If the vector function does not change over time, then it is referred to as a static vector function or static vector field. For instance, we consider a plane region with a static field. The vector function can then be written in the form, $\vec{F} = M(x,y)\vec{i} + N(x,y)\vec{j}$. For example, the equation $\vec{F} = xy^2\vec{i} - xy\vec{j}$ is a vector function because at each point (x,y) can be defined a given vector, $\vec{F}$. A field of forces can also be described as a vector field. The force can be an electric force, a gravitational force or a magnetic force. In the case of the gravitational force, if a particle under such field can move on a curve $\vec{F} = xy^2\vec{i} - xy\vec{j}$, a distance $d\vec{r}$, then the elementary work done by the force on the particle is $dw = \vec{F}\cdot d\vec{r}$. The total work done will be given by the line integral:

$$dw = \vec{F}\cdot d\vec{r} \tag{3.143}$$

3.20.4 Definition

Let $\vec{F}$ be a continuous vector field defined on a smooth curve C given by a vector function $\vec{r}(t)$ and $a \le t \le b$. The line integral of $\vec{F}$ along the curve $a \le t \le b$ is given by

$$\int_C \vec{F}\, d\vec{r} = \int_a^b F\left(\vec{r}(t)\right)\cdot\vec{r}'(t)\, dt = \int_C \vec{F}\cdot\vec{T}\, ds$$

where $\vec{T}$ is the unit tangent vector at point (x,y,z) on the curve C. Notice that with $\vec{r}(t) = x(t)\vec{i} + y(t)\vec{j} + z(t)\vec{k}$, (x,y,z) is given by $\vec{T} = \dfrac{\vec{r}'(t)}{\left|\vec{r}'(t)\right|}$.

In general, the line integral of a vector quantity $\vec{V}$ is represented by

$$\int\vec{V}\cdot d\vec{r} \text{ or } \int\vec{V}\times d\vec{r}. \tag{3.144}$$

For a scalar quantity $\varphi = \varphi(x,y,z)$, the following relation is satisfied:

$$\int\varphi(x,y,z)d\vec{r} = \vec{i}\int\varphi(x,y,z)dx + \vec{j}\int\varphi(x,y,z)dy + \vec{k}\int\varphi(x,y,z)dz \tag{3.145}$$

The line integral of the scalar quantity $\varphi(x,y,z)$ is represented by the integral $\int\varphi(x,y,z)\cdot d\vec{r}$.

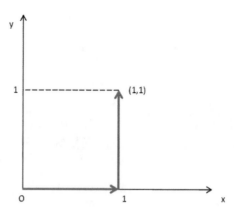

FIGURE 3.27   A path of integration.

Example 24

The work done by a force $\vec{F} = f(x,y,z)\vec{i} + g(x,y,z)\vec{j} + h(x,y,z)\vec{k}$ whose point of application moves by the displacement, $d\vec{r}$, is given in physics by

$$w = \int Fdr = \int f(x,y,z)dx + \int g(x,y,z)dy + \int h(x,y,z)dz \qquad (3.146)$$

Assume the vector force $\vec{F}(x,y,z) = -y\vec{i} + x\vec{j}$;
   The work done by the force is given by

$$w = \int \vec{F}(x,y,z)d\vec{r} = \int -ydx + \int xdy$$

However,

$$\forall x \in [0,1], y = 0 \text{ and } \forall y \in [0,1], x = 1, \ w = \int 0dx + \int 1dy = 1$$

Example 25

Evaluate the line integral $w = \int \left(7y^2\vec{i} - 4xy\vec{j}\right) \cdot d\vec{r}$ over the arc Γ that joins the points $(0,0)$ and $(2,4)$
$\Gamma : \vec{r} = t\vec{i} + 2t\vec{j}$.

Solution

For $\vec{r} = t\vec{i} + 2t\vec{j}$, we know that $x = t$ and $y = 2t$

$$d\vec{r} = dt\vec{i} + 2dt\vec{j}$$

So $w = \int \vec{F}d\vec{r} = \int_0^2 \left(7y^2 - 8xy\right) \cdot dt = \int^2 \left[7(2t)^2 - 8t \cdot 2t\right] \cdot dt$

$$w = \int_0^2 \left[28t^2 - 16t^2\right] \cdot dt = w = \int_0^2 12t^2 \cdot dt = 4t^3\Big|_0^2 = 32$$

3.21   SURFACE INTEGRAL

Surface integrals are generalization of the line integrals over a surface in a 3D space. Surface integrals have applications in continuous media such as solids, fluids and gases. They are also used in the force field such as the electromagnetic field and the gravitational field.

3.21.1 Parametric Surface

Let us consider a vector function $\vec{r}(u,v)$ where u and S are two parameters as follows:

$$\vec{r}(u,v) = x(u,v)\vec{i} + y(u,v)\vec{j} + z(u,v)\vec{k}. \tag{3.147}$$

Assume a domain D formed with the system (u,v). Consider D to be a rectangle which we divide in subrectangles R_{ij} with dimensions Δu and Δv.

Each subrectangle R_{ij} has a surface S_{ij}. Consider a function of three variables f whose domain is D. We will define the surface integral of f over the surface S of u in such a way that, in the case where $f(x,y,z)=1$, the value of the surface integral is equal to the surface area of S. The value of f at each point P_{ij} of the surface S_{ij} is P_{ij} and if the limit exists, we can define the surface integral of f over the surface S by the relation:

$$\iint_S f(x,y,z)\,dS = \lim_{m,n\to\infty} \sum_{i=1}^{m}\sum_{j=1}^{n} f(P_{ij})\Delta S_{ij} \tag{3.148}$$

If we define the tangent vectors at the corner of each S_{ij} as:

$\vec{r}_u = \dfrac{\partial x}{\partial u}\vec{i} + \dfrac{\partial y}{\partial u}\vec{j} + \dfrac{\partial z}{\partial u}\vec{k}$ and $\vec{r}_v = \dfrac{\partial x}{\partial v}\vec{i} + \dfrac{\partial y}{\partial v}\vec{j} + \dfrac{\partial z}{\partial v}\vec{k}$, and use the approximation $\Delta S_{ij} \approx |\vec{r}_u \times \vec{r}_v|\Delta u\Delta v$,

then it can be shown that

$$\iint_S f(x,y,z)\,dS = \iint_D f(\vec{r}(u,v))|\vec{r}_u \times \vec{r}_v|\,dA \tag{3.149}$$

where the components of $\vec{r}_u$ and $\vec{r}_v$ are nonzero and nonparallel in the interior of the domain D.

In the case where $z=g(x,y)$ is a parametric surface with parametric equations: $x=x$, $y=y$, $z=g(x,y)$, we will have

$$\vec{r}_x = \frac{\partial x}{\partial x}\vec{i} + \frac{\partial y}{\partial x}\vec{j} + \frac{\partial z}{\partial x}\vec{k} = \vec{i} + \left(\frac{\partial g}{\partial x}\right)\vec{k} \tag{3.150}$$

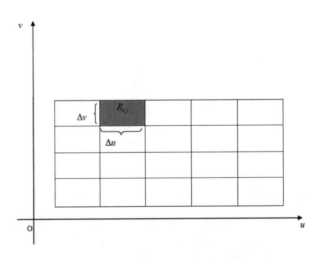

FIGURE 3.28 Domain subdivision subrectangle R_{ij}.

$$\vec{r}_x = \frac{\partial x}{\partial x}\vec{i} + \frac{\partial y}{\partial x}\vec{j} + \frac{\partial z}{\partial x}\vec{k} = \vec{i} + \left(\frac{\partial g}{\partial x}\right)\vec{k} \tag{3.151}$$

and therefore,

$$\vec{r}_x \times \vec{r}_y = -\frac{\partial g}{\partial x}\vec{i} + \frac{\partial g}{\partial y}\vec{j} + \vec{k} \tag{3.152}$$

and

$$\left|\vec{r}_x \times \vec{r}_y\right| = \sqrt{\left(\frac{\partial z}{\partial x}\right)^2 + \left(\frac{\partial z}{\partial y}\right)^2 + 1} \tag{3.153}$$

As a result, formula (3.149) for the surface integral becomes

$$\iint_S f(x,y,z)\,dS = \iint_D f\left(\vec{r}(u,v)\right)\sqrt{\left(\frac{\partial z}{\partial x}\right)^2 + \left(\frac{\partial z}{\partial y}\right)^2 + 1}\,dA \tag{3.154}$$

Example 26

Evaluate $\iint_S y\,dS$, where S is the surface $z = x + y^2$ with $0 \le x \le 1, 0 \le y \le 2$.

Solution

$$\frac{\partial z}{\partial x} = 1, \frac{\partial z}{\partial y} = 2y$$

$$\iint_S y\,dS = \iint_D y\sqrt{\left(\frac{\partial z}{\partial x}\right)^2 + \left(\frac{\partial z}{\partial y}\right)^2 + 1}\,dy\,dx$$

$$= \iint_D y\sqrt{2 + 4y^2}\,dy\,dx$$

$$= \sqrt{2}\int_0^1 dx \int_0^2 y\sqrt{1 + 2y^2}$$

$$= \frac{13\sqrt{2}}{3}$$

3.21.2 ORIENTED SURFACE

Let us start with a surface S that has a tangent plane at each of its point (x, y, z) except at the boundary points. There are two-unit normal vectors, $\vec{n}_1$ and $\vec{n}_2$, associated with S such that $\vec{n}_2 = -\vec{n}_1$.

If it is possible to identify a unit normal vector at (x, y, z) of the surface S so that $\vec{n}$ varies continuously, S is referred to as oriented surface. In this case, it is easy to see that $\vec{n}$ gives an orientation to S.

If the surface is represented by the graph given by $z = g(x, y)$, then the unit normal vector will be written:

$$\vec{n} = \frac{\vec{r}_x \times \vec{r}_y}{\left|\vec{r}_x \times \vec{r}_y\right|} = \frac{-\frac{\partial g}{\partial x}\vec{i} + \frac{\partial g}{\partial y}\vec{j} + \vec{k}}{\sqrt{1 + \left(\frac{\partial g}{\partial x}\right)^2 + \left(\frac{\partial g}{\partial y}\right)^2}}. \tag{3.155}$$

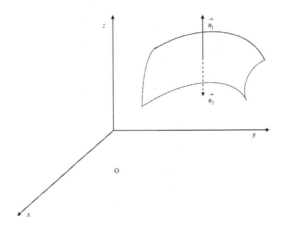

FIGURE 3.29   Oriented surface unit normal vectors, $\vec{n}_1$ and $\vec{n}_2$.

If S is a smooth orientable surface given in the parametric form by a vector function $\vec{r}(u,v)$, then the unit normal vector becomes $\vec{n} = \dfrac{\vec{r}_u \times \vec{r}_v}{\left|\vec{r}_u \times \vec{r}_v\right|}$.

3.21.3   SURFACE INTEGRALS OF VECTOR FIELDS

Assume $\vec{F}$ is a continuous vector field defined on an oriented surface S with unit normal vector $\vec{n}$, then the surface integral of $\vec{F}$ over S is expressed as

$$\varphi_{\vec{F}} = \iint_S \vec{F} \cdot d\vec{S} = \iint_S \left(\vec{F} \cdot \vec{n}\right) dS \tag{3.156}$$

This integral is also called the flux of $\vec{F}$ accros the surface area S.

If the surface $\vec{S}$ is given by the vector function $\vec{r}(u,v)$, then

$\vec{n} = \dfrac{\vec{r}_u \times \vec{r}_v}{\left|\vec{r}_u \times \vec{r}_v\right|}$ and

$$\varphi_{\vec{F}} = \iint_S \vec{F} \cdot d\vec{S}$$

$$= \iint_S \left(\vec{F} \cdot \frac{\vec{r}_u \times \vec{r}_v}{\left|\vec{r}_u \times \vec{r}_v\right|}\right) dS = \iint_D \left(\vec{F}\left(\vec{r}(u,v)\right) \cdot \frac{\vec{r}_u \times \vec{r}_v}{\left|\vec{r}_u \times \vec{r}_v\right|}\right) \cdot \left|\vec{r}_u \times \vec{r}_v\right| dS$$

With D the parameter of the domain. Also, we have

$$\varphi_{\vec{F}} = \iint_S \vec{F} \cdot d\vec{S} = \iint_D \vec{F} \cdot \left(\vec{r}_u \times \vec{r}_v\right) .dS \tag{3.157}$$

In the case, $\vec{F} = P\vec{i} + Q\vec{j} + R\vec{k}$ and the surface S is given by the graph $z = g(x,y)$, the surface integral becomes

$$\varphi_{\vec{F}} = \iint_S \vec{F} \cdot d\vec{S} = \iint_D \left(-P\frac{\partial g}{\partial x} - Q\frac{\partial g}{\partial y} + R\right).dS \tag{3.158}$$

The latest formula assumes the upward orientation of the surface. The opposite will give the downward orientation.

Example 27

Calculate the surface integral $\iint_S \vec{F} \cdot d\vec{S}$ for $\vec{F} = y\vec{i} + x\vec{j} + z\vec{k}$ and the surface S is the boundary of the solid region E enclosed by the paraboloid $z = 1 - x^2 - y^2$ and the plane $z = 0$

Solution

Let us use the convention of positive orientation, that is, outward orientation. The surface S consists of a parabolic top surface S_1 and circular bottom surface S_2. Assume D the projection of S_1 on the xy-plane, that is, the disk $x^2 + y^2 \leq 1$.

We have $P = y$, $Q = x$, $R = z = 1 - x^2 - y^2$, $\dfrac{\partial g}{\partial x} = -2x$ and $\dfrac{\partial g}{\partial y} = -2y$.

$$
\begin{aligned}
\varphi_{\vec{F}} &= \iint_S \vec{F} \cdot d\vec{S} = \iint_D \left(-P\frac{\partial g}{\partial x} - Q\frac{\partial g}{\partial y} + R \right).dS \\
&= \iint_D \left(1 + 4xy - x^2 - y^2 \right).dS \\
&= \int_0^{2\pi} \int_0^1 \left(1 + 4r^2\cos\theta\sin\theta - r^2 \right) r\,dr\,d\theta \\
&= \frac{\pi}{2}
\end{aligned}
$$

3.22 VOLUME INTEGRAL

To define the volume integral, let us start from an example. Consider a cube of side 4. Suppose that one corner of the cube is at the origin of axis and the other adjacent corners on the x, y and z axes. If the density of the cube is $f(x, y, z)$, let us find the mass of the cube. We know that $f(x, y, z)$, that is $f(x, y, z) = \dfrac{dm}{d\tau}$, where $d\tau$ is the volume element, that is, a very small part of the volume of the cube. Basically here, $d\tau = dxdydz$ and the mass of the cube will be

$$
m = \iint_{\text{volume}} f(x, y, z)d\tau \quad \text{with } dx, dy, dz. \tag{3.159}
$$

For $f(x, y, z) = 2z\text{gram} / \text{cm}^3$, we have

$$
\begin{aligned}
m &= \int_0^4 \int_0^4 \int_0^4 2z\,dx\,dy\,dz \\
&= \int_0^4 dx \int_0^4 dy.16 = 4 \times 4 \times 16 = 256. \\
m &= 256 \text{ grams.}
\end{aligned}
$$

In general, for a volume element $d\tau = dxdydz$ in a 3D rectangular space, the volume integral of a function $f(x, y, z)$ is given by

$$
\iint_{\text{Volume}} f(x, y, z)d\tau \tag{3.160}
$$

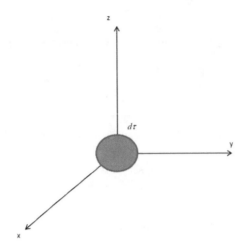

FIGURE 3.30   Volume element, $\vec{a}$ in a 3D rectangular space.

PROBLEM SET 3

3.1   Consider the vectors $\vec{a}, \vec{b}, \vec{c}$ shown below; geometrically represent the following sums: $\vec{a}+\vec{b}; \vec{b}+\vec{c}$.

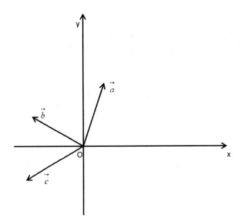

FIGURE 3.31   Vectors $\vec{a}, \vec{x}+\vec{y}, \vec{c}$ in a plane.

3.2   A particle with positive charge $q = 3.20 \times 10^{-19}\text{C}$ moves with a velocity $\vec{v} = \left(2\vec{i} + 3\vec{j} - \vec{k}\right)$m/s through a region where both a uniform magnetic field $\vec{B}$ and a uniform electric field $\vec{E}$ exist.

   a.   Express the vector cross product $\vec{v} \times \vec{B}$.

   b.   Calculate the magnitude of the total force $\vec{F}$ on the moving particle, considering

$$\vec{B} = \left(2\vec{i} + 4\vec{j} + \vec{k}\right)T \text{ and } \vec{E} = \left(4\vec{i} - \vec{j} - 2\vec{k}\right)\text{V/m}.$$

   c.   What angle does the force vector make with the positive x-axis?

3.3   Find the divergence and the curl of each of these vector fields

   a. $\vec{F}\left(x,y,z\right) = 2x^2 y\vec{i} + y^3 z\vec{j} - 3z\vec{k}$

b. $\vec{F}(x,y,z) = 2x\vec{i} + xyz\vec{j} + yz^2\vec{k}$

c. $\vec{F}(x,y) = \dfrac{y}{y^2x^2+1}\vec{i} + \dfrac{x}{y^2x^2+1}\vec{j} + z\vec{k}$

3.4 Find the divergence and the rotational of the vector field

a. $\vec{E}(x,y,z) = x^2\vec{i} + y^2\vec{j} + z^2\vec{k}$

b. $\vec{B}(x,y,z) = -\cos x\vec{i} + \sin y\vec{j} - (z\cos y + z\sin x)\vec{k}$

3.5 Prove that the divergence of the inverse-square field $\vec{F}(x,y,z) = \dfrac{k}{\left(x^2+y^2+z^2\right)^{3/2}}$

$\left(x\vec{i} + y\vec{j} + z\vec{k}\right)$ is zero.

3.6 Evaluate the line integral along the curve C.

$\int_C \left(x^2 - y^2\right)dx + 2xdy$

$C: x = t^{2/3}, y = t, -1 \leq t \leq 1.$

3.7 Consider the vector field $\vec{F}(x,y) = \left(ye^{xy'} - 1\right)\vec{i} + xe^{xy}\vec{j}$

a. Show that $\vec{F}$ is a conservative force
b. Find the potential function for $\vec{F}$
c. Find the work performed by the force field F on a particle that moves along the saw-tooth curve represented by the parametric equations:

3.8 Consider the following three vectors:

$\vec{A} = 2\vec{i} + 3\vec{j} - 5\vec{k}, \vec{B} = -3\vec{i} + 3\vec{j} - \vec{k}, \vec{C} = \vec{i} + \vec{j} + \vec{k}$

Express $\vec{A}\times\vec{B}, \vec{A}\times\vec{C}, \vec{B}\times\vec{B}, \vec{B}\times\vec{C}, \vec{A}\times\left(\vec{B}\times\vec{C}\right), \vec{A}\cdot\left(\vec{B}\times\vec{C}\right).$

3.9 A uniformly charged insulating rod of length 14.0 cm is bent into the shape of a semi-circle, as shown in Figure 3.32. The rod has a total charge of $-7.50\ \mu C$. Find the magnitude and the direction of the electric field at O, the center of the semicircle.

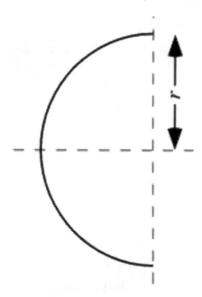

FIGURE 3.32 The uniformly charged semicircle.

3.10   Consider a long cylindrical charge distribution of radius R with a uniform charge density ρ. Find the electric field at distance r from the axis where $r < R$.

3.11   Three charged particles are at the vertices of an isosceles triangle in Figure 3.33. Calculate the electric potential at midpoint of the base, taking $q = -7.00\,\mu C$ $q = -7.00\,\mu C$.

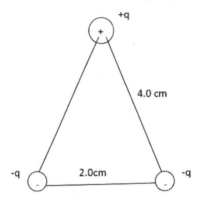

FIGURE 3.33   The three charged particles at the corner of a triangle.

3.12   Consider the following electrical circuit composed of resistors of resistances 8Ω, 1Ω, 5Ω, 4Ω, and two generators of voltages 4V and 12V.

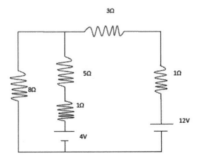

FIGURE 3.34   The electric network.

Write Kirchhoff rules and change the system of equations obtained in matrix form. Solve the system using the matrix inversion technique to determine the current in each branch.

3.13   An infinitely long, thin-walled circular cylinder of radius b is split into half cylinders (see Figure 3.35). The upper half is fixed at potential $V = +V_0$ and the lower half at $V = -V_0$

   a.   Use Laplace equation in spherical coordinates to find the potential inside and outside the cylinder.

     Hint. Prove that these potentials are:

$$\Phi \frac{4V_0}{\pi} \sum_{n=0}^{\infty} \left(\frac{r}{b}\right)^{2n+1} \frac{\sin(2n+1)\varphi}{2n+1}\Bigg|_{int} \quad \text{and}$$

$$\Phi_{ext} = \frac{4V_0}{\pi} \sum_{n=0}^{\infty} \left(\frac{b}{r}\right)^{2n+1} \frac{\sin(2n+1)\varphi}{2n+1}.$$

   b.   Calculate the surface charge density as a function of the angle φ

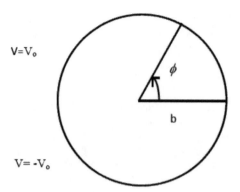

FIGURE 3.35   The uniformly long circular cylinder of radius b.

3.14   Evaluate the line integral along the indicated curve from P and Q

$$\int_{\Gamma} 3y^2 dx + 2xy dy; \ \Gamma : y^2 = 9x; \ P(0,0), \ Q(1,3)$$

3.15   Evaluate the line integral along the indicated curve from P and Q

$$\int_{\Gamma} x dx + y dy + z dz; \ r = ti - tj + 2tk; \ P(0,0,0), \ Q(2,4,6)$$

3.16   Use Green's theorem to evaluate the given line integral over a given curve.

$$\oint (x+y)dx + (x-y)dy$$

C is any circle.

3.17   Use Green's theorem to evaluate the given line integral over a given curve.

$$\oint (y^3 + 1)dx + 3xy^2 dy \quad C : x^2 + y^2 = 10.$$

3.18   Consider the vectors $\vec{u} = (5,-4)$, $\vec{v} = (3,2)$

    a.   Find $\vec{a} = \vec{u} + \vec{v}$; $\vec{b} = \vec{u} - \vec{v}$; $\vec{a} + \vec{b}$.

    b.   Represent $\vec{u}, \vec{a}, \vec{a}, \vec{b}$.

3.19   a.   Determine all vectors $\vec{v}$ that are orthogonal to $\vec{u}$

$$\vec{u} = (3,1); \ \vec{u} = (1,0); \ \vec{u} = (-2,-1,1); \ \vec{u} = (0,1,-2,0)$$

    b.   Determine whether $\vec{u}$ and $\vec{v}$ are orthogonal or parallel

$$\vec{u} = (-2,18); \ \vec{v} = \left(\frac{3}{2}, \frac{1}{6} \right)$$

$$\vec{u} = (-5,-1,0); \ \vec{v} = (1,-5,-2).$$

    c.   Find the angle θ between the vectors

$$\vec{u} = (-1,1) \text{ and } \vec{v} = (2,1)$$

$$\vec{u} = \left(\cos \frac{3\pi}{6}, \sin \frac{3\pi}{6} \right) \text{ and } \vec{v} = \left(\cos \frac{\pi}{6}, \sin \frac{\pi}{6} \right)$$

$$\vec{u} = (-1,0,2,0,1) \text{ and } \vec{v} = (-2,2,1,3).$$

4 Matrix Algebra

The scientist does not study nature because it is useful; he studies it because he delights in it, and he delights in it because it is beautiful.

–Henri Poincaré

INTRODUCTION

Matrix appears in a broad range of physical concepts. A system of equations of several unknowns can be changed into a matrix form, which makes it easy to solve. Matrix is used in special theory of relativity, in quantum and classical mechanics. They are also used in data analysis where many arrays can be transformed in an $m \times n$ matrix. In this chapter, after the basic definitions of matrices, we will develop matrix operations, the transpose of a matrix and matrix form of a system of linear equations, matrix inverse and determinant of a square matrix and diagonal and triangular matrices.

4.1 BASIC DEFINITION

An $m \times n$ (read m by n) matrix A is a rectangular array of real or complex numbers with m rows and n columns. The row is also called a line. The real or complex numbers in the matrix are called entries or coefficients. The entry at the intercept of the ith row and the jth column is designated by a_{ij}. For example, a_{12} is the coefficient located at the intercept of the first row and the second column.

Example 1
The following matrix A has four rows and five columns; It is a 4×5 matrix.

$$A = \begin{bmatrix} 1 & 2 & 3 & 4 & 5 \\ -2 & 4 & 0 & 1 & -3 \\ 0 & 3 & 6 & 4 & 8 \\ -1 & 3 & -5 & 2 & 1 \end{bmatrix} \tag{4.1}$$

4.2 MATRICES OPERATIONS

4.2.1 THE TRANSPOSE OF A MATRIX

The transpose, A^T, of a matrix A is the matrix obtained by changing the rows of A in columns. If A is an $m \times n$ matrix and $B = A^T$, then B is the $n \times m$ matrix with $b_{ij} = a_{ji}$.

Example 2

$$\begin{bmatrix} 0 & 5 & 7 \\ 4 & 1 & -1 \\ 0 & 0 & 12 \end{bmatrix}^T = \begin{bmatrix} 0 & 4 & 0 \\ 5 & 1 & 0 \\ 7 & -1 & 12 \end{bmatrix} \tag{4.2}$$

Theorem 1
Consider two $n \times n$ matrices A and B, then the following qualities are true:
If $A^T = B$, then $B^T = A$, that is, $(A^T)^T = A$

DOI: 10.1201/9781003478812-4

$$(A + B)^T = A^T + B^T$$

$$(AB)^T = B^T A^T$$

4.2.2 SUM AND DIFFERENCE

If matrices A and B have the same dimension, then their sum, $A + B$, is obtained by adding the analogous entries:

$$(A + B)_{ij} = A_{ij} + B_{ij}.$$

If A and B have the same dimension, then their difference, $A \times B$, is obtained by subtracting corresponding entries:

$$(A \times B)_{ij} = A_{ij} \times B_{ij}.$$

Example 3

If $A = \begin{bmatrix} 1 & 2 & 3 \\ 2 & 0 & -1 \\ 0 & 3 & 1 \end{bmatrix}$ and $B = \begin{bmatrix} -1 & 3 & 4 \\ 2 & 1 & 0 \\ 0 & -3 & 11 \end{bmatrix}$ then

$$(AB)^T = B^T A^T.$$

4.2.3 SCALAR MULTIPLE

If A is a matrix and α is a scalar, then the scalar multiple, $(A^T)^T = A$, is obtained by multiplying every entry in A by α. In symbols,

$$\left(\alpha_a \right)_{ij} = \alpha \left(a_{ij} \right).$$

Example 4

   i. For α real,

$$\alpha \begin{bmatrix} 1 & 2 & 3 \\ 2 & 0 & -1 \\ 0 & 3 & 1 \end{bmatrix} = \begin{bmatrix} \alpha & 2\alpha & 3\alpha \\ 2\alpha & 0 & -\alpha \\ 0 & 3\alpha & \alpha \end{bmatrix} \qquad (4.3)$$

   ii. Calculate $-3 \begin{bmatrix} 2 & 5 \\ 3 & 1 \end{bmatrix}$; $5 \begin{bmatrix} -1 & 1 \\ 2 & -1 \end{bmatrix}$

Solution

$$-3 \begin{bmatrix} 2 & 5 \\ 3 & 1 \end{bmatrix} = \begin{bmatrix} -6 & -15 \\ -9 & -3 \end{bmatrix}$$

$$5\begin{bmatrix} -1 & 1 \\ 2 & -1 \end{bmatrix} = \begin{bmatrix} -5 & 5 \\ 10 & -5 \end{bmatrix}$$

4.2.4 PRODUCT OF TWO MATRICES

If A has dimensions $m \times n$ and B has dimensions $n \times p$, then the product AB is defined and has dimensions $m \times p$. The entry $(AB)_{ij}$ is obtained by multiplying row i of A by column j of B, which is done by multiplying corresponding entries together and then adding the results.

Example 5

i. Consider the following 2×2 matrices: $M = \begin{bmatrix} 1 & 2 \\ -1 & 4 \end{bmatrix}$ and $N = \begin{bmatrix} 3 & 5 \\ 1 & 2 \end{bmatrix}$, then the product

of these two matrices is given by

$$MN = \begin{bmatrix} 1 \times 3 + 2 \times 1 & 1 \times 5 + 2 \times 2 \\ -1 \times 3 + 4 \times 1 & -1 \times 5 + 4 \times 2 \end{bmatrix} = \begin{bmatrix} 5 & 9 \\ 1 & 3 \end{bmatrix} \tag{4.4}$$

$$\begin{bmatrix} 2 & 5 \\ 3 & 1 \end{bmatrix} \cdot \begin{bmatrix} -1 & 1 \\ 2 & -1 \end{bmatrix} = \begin{bmatrix} -2 + 10 & 2 - 5 \\ -3 + 2 & 3 - 1 \end{bmatrix} = \begin{bmatrix} 8 & -3 \\ -1 & 2 \end{bmatrix}$$

4.3 ALGEBRA OF MATRIX

The matrix operation has several properties. Among these properties are the associative law, the commutative law and the distributive laws.

4.3.1 ASSOCIATIVE LAW FOR ADDITION

Addition of matrices is commutative. For three matrices A, B, C,

$$(A + B) + C = A + (B + C) \tag{4.5}$$

4.3.2 COMMUTATIVE LAW FOR ADDITION

Addition of matrices is commutative. For two matrices A, B,

$$A + B = B + A \tag{4.6}$$

4.3.3 NEUTRAL ELEMENT

The zero matrix is the neutral element in the set of the matrices. All entries of the zero matrix are equal to zero.

$$A + \Theta = \Theta + A \tag{4.7}$$

4.3.4 SYMMETRY

For each matrix A, there exists a matrix $A\times$ called symmetric of A for addition, such that

$$A + A' = A' + A = \Theta \tag{4.8}$$

This means that the symmetry with respect to addition of a matrix A is $-A$.

4.3.5 DISTRIBUTIVE PROPERTY FROM THE LEFT OF A SCALAR WITH RESPECT TO THE SUM OF MATRICES

For two matrices A and B and for a constant number c:

$$c(A + B) = cA + cB \tag{4.9}$$

4.3.6 DISTRIBUTIVE PROPERTY FROM THE RIGHT OF A SCALAR WITH RESPECT TO THE SUM OF MATRICES

For two matrices A and B, and for a constant number c:

$$(A + B)c = cA + cB \tag{4.10}$$

The one rule that is conspicuously absent from this list is the commutative rule of the matrix product. In general, matrix multiplication is not commutative: AB is not equal to BA in general.

Example 6
The following illustrates the failure of the commutative law for matrix multiplication.

$$A = \begin{bmatrix} 0 & 1 \\ \dfrac{1}{3} & -1 \end{bmatrix} \quad B = \begin{bmatrix} 1 & -1 \\ \dfrac{2}{3} & -2 \end{bmatrix}$$

$$AB = \begin{bmatrix} \dfrac{2}{3} & -2 \\ -\dfrac{1}{3} & 5/3 \end{bmatrix} \quad BA = \begin{bmatrix} 1/3 & 2 \\ -\dfrac{2}{3} & 8/3 \end{bmatrix}$$

As we can see, $AB \neq BA$.

4.3.7 ADJOINT OF A MATRIX

The adjoint of the matrix A is defined by the matrix A^+ whose entries are given by $A_{ij}^+ = a_{ji}^*$.

In practice we can see that the adjoint of matrix A is the conjugate of the transpose of A, that is, $A^+ = A^{T*}$.

Example 7 Find the adjoint of the matrix $M = \begin{bmatrix} 1 & 1 & -1 \\ 3 & 1 & -1 \\ 1 & 1 & -2 \\ 3 & 2 & -1 \end{bmatrix}$.

Solution: The adjoint of matrix M is given by $M^+ = \begin{bmatrix} 1 & 3 & 1 & 3 \\ 1 & 1 & 1 & 2 \\ -1 & -1 & -2 & -1 \end{bmatrix}$.

Theorem 2

(a) If $A^+ = B$, then $(A + B)^+ = A^+ + B^+$, that is, $(A^+)^+ = A$

(b) $(A + B)^+ = A^+ + B^+$

(c) $(AB)^+ = B^+ A^+$

4.4   MATRIX FORM OF A SYSTEM OF LINEAR EQUATIONS

4.4.1   DEFINITION OF THE MATRIX FORM

To solve systems of linear equations, it can be useful to write them in matrix forms. Consider the system of linear equations:

$$\begin{cases} a_{11}x_1 + a_{12}x_2 + a_{13}x_3 + \ldots + a_{1n}x_n = b_1 \\ a_{21}x_1 + a_{22}x_2 + a_{23}x_3 + \ldots + a_{2n}x_n = b_2 \\ \ldots\ldots\ldots\ldots \\ a_{m1}x_1 + a_{m2}x_2 + a_{m3}x_3 + \ldots + a_{mn}x_n = b_m \end{cases} \tag{4.11}$$

The system (4.11) can be rewritten in the form of a matrix equation:

$$AX = B \tag{4.12}$$

$$X = [x_1, x_2, \ldots, x_n]^T \tag{4.13}$$

and

$$B = [b_1, b_2, \ldots, b_m]^T \tag{4.14}$$

Example 7

$$\text{The system} \begin{cases} x + y - z = 4 \\ 3x + y - z = 6 \\ x + y - 2z = 4 \\ 3x + 2y - z = 9 \end{cases} \tag{4.15}$$

has the matrix form:

$$\begin{bmatrix} 1 & 1 & -1 \\ 3 & 1 & -1 \\ 1 & 1 & -2 \\ 3 & 2 & -1 \end{bmatrix} \begin{bmatrix} x \\ y \\ z \end{bmatrix} = \begin{bmatrix} 4 \\ 6 \\ 4 \\ 9 \end{bmatrix} \tag{4.16}$$

4.4.2 AUGMENTED MATRIX

Definition 1

Consider the system of equations $\begin{cases} a_{11}x + a_{12}y = b_1 \\ a_{21}x + a_{22}y = b_2 \end{cases}$.

The augmented matrix associated with this matrix is given by the matrix:

$$\left[\begin{array}{cc|c} a_{11} & a_{12} & b_1 \\ a_{21} & a_{22} & b_2 \end{array}\right].$$ (4.17)

Example 8

Find the augmented matrix of the system $\begin{cases} 2x + 3y = 1 \\ 5x + 6y = 3 \end{cases}$.

Solution

The augmented matrix of the system $\begin{cases} 2x + 3y = 1 \\ 5x + 6y = 3 \end{cases}$ is $\left[\begin{array}{cc|c} 2 & 3 & 1 \\ 5 & 6 & 3 \end{array}\right]$.

Example 9

Find the augmented matrix of the system $\begin{cases} 3x - 4y = -6 \\ 2x - 3y = -5 \end{cases}$.

Solution

The augmented matrix of the system $\begin{cases} 2x - y + z = 0 \\ x + z - 1 = 0 \\ x + 2y - 8 = 0 \end{cases}$ is $\left[\begin{array}{cc|c} 3 & -4 & -6 \\ 2 & -3 & -5 \end{array}\right]$.

Example 10

Find the system of equations associated with the augmented matrix:

$$\left[\begin{array}{cc|c} 1 & -2 & -1 \\ 3 & 5 & 10 \end{array}\right].$$

Solution

The system of equations associated with the augmented matrix $\left[\begin{array}{cc|c} 1 & -2 & -1 \\ 3 & 5 & 10 \end{array}\right]$ is $\begin{cases} x - 2y = -1 \\ 3x + 5y = 10 \end{cases}$.

Example 11

Find the augmented matrix of the system of three equations,

$$\begin{cases} 2x - y + z = 0 \\ x + z - 1 = 0 \\ x + 2y - 8 = 0 \end{cases}$$

Solution

The augmented matrix of the system of equations is as follows:

$$\begin{cases} 2x - y + z = 0 \\ x + z - 1 = 0 \\ x + 2y - 8 = 0 \end{cases} \text{ is } \begin{bmatrix} 2 & -1 & 1 & | & -1 \\ 1 & 0 & 1 & | & 1 \\ 1 & 2 & 0 & | & 8 \end{bmatrix}$$

Thus, as we can see, from a system of equations, an associated augmented matrix can be written. Similarly, for an augmented matrix, the associated system of equations can be written.

4.4.3 ROW OPERATIONS

The row operations consist of multiplying rows by a given real number and adding (or subtracting) them in another row. For example, consider the augmented matrix in Solution 1, that is, $\begin{bmatrix} 2 & 3 & | & 1 \\ 5 & 6 & | & 3 \end{bmatrix}$. We have two rows: r_2 and r_2, which we can call old rows: r_1 consists of 2, 3 and 1. r_2 consists of 5, 6 and 3. If α and β are two real numbers, the combinations $R_2 = \alpha r_1 + \beta r_2$ mean that we obtain a new augmented matrix whose first row is unchanged, but whose second row is obtained by multiplying the first old row by α and the second old row by β. The new matrix will be $\begin{bmatrix} 2 & 3 & | & 1 \\ 2\alpha + 5\beta & 3\alpha + 6\beta & | & \alpha + 3\beta \end{bmatrix}$.

It appears that this augmented matrix is associated to a system of equations which will have the same solution as the original system of equations $\begin{cases} 2x + 3y = 1 \\ 5x + 6y = 3 \end{cases}$. This method is used to obtain some zeros in the augmented matrix in order to solve a system of equations. It is also used to obtain the row echelon matrix, which will be defined in the next section.

4.4.4 ROW ECHELON MATRIX

A matrix is in row echelon form when the following conditions are met:

1. The entry in row 1 and column 1 is 1 and only 0 appears below it.
2. The entry in row 2 and column 2 is 1 and only 0 appears below it.
3. The entry in row 3 and column 3 is 1.

The general form for a 3×3 row echelon matrix is $\begin{bmatrix} 1 & a & b \\ 0 & 1 & c \\ 0 & 0 & 1 \end{bmatrix}$.

Thus, the principal diagonal of the matrix is reduced to unity.

The corresponding augmented matrix will be $\begin{bmatrix} 1 & a & b & | & d \\ 0 & 1 & c & | & e \\ 0 & 0 & 1 & | & f \end{bmatrix}$.

This method helps solve the system of equations associated with the augmented matrix. In the process of solving this system we can identify $z = f$. The whole process which consists of writing a matrix in a reduced echelon form is also called the Gauss–Jordan elimination method for solving a system of equations.

Example 12

Use the Gauss–Jordan elimination method to find the solutions of the following system of equations

$$\begin{cases} x+y+z=1 \\ 2x-y-z=2 \\ 3x+z=4 \end{cases}.$$

Solution

The first thing to do here is to write the augmented corresponding to the system.

The augmented matrix is written as $\begin{vmatrix} 1 & 1 & 1 & 1 \\ 2 & -1 & -1 & 2 \\ 3 & 0 & 1 & 4 \end{vmatrix}$.

Now, we need to perform some row operations. The first-row operation will be given by $R_2 = r_1 + r_2$, which gives the following augmented matrix:

$$\begin{vmatrix} 1 & 1 & 1 & 1 \\ 3 & 0 & 0 & 3 \\ 3 & 0 & 1 & 4 \end{vmatrix}.$$

The second-row operation is given by $R_3 = r_2 - r_3$, which gives the following augmented matrix:

$$\begin{vmatrix} 1 & 1 & 1 & 1 \\ 3 & 0 & 0 & 3 \\ 0 & 0 & -1 & -1 \end{vmatrix}.$$

The third-row operation is given by $R_2 = 3r_1 - r_2$, which will give

$$\begin{vmatrix} 1 & 1 & 1 & 1 \\ 0 & 3 & 3 & 0 \\ 0 & 0 & -1 & -1 \end{vmatrix}.$$

We can write the equivalent system of equations as follows:

$$\begin{cases} x+y+z=1 \\ 3y+3z=0 \\ -z=-1 \end{cases}.$$

The solutions to the system of equations will be $x = 1, y = -1, x = 1$. The set of solutions is $\{(1,-1,1)\}$.

4.5 MATRIX INVERSE

If A is a square matrix with a nonzero determinant and has the same number of rows and columns, it is possible to take a matrix equation such as $AX = B$ and solve for X by "dividing by A." Precisely, a square matrix A, with a nonzero determinant, may have an inverse, written A^{-1}, with the property that

$$AA^{-1} = A^{-1}A = I \qquad\qquad . (4.18)$$

If matrix A has an inverse it is said to be invertible, otherwise A is said to be singular. When matrix A is invertible we can solve the equation: $AX = B$, by multiplying both sides by $A^{\times 1}$, which gives us

$$X = A^{\times 1}B. \tag{4.19}$$

Example 13

The system of equations $\begin{bmatrix} 1 & 2 & 4 \\ 2 & 4 & 6 \\ 4 & 6 & 8 \end{bmatrix} \begin{bmatrix} x \\ y \\ z \end{bmatrix} = \begin{bmatrix} 1 \\ 1 \\ -1 \end{bmatrix}$ has the solutions:

$$\begin{bmatrix} x \\ y \\ z \end{bmatrix} = \begin{bmatrix} 1 & 2 & 4 \\ 2 & 4 & 6 \\ 4 & 6 & 8 \end{bmatrix}^{-1} \begin{bmatrix} 1 \\ 1 \\ -1 \end{bmatrix} = \begin{bmatrix} 1 & -2 & 1 \\ -2 & 2 & -1/2 \\ 1 & -1/2 & 0 \end{bmatrix}^{-1} \begin{bmatrix} 1 \\ 1 \\ -1 \end{bmatrix} = \begin{bmatrix} -2 \\ 1/2 \\ 1/2 \end{bmatrix}$$

4.5.1 Determining Whether a Matrix Is Invertible

In order to determine whether an $n \times n$ matrix A is invertible or not, and to find $A^{\times 1}$ if it does exist, write down the $n \times (2n)$ matrix $[A \mid I]$ (this is A with the $n \times n$ identity matrix set next to it). Reduce this matrix in the form $[I \mid B]$ with the identity matrix on the left; then, A is invertible and $B = A^{\times 1}$. If you cannot obtain I in the left part, then A is singular.

Example 14

The matrix $\begin{bmatrix} 1 & 2 & 4 \\ 2 & 4 & 6 \\ 4 & 6 & 8 \end{bmatrix}$ is invertible.

Proof

$$[A|I] = \begin{bmatrix} 1 & 2 & 4 & 1 & 0 & 0 \\ 2 & 4 & 6 & 0 & 1 & 0 \\ 4 & 6 & 8 & 0 & 0 & 1 \end{bmatrix}$$

$$R_2 = 2r_2 - r_3 \rightarrow \begin{bmatrix} 1 & 2 & 4 & 1 & 0 & 0 \\ 0 & 2 & 4 & 0 & 2 & -1 \\ 4 & 6 & 8 & 0 & 0 & 1 \end{bmatrix}$$

$$R_3 = 4r_1 - r_3 \rightarrow \begin{bmatrix} 1 & 2 & 4 & 1 & 0 & 0 \\ 0 & 2 & 4 & 0 & 2 & -1 \\ 0 & 2 & 8 & 4 & 0 & -1 \end{bmatrix}$$

$$R_1 = r_3 - r_1 \rightarrow \begin{bmatrix} -1 & 0 & 4 & 3 & 0 & -1 \\ 0 & 2 & 4 & 0 & 2 & -1 \\ 0 & 2 & 8 & 4 & 0 & -1 \end{bmatrix}$$

$$R_3 = r_3 - r_2 \rightarrow \begin{bmatrix} -1 & 0 & 4 & 3 & 0 & -1 \\ 0 & 2 & 4 & 0 & 2 & -1 \\ 0 & 0 & 4 & 4 & -2 & 0 \end{bmatrix}$$

$$\left.\begin{array}{l} R_1 = r_3 - r_1 \\ R_2 = r_3 - r_2 \end{array}\right\} \rightarrow \begin{bmatrix} 1 & 0 & 0 & 1-2 & 1 \\ 0 & -2 & 0 & 4-4 & 1 \\ 0 & 0 & 4 & 4-2 & 0 \end{bmatrix}$$

$$\left.\begin{array}{l} R_1 = r_1 \\ R_2 = -r_2/2 \\ R_3 = r_3/4 \end{array}\right\} \rightarrow \begin{bmatrix} 1 & 0 & 0 & 1 & -2 & 1 \\ 0 & 1 & 0 & -2 & 2-1/2 \\ 0 & 0 & 1 & 1-1/2 & 0 \end{bmatrix} = \left[I_3 \middle| A^{-1}\right]$$

So, the inverse matrix of A would be

$$A^{-1} = \begin{bmatrix} 1 & -2 & 1 \\ -2 & 2 & -1/2 \\ 1 & -1/2 & 0 \end{bmatrix}$$

To double check, it is useful to calculate

$$A \cdot A^{-1} = \begin{bmatrix} 1 & 2 & 4 \\ 2 & 4 & 6 \\ 4 & 6 & 8 \end{bmatrix} \begin{bmatrix} 1 & -2 & 1 \\ -2 & 2 & -1/2 \\ 1 & -1/2 & 0 \end{bmatrix} = \begin{bmatrix} 1 & 0 & 0 \\ 0 & 1 & 0 \\ 0 & 0 & 1 \end{bmatrix} = I_3$$

which proves that the inverse A^{-1} obtained was true.

4.5.2 INVERSE OF A MATRIX USING THE ROW OPERATION

Consider the 2×2 matrix: $= \begin{bmatrix} 1 & 1 \\ 3 & 2 \end{bmatrix}$. Let us form the augmented matrix $\left[A \middle| I_2\right]$, where I_2 is the 2×2

unit matrix $\begin{bmatrix} 1 & 0 \\ 0 & 1 \end{bmatrix}$. We have: $\left[A \mid I_2\right] = \begin{bmatrix} 1 & 1 \\ 3 & 2 \end{bmatrix} \begin{bmatrix} 1 & 0 \\ 0 & 1 \end{bmatrix}$.

Using the operation, $R_1 = r_2 - 2r_1$, we obtain $\begin{bmatrix} 1 & 0 & -2 & 1 \\ 3 & 2 & 0 & 1 \end{bmatrix}$.

Now let's calculate $R_2 = 3r_1 - r_2 x$; we get $\begin{bmatrix} 1 & 0 & -2 & 1 \\ 0 & 1 & 3 & -1 \end{bmatrix}$.

Moreover, $R_2 = -r_2/2$ gives $\begin{bmatrix} 1 & 0 & -2 & 1 \\ 0 & 1 & 3 & -1 \end{bmatrix}$

$A^{-1} = \begin{bmatrix} -2 & 1 \\ 3 & -1 \end{bmatrix}$ is the inverse of the matrix $A = \begin{bmatrix} 1 & 1 \\ 3 & 2 \end{bmatrix}$. To double check, let us multiply them.

$$\begin{bmatrix} 1 & 1 \\ 3 & 2 \end{bmatrix} \begin{bmatrix} -2 & 1 \\ 3 & -1 \end{bmatrix} = \begin{bmatrix} 1 & 0 \\ 0 & 1 \end{bmatrix} = I_2$$

I_2 is a 2 × 2 identity matrix. As a result, we conclude that the inverse calculated is justified.

4.6   DETERMINANT OF A SQUARE MATRIX

A determinant is a real number associated with every square matrix. The determinant of a square matrix A is denoted by

"det (A)" or $| A |$.

4.6.1   DETERMINANT OF A 2 × 2 MATRIX

The determinant of a 2 × 2 matrix is found much like a pivot operation. It is the product of the entries on the main diagonal minus the product of the entries off the main diagonal.

$$\det(A) = \begin{vmatrix} a & b \\ c & d \end{vmatrix} = ad - bc \tag{4.20}$$

4.6.1.1   Inverse of a 2 × 2 Matrix

The 2 × 2 matrix $A = \begin{bmatrix} 1 & 1 \\ 3 & 2 \end{bmatrix}$ is invertible if $ad \times bc$ is nonzero and is singular if $ad \times bc = 0$. The

number $ad \times bc$ is called the determinant of the matrix. When the matrix is invertible its inverse is given by

$$A^{-1} = \frac{1}{ad - bc} \begin{bmatrix} d & -b \\ -c & a \end{bmatrix} \tag{4.21}$$

Example 15

If $A = \begin{bmatrix} 1 & 1 \\ 3 & 2 \end{bmatrix}$, the inverse of matrix A is given by

$$A^{-1} = \frac{1}{2-3} \begin{bmatrix} 2 & -1 \\ -3 & 1 \end{bmatrix} = \begin{bmatrix} -2 & 1 \\ 3 & -1 \end{bmatrix}.$$

Example 16

The determinant of two vectors $u = \begin{pmatrix} 2 \\ 1 \end{pmatrix}$ and $v = \begin{pmatrix} 3 \\ 2 \end{pmatrix}$ is the two-dimensional array $\begin{vmatrix} 2 & 3 \\ 1 & 2 \end{vmatrix} =$

$2 \times 2 - 3 \times 1 = 4 - 3 = 1 \cdot$

4.6.1.2   Determinant of Three Vectors of Three Components

The determinant of three vectors $u_1 = \begin{pmatrix} a_1 \\ b_1 \\ c_1 \end{pmatrix}, u_2 = \begin{pmatrix} a_2 \\ b_2 \\ c_2 \end{pmatrix}$ and

$$u_3 = \begin{pmatrix} a_3 \\ b_3 \\ c_3 \end{pmatrix} \text{ is the number denoted by } \begin{vmatrix} a_1 a_2 a_3 \\ b_1 b_2 b_3 \\ c_1 c_2 c_3 \end{vmatrix} = a_1 \begin{vmatrix} b_2 b_3 \\ c_2 c_3 \end{vmatrix} - a_2 \begin{vmatrix} b_1 b_3 \\ c_1 c_3 \end{vmatrix} + a_3 \begin{vmatrix} b_1 b_2 \\ c_1 c_2 \end{vmatrix}, \text{ where } \begin{vmatrix} b_2 b_3 \\ c_2 c_3 \end{vmatrix} \text{ is the deter-}$$

minant obtained by crossing out the first row and the first column of $\begin{vmatrix} a_1 a_2 a_3 \\ b_1 b_2 b_3 \\ c_1 c_2 c_3 \end{vmatrix}$.

$\begin{vmatrix} b_1 b_3 \\ c_1 c_3 \end{vmatrix}$ is the determinant obtained by crossing out the first row and the second column of $\begin{vmatrix} a_1 a_2 a_3 \\ b_1 b_2 b_3 \\ c_1 c_2 c_3 \end{vmatrix}$.

$\begin{vmatrix} b_1 b_2 \\ c_1 c_2 \end{vmatrix}$ is the determinant obtained by crossing out the first row and the third column of $\begin{vmatrix} a_1 a_2 a_3 \\ b_1 b_2 b_3 \\ c_1 c_2 c_3 \end{vmatrix}$.

4.6.1.3   Expansion Using Minors and Cofactors

The definition of determinant that we have so far is only for a 2×2 matrix. There is a shortcut of a 3×3 matrix, but we firmly believe you should learn the way that will work for all sizes, not just a special case for a 3×3 matrix. The method is called expansion using minors and cofactors. Before we can use these terms, we need to define them.

4.6.1.4   Minors

The minor of an entry a_{ij} is the determinant of the matrix obtained by crossing out the ith row and the jth column. The notation M_{ij} is used to stand for the minor of the entry in row i and column j. So M_{23} would mean the minor for the entry in row 2 and column 3.

Consider now the following 3×3 matrix:

$$A = \begin{bmatrix} a_{11} & a_{12} & a_{13} \\ a_{21} & a_{22} & a_{23} \\ a_{31} & a_{32} & a_{33} \end{bmatrix} \tag{4.22}$$

The minor corresponding to the entry a_{ij} is the determinant of the matrix $\begin{bmatrix} a_{22} & a_{23} \\ a_{32} & a_{33} \end{bmatrix}$.

The determinant of the 3×3 matrix is written as follows:

$$\alpha \begin{bmatrix} 1 & 2 & 3 \\ 2 & 0 & -1 \\ 0 & 3 & 1 \end{bmatrix} = \begin{bmatrix} \alpha & 2\alpha & 3\alpha \\ 2\alpha & 0 & -\alpha \\ 0 & 3\alpha & \alpha \end{bmatrix} \tag{4.23}$$

In this determinant, the following minors are used: $\begin{bmatrix} a_{22} & a_{23} \\ a_{32} & a_{33} \end{bmatrix}$ for the entry a_{11}; $\begin{vmatrix} a_{21} & a_{23} \\ a_{31} & a_{33} \end{vmatrix}$ for the entry a_{12}; $\begin{vmatrix} a_{21} & a_{22} \\ a_{31} & a_{32} \end{vmatrix}$ for the entry a_{13}.

4.6.1.5 Cofactors

A cofactor for any entry is either the minor or the opposite of the minor, depending on where the entry is in the original determinant. If the rank of the row and the rank of the column of the entry add up to be an even number, then the cofactor is the same as the minor. If the rank of the row and the rank of the column of the entry add up to be an odd number, then the cofactor is the opposite of the minor. For example, if M_{ij} is the minor corresponding to the entry a_{ij}, the cofactor is:

$$C_{ij} = (-1)^{i+j} M_{ij}.$$

4.6.1.6 Expansion of a determinant by cofactors

In general, the determinant of a $C_{ij} = (-1)^{i+j} M_{ij}$ matrix A can be expressed as a cofactor expansion using any row or any column of A.

Theorem 3

Let A be a square matrix, then

$$det(A) = \sum_{j=1}^{n} a_{ij} C_{ij} = a_{i1} C_{i1} + a_{i2} C_{i2} + \ldots + a_{in} C_{in}$$

or

$$det(A) = \sum_{i=1}^{n} a_{ij} C_{ij} = a_{1j} C_{1j} + a_{2j} C_{2j} + \ldots + a_{nj} C_{nj}$$

where C_{ij} are cofactors of entries a_{ij}.

Example 17

Find the determinant of matrix $A = \begin{bmatrix} 1-2 & 3 & 0 \\ -1 & 1 & 0 & 2 \\ 0 & 2 & 0 & 3 \\ 3 & 4 & 0-2 \end{bmatrix}$

Solution

Column 3 of matrix A has three zeros; so, we can use it to save time. The entry a_{13} is the only one different from zero in column 3. So, let us find the minor of the entry a_{13}.

$$M_{13} = \begin{vmatrix} -1 & 1 & 2 \\ 0 & 2 & 3 \\ 3 & 4 & -2 \end{vmatrix} = -1(-4-12)+3(3-4) = 13 \cdot$$

The cofactor of the entry a_{13} is $C_{13} = (-1)^{1+3} M_{13} = 13$. Then, the determinant of A will be:

$$det(A) = \sum_{i=1}^{4} a_{ij} C_{ij} = a_{13} C_{13} + a_{23} C_{23} + a_{33} C_{33} + a_{43} C_{43}$$

$$M_{13} = \begin{vmatrix} -1 & 1 & 2 \\ 0 & 2 & 3 \\ 3 & 4 & -2 \end{vmatrix} = -1(-4-12)+3(3-4) = 13$$

4.6.1.7 Alternative Method

An alternative method used to evaluate a determinant of a 3×3 matrix A can be done by copying the first and the second columns of A, respectively, and pasting them on the fourth and fifth columns. Next, obtain the determinant of the matrix A by adding (or subtracting) the product of the six diagonals as shown in the diagram below.

4.6.1.8 Subtract These Products

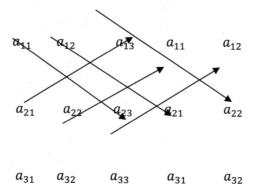

By adding these products, we obtain the determinant,

$$\det(A) = |A| = a_{11}a_{22}a_{33} + a_{12}a_{23}a_{31} + a_{13}a_{21}a_{32}$$
$$- a_{31}a_{22}a_{13} - a_{32}a_{23}a_{11} - a_{33}a_{21}a_{12}$$

Example 18

Find the determinant of $A = \begin{bmatrix} 0 & 2 & 1 \\ 3 & -1 & 2 \\ 4 & -4 & 1 \end{bmatrix}$

Solution

Using the same method, we have:

So, the determinant of matrix A is

$$|A| = (0 + 16 - 12) - (-4 + 0 + 6) = 2$$

Remark 1

The diagonal process in the alternative method is valid only for a square matrix of order 3. For matrices of higher order, it is better to use another method.

4.6.2   PROPERTIES OF DETERMINANTS

The determinant is a real or a complex number; it is not a matrix. The determinant exists only for square matrices that is 2×2, or 3×3, $\times$, or $n \times n$ matrices. The determinant of a 1×1 matrix is 1. The inverse of a matrix will exist only if the determinant is not zero.

4.6.2.1   Inverse of a Matrix

Let's consider the following matrix $\begin{bmatrix} 1 & 3 & 2 \\ 4 & 1 & 3 \\ 2 & 5 & 2 \end{bmatrix}$. The matrix of minors as explained above is given

by $\begin{bmatrix} -13 & 2 & 18 \\ -4 & -2 & -1 \\ 7 & -5 & -11 \end{bmatrix}$

Let's now turn it into a matrix of cofactors by changing the signs on the appropriate entries based on the sign chart. We obtain:

$$\begin{bmatrix} -13 & -2 & 18 \\ 4 & -2 & 1 \\ 7 & 5 & -11 \end{bmatrix}$$

We will now find the adjoint by transposing the matrix of cofactors. To transpose a matrix, we switch the rows and columns. That is, the rows become columns and the columns become rows. The transpose of a matrix A is denoted by $[A]^T$. In the case of the matrix above, the transpose is given by

$$\begin{bmatrix} -13 & 4 & 7 \\ -2 & -2 & 5 \\ 18 & 1 & -11 \end{bmatrix}$$

Finally, we divide the adjoint of the matrix by the determinant of the matrix. In this problem, the determinant is 17; so, we'll divide each entry by 17. The resulting matrix is the inverse of the original matrix, that is:

$$\begin{bmatrix} -13/17 & 4/17 & 7/17 \\ -2/17 & -2/17 & 5/17 \\ 18/17 & 1/17 & -11/17 \end{bmatrix}$$

The inverse of a matrix is found by dividing the adjoint of the matrix by the determinant of the matrix:

$$Inverse\ of\ A = \frac{1}{\det A} \times \left(Adjoint\ Matrix\ of\ cofactors \right)$$

Example 19

Prove that the matrix $A = \begin{bmatrix} 1 & 2 & 4 \\ 2 & 4 & 6 \\ 2 & 4 & 7 \end{bmatrix}$ is not invertible.

Solution

It suffices to calculate the determinant of the matrix and to prove that the determinant is zero.

$$\det(A) = 1.\begin{vmatrix} 4 & 6 \\ 4 & 7 \end{vmatrix} - 2\begin{vmatrix} 2 & 6 \\ 2 & 7 \end{vmatrix} + 4\begin{vmatrix} 2 & 4 \\ 2 & 4 \end{vmatrix}$$

$$= (28-24) - 2(14-12) + 4(8-8)$$

$$= 4 - 4 + 0$$

$$= 0$$

Because $\det(A)=0$, the above matrix A is not invertible.

4.6.3 EIGENVALUE AND EIGENVECTOR

Eigenvalue and eigenvector problem is of great interest in a wide range of domains such as in solving systems of differential equations, analyzing population dynamics.

Definition 2
Let A be a square matrix. A non-zero vector A is called eigenvector of A if and only if there exists a real or complex number λ_i, such that $A\vec{V}_i = \lambda_i \vec{V}_i$.

Example 20
Find the eigenvector and the eigenvalue of the matrix

$$A = \begin{bmatrix} 7 & 4 \\ 3 & 6 \end{bmatrix}.$$

Solution
If we consider $\vec{V}$ the eigenvector of A with the eigenvalue λ, then we would need to solve the equation $A\vec{V} = \lambda\vec{V}v$. This is a two-dimensional problem. Assume $\lambda^2 - 13\lambda + 30 = 0$ has two components, $\begin{bmatrix} x \\ y \end{bmatrix}$, then we have the following matrix equation:

$$\begin{bmatrix} 7 & 4 \\ 3 & 6 \end{bmatrix}\begin{bmatrix} x \\ y \end{bmatrix} = \lambda\begin{bmatrix} x \\ y \end{bmatrix}.$$

The previous equation can be written as follows:

$$\begin{bmatrix} 7-\lambda & 4 \\ 3 & 6-\lambda \end{bmatrix}\begin{bmatrix} x \\ y \end{bmatrix} = 0$$

The solution is the null vector unless the determinant is zero, that is:

$$\begin{vmatrix} 7-\lambda & 4 \\ 3 & 6-\lambda \end{vmatrix} = 0$$

which gives the quadratic equation, $\lambda^2 - 13\lambda + 30 = 0$, known as the characteristic equation of A.

The roots of the characteristics equation, $\begin{bmatrix} 7-\lambda & 4 \\ 3 & 6-\lambda \end{bmatrix}\begin{bmatrix} x \\ y \end{bmatrix} = 0$ and $\lambda_2 = 10$ are the eigenvalues of A.

For $\lambda_1 = 3$, the matrix equation gives the following system of linearly dependent equations:

$$\begin{cases} 7x + 4y = 3x \\ 3x + 6y = 3y \end{cases}$$

the solution of which gives the eigenvector

$V_1 = \begin{bmatrix} c \\ -c \end{bmatrix} = c \begin{bmatrix} 1 \\ -1 \end{bmatrix}$. However, if a vector is eigenvector, every linear combination of this vector

is also eigenvector. Then, it is useful to consider $\lambda^2 - 13\lambda + 30 = 0$ as the eigenvector. Furthermore, it is often useful to normalize the eigenvector vectors. Hence, to find the normalized eigenvector

of the previous one, let us proceed this way: $V_1 = \sqrt{2} \begin{bmatrix} \dfrac{1}{\sqrt{2}} \\ -\dfrac{1}{\sqrt{2}} \end{bmatrix}$. Also, since V_1 and $\begin{bmatrix} \dfrac{1}{\sqrt{2}} \\ -\dfrac{1}{\sqrt{2}} \end{bmatrix}$ are

proportional, we will consider the normalized eigenvector as $v_1 = \begin{bmatrix} \dfrac{1}{\sqrt{2}} \\ -\dfrac{1}{\sqrt{2}} \end{bmatrix}$.

Similarly, the normalized eigenvector corresponding to the second eigenvalue, $\lambda_2 = 10$ can be

found to be $v_2 = \begin{bmatrix} \dfrac{4}{5} \\ \dfrac{3}{5} \end{bmatrix}$.

In general, in the theory of eigenvalue and eigenvector, if A is an $n \times n$ matrix, the eigenvalue equation is written, $A\vec{V} = \lambda \vec{V}$, and has the solution $\vec{V} = \vec{0}$, unless $\det(A - \lambda I) = 0$, where I is the n by n identity matrix. $\det(A - \lambda I) = 0$ leads to the characteristics equation, $f(\lambda) = 0$, where $f(\lambda)$ is a polynomial of degree n. The zeros of $f(\lambda) = 0$, $\lambda_1, \lambda_2, ..., \lambda_n$ are the eigenvalues of matrix A. When each eigenvalue is plugged back into the eigenvalue equation, it leads to the corresponding eigenvector which is usually normalized. A normalized eigenvector of a given matrix is an eigenvector of the matrix. Indeed, an eigenvector remains eigenvector upon multiplication by any nonzero scalar. Suppose $\vec{V}$ is an eigenvector of a square matrix A with eigenvalue λ. We can therefore write the eigenvalues equation, $A\vec{V} = \lambda \vec{V}$. Moreover, for all α, real or complex, we can multiply both sides of the eigenvalues' equation by α. Therefore, we have $\alpha A\vec{V} = \alpha\lambda\vec{V}$. Furthermore, by the commutative property, $\alpha\vec{V}$ and $\alpha\lambda = \lambda\alpha$, we can write $A(\alpha\vec{V}) = \lambda(\alpha\vec{V})$, which proves that $\alpha\vec{V}$ is also eigenvector of matrix A.

4.7   TRIANGULAR MATRICES

4.7.1   UPPER TRIANGULAR MATRIX

A matrix in which all the non-zero entries are either on or above the main diagonal is called upper triangular matrix. That is, all the non-zero entries are in the upper triangle. Every entry below the main diagonal is a zero.

An upper triangular matrix can be represented as follows:

$$\begin{bmatrix} a_{11} & a_{12} & a_{13} & \cdots\cdots\cdots & a_{1n} \\ 0 & a_{22} & a_{23} & \cdots\cdots\cdots & a_{2n} \\ 0 & 0 & a_{33} & a_{34}\cdots\cdots\cdots & a_{3n} \\ & & & & \\ & & & & \\ 0 & 0 & 0 & \cdots\cdots\cdots & a_{nn} \end{bmatrix}$$

4.7.2 LOWER TRIANGULAR MATRIX

A lower triangular matrix is a matrix in which all the non-zero entries are either on or below the main diagonal. That is, all the non-zero values are in the lower triangle. Every entry above the diagonal is zero. Such matrix can be represented as follows:

$$\begin{bmatrix} a_{11}0 & 0\ldots\ldots\ldots\ldots\ldots\ldots\ldots.0 \\ a_{21}\,a_{22}\,0\ldots\ldots\ldots\ldots\ldots\ldots.0 \\ a_{31}\,a_{32}\,a_{33}0\ldots\ldots\ldots\ldots\ldots.0 \\ \cdot \\ \cdot \\ \cdot \\ a_{n1}\,a_{n2}\ldots\ldots\ldots\ldots\ldots\ldots\ldots.a_{nn} \end{bmatrix}.$$

To find the determinant of a triangular matrix, simply form the product of the entries of the main diagonal.

Example 21

Find the determinant of the following triangular matrix:

$$A = \begin{bmatrix} 2 & 3 & -1 \\ 0 & -1 & 2 \\ 0 & 0 & 3 \end{bmatrix}.$$

Solution

$$|A| = 2(-1)(3) = -6.$$

We can double check that this determinant is the same as the one obtained by using the method of crossing out the first row and the first column:

$$|A| = 2\begin{vmatrix} -1 & 2 \\ 0 & 3 \end{vmatrix} = 2(-3-0) = -6.$$

Theorem 4 (determinant of a triangular matrix)

If A is a triangular matrix of order n, then its determinant is the product of entries on the main diagonal, that is:

$$|A| = \det(A) = a_{11}a_{22}\ldots a_{nn}$$

Proof

Let us prove the above theorem by using the case of an upper triangular matrix and the mathematical induction. For a matrix of order 1,

$$[A] = [a_{11}] \text{ and } |A| = a_{11}.$$

The only entry of matrix A is also its main diagonal.

Assume now that the theorem is true for any upper triangular matrix of order $(k-1)$ and consider an upper triangular matrix of order k and use the $(k-1)$ row expansion.

$$|A| = 0C_{k1} + 0C_{k2} + 0C_{k3} + \ldots + 0C_{kk-1} + a_{kk}C_{kk}$$
$$= a_{kk}C_{kk}$$
$$= a_{kk}(-1)^{2k}M_{kk} = a_{kk}M_{kk}.$$

Remember that M_{kk} is the determinant of the matrix obtained by deleting the kth row and the kth column. Such a matrix is of order $k-1$ with determinant $\mathrm{col}_1(P), \mathrm{col}_2(P), \ldots, \mathrm{col}_n(P)\ a_{11}a_{22}\ldots C_{k-1k-1}$.

So, $|A| = a_{kk}M_{kk} = a_{kk}(a_{11}a_{22}\ldots a_{k-1k-1}) == a_{11}a_{22}\ldots a_{k-1k-1}a_{kk}$.

Example 22

Find the determinant of the following matrix

$$B = \begin{bmatrix} 2 & 3 & 0 & 0 \\ 4 & -2 & 0 & 0 \\ -5 & 6 & 1 & 0 \\ 1 & 5 & 3 & 3 \end{bmatrix}.$$

Solution

$$|B| = 2(-2)(1)(3) = -12$$

4.7.3 DETERMINANTS THAT ARE ZERO

The determinant of a matrix will be zero if

1. An entire row is zero.
2. Two rows or columns are equal.
3. A row or column is a constant multiple of another row or column.

Remember, that a matrix is invertible, non-singular, if and only if the determinant is not zero. So, if the determinant is zero, the matrix is singular and does not have an inverse.

4.8 DIAGONALIZATION OF A MATRIX

A matrix in which all the non-zero entries are on the main diagonal is called diagonal matrix. Everything off the main diagonal is a zero. The determinant of a triangular matrix or a diagonal matrix is the product of the entries on the main diagonal. A matrix A is diagonalizable if there exists a nonsingular matrix P and a diagonal matrix D such that

$$D = P^{-1}AP.$$

Example 23

Let us use an example to illustrate the diagonalization of a matrix. Consider the matrix,

$$\text{col}_1(P), \text{col}_2(P), \ldots, \text{col}_n(P)$$

and let us calculate its determinant.

Solution

$$|A| = 2 \times 2 - 0 \times 1 = 4$$

The determinant is different from zero; therefore, the matrix is a nonsingular matrix.

Let us write the secular equation.

$$|A - \lambda I_2| = \begin{vmatrix} 2-\lambda & 0 \\ 1 & 2-\lambda \end{vmatrix} = (2-\lambda)^2 = 0.$$

So we only have one eigenvalue: $\lambda = 2$. If we try to find the eigenvectors, we would have

$$\begin{bmatrix} 2 & 0 \\ 1 & 2 \end{bmatrix} \begin{bmatrix} x \\ y \end{bmatrix} = 2 \begin{bmatrix} x \\ y \end{bmatrix} \Rightarrow \begin{cases} 2x = 2x \\ x + 2y = 2y \end{cases}.$$

Therefore $\begin{cases} x = 0 \\ y = y \end{cases}$

The single vector, $\begin{pmatrix} x \\ y \end{pmatrix} = \begin{pmatrix} 0 \\ y \end{pmatrix} = y \begin{pmatrix} 0 \\ 1 \end{pmatrix}$. Therefore, $\begin{pmatrix} 0 \\ 1 \end{pmatrix}$, can be chosen as an eigenvector. Since

for the 2×2 matrix we only have a single eigenvalue, we can infer that the upper matrix is not diagonalizable.

Example 24

Find the diagonal matrix related to the matrix $A = \begin{bmatrix} 1 & 3 \\ 0 & 4 \end{bmatrix}$.

Solution

We need to find the eigenvalues and eigenvectors of matrix A. If λ is the eigenvalue associated with the eigenvector V, then the eigenvalue equation is written as

$$AV = \lambda V,$$

which has solutions if $\det(A - \lambda V) = 0$, that is:

$$\begin{vmatrix} 1-\lambda & 3 \\ 0 & 4-\lambda \end{vmatrix} = 0.$$

The characteristics equation is $(1-\lambda)(4-\lambda) = 0$; and the eigenvalues are $\lambda = 1$ and $\lambda = 4$.

We now need to look for the eigenvectors.

For $\lambda = 1$, consider the vector $V = \begin{bmatrix} x \\ y \end{bmatrix}$ as the associated eigenvector, then

$AV = \lambda V$ gives $\begin{bmatrix} 1 & 3 \\ 0 & 4 \end{bmatrix}\begin{bmatrix} x \\ y \end{bmatrix} = 1\begin{bmatrix} x \\ y \end{bmatrix}$, which results in the system of equations:

$$\begin{cases} x + 3y = x \\ 4y = y \end{cases} \text{ or } \begin{cases} 0x = 0 \\ y = 0 \end{cases}$$

So, $x \in \mathbb{R}$, $V = \begin{bmatrix} x \\ 0 \end{bmatrix} = x\begin{bmatrix} 1 \\ 0 \end{bmatrix}$ and $V_1 = \begin{bmatrix} 1 \\ 0 \end{bmatrix}$ can be chosen as the first eigenvector since every linear combination of an eigenvector is also eigenvector.

Similarly, for $\lambda = 1$, consider the vector $V = \begin{bmatrix} x \\ y \end{bmatrix}$ as the associated eigenvector, then

$col_j(P) = x_j$) gives $\begin{bmatrix} 1 & 3 \\ 0 & 4 \end{bmatrix}\begin{bmatrix} x \\ y \end{bmatrix} = 4\begin{bmatrix} x \\ y \end{bmatrix}$, which results in the system of equations:

$$\begin{cases} x + 3y = 4x \\ 4y = 4y \end{cases} \text{ or } \begin{cases} x - y = 0 \\ 0y = 0 \end{cases}.$$

So, $y \in \mathbb{R}$, $V = \begin{bmatrix} x \\ x \end{bmatrix} = x\begin{bmatrix} 1 \\ 1 \end{bmatrix}$ and $V_2 = \begin{bmatrix} 1 \\ 1 \end{bmatrix}$ can be chosen as the second eigenvector. The transformation matrix is $P = \begin{bmatrix} 1 & 1 \\ 0 & 1 \end{bmatrix}$ and its inverse is $P^{-1} = \begin{bmatrix} 1 & -1 \\ 0 & 1 \end{bmatrix}$. The diagonal matrix is

$$D = \begin{bmatrix} 1 & 0 \\ 0 & 4 \end{bmatrix}.$$

Theorem 5
An $n \times n$ matrix A is diagonalizable if and only if it has n linearly independent eigenvectors.

Proof
Assume that A is diagonalizable. Then, there exists a nonsingular matrix P and a diagonal matrix

$$D = \begin{bmatrix} \lambda_1 & 0 & \dots & 0 \\ 0 & \lambda_2 & \dots & 0 \\ & & \cdot & \\ & & \cdot & \\ & & \cdot & \\ 0 & 0 & \dots & \infty\lambda_n \end{bmatrix} \qquad (4.27)$$

such that

$$D = P^{-1}AP \Leftrightarrow AP = PD$$

$$\Leftrightarrow A\left[col_1(P) col_2(P) \dots col_n(P) \right]$$

$$\Leftrightarrow \left[col_1(P) col_2(P) ... col_n(P) \right] \begin{bmatrix} \lambda_1 & 0 & ... & 0 \\ 0 & \lambda_2 & ... & 0 \\ & & \cdot & \\ & & \cdot & \\ & & \cdot & \\ 0 & 0 & ... & \lambda_n \end{bmatrix} \tag{4.28}$$

Then,

$$\Leftrightarrow A \left[col_1(P) col_2(P) ... col_n(P) \right] \tag{4.29}$$

This means that, $col_1(P), col_2(P), ..., col_n(P)$ are eigenvectors associated with the eigenvalues $\lambda_1, \lambda_2, ..., \lambda_n$. Since P is a nonsingular matrix, that is a matrix whose determinant is different from zero, it can be said that the eigenvectors, $col_1(P), col_2(P), ..., col_n(P)$ are linearly independent.

Let $x_1, x_2, ..., x_n$ be n linearly independent eigenvectors of A associated with the eigenvalues $\lambda_1, \lambda_2, ..., \lambda_n$. That is, $Ax_j = \lambda_j x_j, j = 1, 2, ..., n$.

Now, let $P = \begin{bmatrix} x_1 & x_2 & \cdots & x_n \end{bmatrix}$ (i.e., $col_j(P) = x_j$) and,

$$D = \begin{bmatrix} \lambda_1 & 0 & ..... & 0 \\ 0 & \lambda_2 & ... & .0 \\ & & \cdot & \\ & & \cdot & \\ & & \cdot & \\ 0 & 0 & ..... & \lambda_n \end{bmatrix}.$$

Since $Ax_j = \lambda_j x_j$,

$$AP = A \begin{bmatrix} x_1 & x_2 & \cdots & x_n \end{bmatrix} = \begin{bmatrix} x_1 & x_2 & \cdots & x_n \end{bmatrix} \begin{bmatrix} \lambda_1 & 0 & \cdots & 0 \\ 0 & \lambda_2 & \cdots & 0 \\ \vdots & \vdots & \ddots & \vdots \\ 0 & 0 & \cdots & \lambda_n \end{bmatrix} = PD .$$

Thus,

$$P^{-1}AP = P^{-1}PD = D, \tag{4.30}$$

$P^{\times 1}$ exists because $x_1, x_2, ..., x_n$ are linearly independent and thus P is nonsingular.

Important Remark

An $n \times n$ matrix A is diagonalizable if all the roots of its characteristic equation are real and distinct.

Example 25

Consider the matrix $A = \begin{bmatrix} -4 & -6 \\ 3 & 5 \end{bmatrix}$.

Find the nonsingular matrix P and the diagonal matrix D such that, $= P^{-1}AP$, and find A^n, n is any positive integer.

Solution

We need to find the eigenvalues and eigenvectors of A first. The characteristic equation of A is the following:

$$\det\left(\lambda I - A\right) = \begin{vmatrix} \lambda+4 & 6 \\ -3 & \lambda-5 \end{vmatrix} = \left(\lambda+1\right)\left(\lambda-2\right) = 0$$

which gives $\lambda = -1$ or $\lambda = 2$.

By the above important result, A is diagonalizable; then,

1. As $\lambda = 2$, $Ax = 2x \Leftrightarrow \left(2I - A\right)x = 0 \Leftrightarrow x = r\begin{bmatrix} -1 \\ 1 \end{bmatrix}$, $r \in \mathbb{R}$

2. As $\lambda = -1$,

$$Ax = 2x \Leftrightarrow \left(-I - A\right)x = 0 \Leftrightarrow x = t\begin{bmatrix} -2 \\ 1 \end{bmatrix}, \ t \in \mathbb{R}$$

So we have two linearly independent eigenvectors of the matrix A, $\begin{bmatrix} -1 \\ 1 \end{bmatrix}$ and $\begin{bmatrix} -2 \\ 1 \end{bmatrix}$.

Now let $D^k = \begin{bmatrix} \lambda_1^k & 0 & \cdots & 0 \\ 0 & \lambda_2^k & \cdots & 0 \\ \vdots & \vdots & \ddots & \vdots \\ 0 & 0 & \cdots & \lambda_n^k \end{bmatrix}$ and $D = \begin{bmatrix} 2 & 0 \\ 0 & -1 \end{bmatrix}$.

Then, from Section 4.8, $= P^{-1}AP$.

To find A^n, let us calculate:

$$D^n = \begin{bmatrix} 2^n & 0 \\ 0 & (-1)^n \end{bmatrix} = \left(P^{-1}AP\right)\left(P^{-1}AP\right)\ldots\left(P^{-1}AP\right) = P^{-1}A^nP.$$

When multiplied by P and $P^{\times 1}$ on both sides, we get

$$PD^nP^{-1} = PP^{-1}A^nPP^{-1} = A^n = \begin{bmatrix} -1 & -2 \\ 1 & 1 \end{bmatrix}\begin{bmatrix} 2^n & 0 \\ 0 & (-1)^n \end{bmatrix}\begin{bmatrix} -1 & -2 \\ 1 & 1 \end{bmatrix}^{-1}$$

$$= \begin{bmatrix} & -\left[2^n + 2\cdot(-1)^{n+1}\right] \\ 2^n + (-1)^{n+1} & -\left[2^{n+1} + 2\cdot(-1)^{n+1}\right] \\ & 2^{n+1} + (-1)^{n+1} \end{bmatrix}$$

Remark

If A is an $n \times n$ diagonalizable matrix, then there exists a nonsingular matrix P such that

$D = P^{-1}AP$, where $\text{col}_1(P), \text{col}_2(P), \ldots, \text{col}_n(P)$ are n linearly independent eigenvectors of A and the diagonal entries of the diagonal matrix D are the eigenvalues of A associated with these eigenvectors.

Note

For any $n \times n$ diagonalizable matrix A, if $\lambda = 2$ then

$$A^k = PD^k P^{-1}, k = 1, 2, \ldots \tag{4.31}$$

where

$$D^k = \begin{bmatrix} \lambda_1^k & 0 & \cdots & 0 \\ 0 & \lambda_2^k & \cdots & 0 \\ \vdots & \vdots & \ddots & \vdots \\ 0 & 0 & \cdots & \lambda_n^k \end{bmatrix}. \tag{4.32}$$

Example 26

Is the matrix, $A = \begin{bmatrix} 5 & -3 \\ 3 & -1 \end{bmatrix}$ diagonalizable?

Solution

$$\det(\lambda I - A) = \begin{vmatrix} \lambda - 5 & 3 \\ -3 & \lambda + 1 \end{vmatrix} = (\lambda - 2)^2 = 0.$$

Then, $\lambda = 2, 2$; As $\lambda = 2$

$$(2I - A)x = 0 \Leftrightarrow x = t\begin{bmatrix} 1 \\ 1 \end{bmatrix}, \ t \in R.$$

Therefore, all the eigenvectors are spanned by $\begin{bmatrix} 1 \\ 1 \end{bmatrix}$. We do not have two linearly independent eigen-

vectors. By the previous theorem, A is not diagonalizable.

Note 1

An $n \times n$ matrix may fail to be diagonalizable since not all roots of its characteristic equation are real numbers, or it does not have n linearly independent eigenvectors.

Note 2

The set S_j consisting of all eigenvectors of an $n \times n$ matrix A associated with eigenvalue λ_j and zero vector $\vec{0}$ is a subspace of $\mathbb{R}^n$. S_j is called the eigenspace associated with λ_j.

Example 27 (diagonalization of a 3×3 matrix)

Diagonalize the matrix: $A = \begin{bmatrix} 1 & 0 & 0 \\ 0 & 0 & 1 \\ 1 & 1 & 0 \end{bmatrix}$

Solution
Let us find the eigenvalues:

$$\begin{vmatrix} 1-\lambda & 0 & 0 \\ 0 & -\lambda & 1 \\ 1 & 1 & -\lambda \end{vmatrix} = \left(1-\lambda\right)\left(\lambda^2 -1\right) = -\left(1-\lambda\right)^2\left(1+\lambda\right) = 0$$

There are two eigenvalues: the first one $\lambda = 1$ is degenerate with a multiplicity 2; the second one is $\lambda = -1$. For each of these eigenvalues, let us find the associated eigenvectors.

Eigenvector for $\lambda = -1$.
The eigenvalue equation is $AV = \lambda V$ where V is the eigenvector associated with the eigenvalue λ.

$$\begin{bmatrix} 1 & 0 & 0 \\ 0 & 0 & 1 \\ 1 & 1 & 0 \end{bmatrix}\begin{bmatrix} x \\ y \\ z \end{bmatrix} = -1\begin{bmatrix} x \\ y \\ z \end{bmatrix} \text{ gives } \begin{cases} x = -x \\ z = -y \\ x+y = -z \end{cases} \text{ which gives}$$

$$\begin{cases} x = 0 \\ z = -y \\ 0+y = -z \end{cases}$$

So, vector V can be expressed as follows:

$$\begin{bmatrix} x \\ y \\ z \end{bmatrix} = \begin{bmatrix} 0 \\ y \\ -y \end{bmatrix} = y\begin{bmatrix} 0 \\ 1 \\ -1 \end{bmatrix}$$

We can choose $V_1 = \begin{bmatrix} 0 \\ 1 \\ -1 \end{bmatrix}$ as the first eigenvector.

Eigenvector for $\lambda = 1$.
Let us now find the second and the third eigenvector:
    For $\lambda = 1$, the eigenvalue equation is written:

$$\begin{bmatrix} 1 & 0 & 0 \\ 0 & 0 & 1 \\ 1 & 1 & 0 \end{bmatrix}\begin{bmatrix} x \\ y \\ z \end{bmatrix} = 1\begin{bmatrix} x \\ y \\ z \end{bmatrix} \text{ which gives the equations}$$

$$\begin{cases} x = x \\ z = y \\ x+y = z \end{cases} \text{ and } \begin{cases} 0x = 0 \\ z = y \\ x+y = z \end{cases}$$

The second equation of the previous system $z = y$ plugged into the third equation gives $A = \begin{bmatrix} 1 & 0 & 0 \\ 0 & 0 & 1 \\ 1 & 1 & 0 \end{bmatrix}$. Therefore, we have:

$$\begin{bmatrix} x \\ y \\ z \end{bmatrix} = \begin{bmatrix} 0 \\ y \\ y \end{bmatrix} = y \begin{bmatrix} 0 \\ 1 \\ 1 \end{bmatrix}.$$

There is a second eigenvector, $A = \begin{bmatrix} 1 & 0 & 0 \\ 0 & 0 & 1 \\ 1 & 1 & 0 \end{bmatrix}$.

We only have here two eigenvectors for the 3×3 matrix $= \begin{bmatrix} 1 & 0 & 0 \\ 0 & 0 & 1 \\ 1 & 1 & 0 \end{bmatrix}$; according to Theorem 5,

the matrix $A = \begin{bmatrix} 1 & 0 & 0 \\ 0 & 0 & 1 \\ 1 & 1 & 0 \end{bmatrix}$ is not diagonalizable because it does not have three linearly independent eigenvectors.

4.9 LINEAR TRANSFORMATION

Definition 3

Let V and W be two vector spaces and L a function defined on V with values in W by: $L : V \to W$.
L is called a linear transformation if the following conditions are satisfied:

i. $L(x + y) = L(x) + L(y)$ for x and y elements of V.

ii. $L(\alpha x) = \alpha L(x)$ for α a real number.

The previous two properties of the linear transformation can be condensed in one that is L is called a linear transformation if $L(\alpha x + \beta y) = \alpha L(x) + \beta L(y)$.

Example 28

Consider

$$\begin{cases} x = -x \\ z = -y \\ x + y = -z \end{cases} \quad \mathbb{R}^3 \to \mathbb{R}^2 \text{ defined by } L(x_1, x_2, x_3) = (x_3 - x_1, x_1 + x_2)$$

a. Compute $L(e_1)$, $L(e_2)$, $L(e_3)$

b. Show that L is a linear transformation

c. Show that $L(x_1, x_2, x_3) = x_1 L(e_1) + x_2 L(e_2) + x_3 L(e_3)$

Solution

a. $e_1 = (1,0,0), e_2 = (0,1,0), e_3 = (0,0,1)$

$L(e_1) = L(1,0,0) = (\times 1, 1)$
$L(e_2) = L(0,1,0) = (0,1)$
$L(e_3) = L(0,0,1) = (1,0)$

b. Consider α, β elements of $\mathbb{R}$ and vectors (x_1, y_1, z_1) and (x_2, y_2, z_2) elements of $\mathbb{R}^3$.

$$L(x_1 + x_2, y_1 + y_2, z_1 + z_2) = (z_1 + z_2 - x_1 - x_2, x_1 + x_2 + y_1 + y_2)$$
$$= (z_1 - x_1, x_1 - y_1) + (z_2 - x_2, x_2 + y_2)$$
$$= L(x_1, y_1, z_1) + L(x_2, y_2, z_2)$$

Also,

$$L[\alpha(x, y, z)] = L(\alpha x, \alpha y, \alpha z)$$
$$= (\alpha z - \alpha x, \alpha x + \alpha y)$$
$$= \alpha(z - x, x + y)$$
$$= \alpha L(x, y, z)$$

So, L is a linear transformation.

$$x_1 L(e_1) + x_2 L(e_2) + x_3 L(e_3) = x_1(-1,1) + x_2(0,1) + x_3(1,0) = (x_3 - x_1, x_1 + x_2) = L(x_1, x_2, x_3).$$

Theorem 6

If L is a linear transformation from a vector space V to a vector space W, then

$$L(o) = o$$

$$L(-x) = -L(x)$$

$$L\left(\sum_{i=1}^{n} \alpha_i x_i\right) = \sum_{i=1}^{n} \alpha_i L(x_i)$$

Proof

1. Since L is a linear transformation, $L(x + y) = L(x) + L(y)$

Assume $\rightarrow$; then

$$L(o + o) = L(o) + L(o)$$

$$L(o) = 2L(o)$$

Therefore, $L(o) = o$

2. $L(x - x) = L(x) + L(-x) = o$; so, $L(-x) = -L(x)$

3. Since L is a linear transformation, we have

$$L(\alpha_1 x_1 + \alpha_2 x_2) = \alpha_1 L(x_1) + \alpha_2 L(x_2)$$

So, for $n = 2$, $L\left(\sum_{i=1}^{2} \alpha_i x_i\right) = \sum_{i=1}^{2} \alpha_i L(x_i)$

Also, $L(\alpha_1 x_1 + \alpha_2 x_2 + \alpha_3 x_3) = \alpha_1 L(x_1) + \alpha_2 L(x_2) + \alpha_3 L(x_3)$

So, for $n = 3$, $L\left(\sum_{i=1}^{3} \alpha_i x_i\right) = \sum_{i=1}^{3} \alpha_i L(x_i)$.

Let us use the induction method. Assume now that the property is true at the nth order; that is,

$$L\left(\sum_{i=1}^{3} \alpha_i x_i\right) = \sum_{i=1}^{3} \alpha_i L(x_i).$$

But at $n+1$ order, we have:

$$L\left(\sum_{i=1}^{n+1} \alpha_i x_i\right) = \sum_{i=1}^{n} L(\alpha_i x_i) + L(\alpha_{n+1} x_{n+1})$$

$$= \sum_{i=1}^{n} \alpha_i L(x_i) + \alpha_{n+1} L(x_{n+1})$$

$$= \sum_{i=1}^{n+1} \alpha_i L(x_i)$$

which proves that the property is true at the higher order $n+1$.

4.9.1 CONCLUSION

For all real number α_i and vector x_i, if L is a linear transformation, then

$$= \sum_{i=1}^{n} \alpha_i L(x_i) + \alpha_{n+1} L(x_{n+1})$$

4.10   MATRIX REPRESENTATION OF A LINEAR TRANSFORMATION

Definition 4
A linear transformation

$L : \mathbb{R}^n \to \mathbb{R}^m$ with $L(e_k) = (a_{1k}, a_{2k}, ..., a_{mk})$, where α_i is the standard basis of R^n, can be represented by a matrix $A = [a_{jk}]$ which is an $m \times n$ matrix whose entries in the kth column are the coordinates of $L(e_k)$.

Example 29 Consider the linear transformation:
$L(-x) = -L(x)\ \mathbb{R}^n \to \mathbb{R}^m$

$$(x_1, x_2, x_3) \to (-6x_2 + 2x_3, x_1 - x_2 + x_3, -x_1 + x_2 - 6x_3, 3x_1 - x_2 + 4x_3)$$

Find the matrix representation of L.

Solution
We can express:

$$L(e_1) = L(1,0,0) = (0,1,-1,3)$$

$$\left(e_2\right) = L\left(0,1,0\right) = \left(-6,-1,1,-1\right)$$

$$L\left(e_3\right) = L\left(0,0,1\right) = \left(2,1,-6,4\right)$$

So, the matrix representation of L is

$$A = \begin{bmatrix} 0 & -6 & 2 \\ 1 & -1 & 1 \\ -1 & 1 & -6 \\ 3 & -1 & 4 \end{bmatrix}$$

4.10.1 KERNEL AND RANGE OF A LINEAR TRANSFORMATION

Consider a linear transformation L, mapping vector space V into vector space W. There are two subspaces associated with L. One is a subspace of V called the kernel of L; the second is a subspace of W called the range of L.

Definition 5 of the Kernel of L

If $\begin{cases} 0x = 0 \\ z = y \\ x + y = z \end{cases}$ is a linear transformation mapping V into W, the kernel of L is the set of vectors x in V for which $L\left(x\right) = 0$.

So, Ker(L)=$\{x \in V : L\left(x\right) = 0\}$.

Example 30

Consider the matrix representation $A = \begin{bmatrix} 2 & -6 & 4 \\ 1 & -1 & 2 \end{bmatrix}$ of a linear transformation L. Find the kernel of L.

Solution

$$L\left(e_1\right) = 2e_1 + e_2$$

$$L\left(e_2\right) = -6e_1 - e_2$$

$$L\left(e_3\right) = 4e_1 + 2e_2$$

So, L is a linear transformation mapping a three-dimensional vector space V into a two-dimensional vector space W.

Consider $X = \begin{bmatrix} x_1 \\ x_2 \\ x_3 \end{bmatrix} \in V$

$$L\left(X\right) = AX$$

$L(X) = 0$ implies $AX = 0$; that is: $\begin{bmatrix} 2 & -6 & 4 \\ 1 & -1 & 2 \end{bmatrix} \begin{bmatrix} x_1 \\ x_2 \\ x_3 \end{bmatrix} = \begin{bmatrix} 0 \\ 0 \end{bmatrix}$

We then have the system of equations: $\begin{cases} x_1 - 3x_2 + 2x_3 = 0 \\ x_1 - x_2 + 2x_3 = 0 \end{cases}$, which gives the relation, $x_1 = -2x_3$. As

a result, we can choose the vector $(-2, 0, 1)$ as the basis of the kernel, which implies that the kernel

is a one-dimensional vector space with basis $A = \begin{bmatrix} 1 & 1 \\ 0 & 1 \end{bmatrix}$.

Example 31

The matrix representation of a linear transformation L is $A = \begin{bmatrix} 1 & 1 \\ 0 & 1 \end{bmatrix}$. Find the kernel of L.

Solution

Consider $u = \begin{bmatrix} x \\ y \end{bmatrix}$.

$L(u) = Au = 0$ if and only if $\begin{bmatrix} 1 & 1 \\ 0 & 1 \end{bmatrix}\begin{bmatrix} x \\ y \end{bmatrix} = \begin{bmatrix} 0 \\ 0 \end{bmatrix}$.

We have the system $\begin{cases} x + y = 0 \\ y = 0 \end{cases}$

So, $x = y = 0$ and the kernel is the null vector: $Ker(L) = \{0\}$.

4.10.2 Definition of the Range of a Linear Transformation

Let L be a linear transformation mapping V into W. Then the range of L is the set of all vectors y of W such that there exists a vector x of V and $L(x) = y$. That is:

$$R_g(L) = \{y \in W : L(x) = y; x \in V\} \tag{4.33}$$

Example 32

Find the range of the linear transformation L whose matrix representation is $A = \begin{bmatrix} 2 & -6 & 4 \\ 1 & -1 & 2 \end{bmatrix}$

Solution

$$AX = Y$$

$$A = \begin{bmatrix} 2 & -6 & 4 \\ 1 & -1 & 2 \end{bmatrix}$$

which gives the system

$$(S)\begin{cases} x - 3y + 2z = x'/2 \\ x - y + 2z = y' \end{cases}$$

and $\text{col}_1(P), \text{col}_2(P), \ldots, \text{col}_n(P); \text{col}_1(P), \text{col}_2(P), \ldots, \text{col}_n(P); z = z$

Because the system (S) has a solution, the range is $R_g(L) = \mathbb{R}^2$; x' and y' are both real numbers.

Definition 6

Let $L : V \rightarrow W$ be a linear transformation.

L is said to be one-to-one if $L(v) = L(w) \Rightarrow v = w$.

L is said to be unto if $R_g(L) = W$.

Also,

$$dim\left[Ker(L) \right] + dim\left[R_g(L) \right] = dim\left[V \right].$$ (4.34)

PROBLEM SET 4

4.1   Consider the following matrix

$$A = \begin{bmatrix} 3 & 3 & 1 \\ 1 & 2 & 1 \\ 2 & -1 & 1 \end{bmatrix}$$

    a.   Find the determinant of A

    b.   Is A invertible? Why?

    c.   Use the row operations to find the inverse $A^{\times 1}$ of A.

    d.   Calculate the product $A.A^{\times 1}$

4.2   Calculate the product AB and BA of the following matrices:

$$col_1(P), col_2(P), \ldots, col_n(P)$$

    a. $A = \begin{bmatrix} 1 & 0 & 2 \\ 2 & 4 & -2 \\ -1 & 2 & 3 \end{bmatrix}$, $B = \begin{bmatrix} 8 & 5 & 7 \\ 2 & -5 & 0 \\ 4 & 1 & 2 \end{bmatrix}$

    b. $A = \begin{bmatrix} 3 & 0 & 1 \\ 3 & -4 & 2 \\ 1 & -2 & -3 \end{bmatrix}$, $B = \begin{bmatrix} 2 & 1 & 3 \\ 3 & 6 & -4 \\ 5 & 3 & 2 \end{bmatrix}$

    c. $A = \begin{bmatrix} 4 & 1 \\ 7 & 2 \end{bmatrix}$, $B = \begin{bmatrix} 2 & -1 \\ -7 & 4 \end{bmatrix}$

4.3   Compute if possible, AB, BA, $6A$ and $-4B$

$$A = \begin{bmatrix} 1 & 0 \\ 2 & 4 \\ -1 & 2 \end{bmatrix} B = \begin{bmatrix} 4 & -3 & 0 \\ 1 & 1 & -2 \end{bmatrix}$$

4.4   a.   Use Cramer's to solve the system of equations. Remember to find D_x, Dy, D.

$$\begin{cases} 3x + 3y + z = -4 \\ x + 2y + z = -15 \\ 2x - y + z = 12 \end{cases}$$

    b.   Use the Gauss Jordan elimination to find the solutions to the same system of linear equations.

    c.   Use the method of the inverse matrix to solve the same (above) system.

4.5   Consider the system of linear equations:

$$(S) \begin{cases} x+y+z = 4 \\ x-y-z = -4 \\ 2x+3y+2z = 8 \end{cases}$$

b_1. Find the determinant of matrix A associated with system S
b_2. Find the inverse of A
b_3. Use the inverse matrix to solve the system of equations S
b_4. Write the augmented matrix of system S
b_5. Use the row operations to solve system S

4.6   Find the inverse of the matrix.

a. $A = \begin{bmatrix} 1 & -1 & 2 \\ -2 & 2 & 1 \\ -1 & 3 & 5 \end{bmatrix}$

b. $A = \begin{bmatrix} 1 & 0 & 2 \\ 2 & 4 & -2 \\ -1 & 2 & 3 \end{bmatrix}$

c. $A = \begin{bmatrix} 2 & 4 & 5 \\ 0 & 1 & 9 \\ 6 & 7 & 12 \end{bmatrix}$

d. $A = \begin{bmatrix} 1 & 1 & -1 \\ 3 & -1 & -2 \\ 5 & 1 & 8 \end{bmatrix}$

4.7   Find the eigenvalues and eigenvectors of the matrix $A = \begin{bmatrix} 1 & 0 \\ 2 & 3 \end{bmatrix}$.

4.8   Find the eigenvalues and eigenvectors of the matrix $A = \begin{bmatrix} 4 & 1 \\ 7 & 2 \end{bmatrix}$, $B = \begin{bmatrix} 2 & -1 \\ -7 & 4 \end{bmatrix}$.

4.9   You are given the following matrices:

$$A = \begin{bmatrix} 3/2 & -i/2 \\ i/2 & 3/2 \end{bmatrix}, B = \begin{bmatrix} 1 & 0 \\ 0 & 1 \end{bmatrix}, C = \begin{bmatrix} 0 & -i \\ i & 0 \end{bmatrix} \text{ and } D = \begin{bmatrix} -1/2 & 3i/2 \\ -3i/2 & -1/2 \end{bmatrix}$$

a.   Prove that they all commute with each other.
b.   Find the eigenvalues of each.
c.   Construct a common orthonormal set of eigenvectors for A, B, C and D.
d.   Construct the unitary matrix that will diagonalize A, B, C and D simultaneously.
e.   Write down the diagonal forms of A, B, C and D.

4.10 Write the augmented matrix and use Gauss–Jordan elimination to solve the system:

a. $\begin{cases} 3x-4y-z = 6 \\ 2x-y+z = -1 \\ 4x-7y-3z = 13 \end{cases}$

b.   Evaluate the determinant

$$b_1 = \begin{vmatrix} -2 & -\sqrt{5} \\ -\sqrt{5} & 3 \end{vmatrix}; \quad b_2 = \begin{vmatrix} 4 & 5 \\ -6 & -3 \end{vmatrix}. \mathrm{col}_1(P), \mathrm{col}_2(P), \ldots, \mathrm{col}_n(P)$$

4.11   Use the Gauss–Jordan method to find the inverse of the following matrix:

$$A = \begin{bmatrix} 1 & 2 & -1 \\ 3 & 5 & 3 \\ 2 & 4 & 3 \end{bmatrix}$$

4.12   Use Gauss–Jordan elimination (the Row echelon matrix) to find the solutions of the following system of equations:

$$\begin{cases} 2x + 8y - 4z = 4 \\ 3x + 8y + 5z = -11 \\ -2x + y + 12z = -17 \end{cases}$$

4.13   The Maxwell stress tensor in electrostatic is as follows:

$$T = \varepsilon \begin{bmatrix} \frac{1}{2}\left(E_x^2 - E_y^2 - E_z^2\right) & E_x E_y & E_x E_z \\ E_y E_x & \frac{1}{2}\left(E_y^2 - E_x^2 - E_z^2\right) & E_y E_z \\ E_z E_x & E_z E_y & \frac{1}{2}\left(E_z^2 - E_x^2 - E_y^2\right) \end{bmatrix}$$

where ε is a constant and the electric vector is $\vec{E} = E_x \vec{i} + E_y \vec{j} + E_z \vec{k}$.

Show that the eigenvalues of T (which are referred to as the principle stresses) are $\frac{1}{2}\varepsilon E^2$,

$-\frac{1}{2}\varepsilon E^2$ and $-\frac{1}{2}\varepsilon E^2$ where $E^2 = E_x^2 + E_y^2 + E_z^2$ in a three-dimensional (x, y, z) system.

Show that one of the corresponding principal axes is parallel to the electric field $\vec{E}$.

4.14   Find the eigenvalues and eigenvectors of the matrix $A = \begin{bmatrix} 3 & 3 & 1 \\ 1 & 2 & 1 \\ 2 & -1 & 1 \end{bmatrix}$; normalize the

eigenvectors and form a complete set of eigenvectors.

4.15   Find the eigenvalues and eigenvectors of the matrix $= \begin{bmatrix} 1 & -1 & 2 \\ -2 & 2 & 1 \\ -1 & 3 & 5 \end{bmatrix}$; normalize the

eigenvectors and form a complete set of eigenvectors.

4.16   Let $L: \mathbb{R}^n \rightarrow \mathbb{R}^m$ be a linear transformation.

   a.   Show that if $\dfrac{1}{2}\varepsilon E^2$, then L must have a non-trivial kernel, i.e., $dim[Ker(L)] > 0$.

   b.   If $n \leq m$ does L have to be one-to-one?

4.17   Let

$$L\left(x_1, x_2, x_3\right) = \left(x_2 + x_3, 6x_1 - x_2 + 3x_3, 2x_1 + 3x_2 - 7x_3, 2x_1 + 6x_3, 2x_1 + 6x_3\right)$$

   a.   Compute $L\left(e_k\right)$ with $k = 1, 2, 3$
   b.   Find the matrix representation A of L with respect to the standard bases in $\mathbb{R}^4$ and $\mathbb{R}^2$.

4.18   You are given the equation

$$\begin{vmatrix} x & 0 & c \\ -1 & x & b \\ 0 & -1 & a \end{vmatrix} = ax^2 + bx + c$$

   a.   Verify the equation.
   b.   Use the equation as a model to find a determinant that is equal to $ax^3 + bx^2 + cx + d$.

5 Tensor Analysis

A good deal of my research in physics has consisted in not setting out to solve some particular problem, but simply examining mathematical equations of a kind that physicists use and trying to fit them together in an interesting way, regardless of any application that the work may have. It is simply a search for pretty mathematics. It may turn out later to have an application. Then one has good luck.

–Paul A. M. Dirac

INTRODUCTION

Tensor analysis is an important topic in physics especially in theoretical physics and chemistry. In rheology, the consideration of direction is of great importance, and tensor analysis is to play a key role. Tensor analysis has also applications in special and general theories of relativity. In an anisotropic medium, where the physical properties are not the same in all directions, tensors are widely used in equations involving the description of the medium. Tensor analysis deals with repeated indices.

5.1 DEFINITION OF THE TENSOR

To define the tensors, we will consider two vector spaces: E_{n_j} with basis $\left(u_1, u_2, u_3, \ldots, u_{m_1}\right)$ and E_{m_1} with the basis $c^i = \left\{0, 0, \dfrac{1}{h_3}\right\}$. We can define the $n_1 n_2$-dimensional tensor product space, $E_{n_1} \otimes E_{n_2}$ with the basis $\left(u_1 \otimes v_1, u_2 \otimes v_2, u_3 \otimes v_3, \ldots, u_{m_1} \otimes u_{n_2}\right)$ or in general, $u_i \otimes v_3$. Tensor can be regarded as the operation $\otimes$ satisfying the following properties:

$\otimes$ **is distributive with respect to addition:**

For three vectors, u_i, u_j and E_{m_1}, we have the following relation: $\left(u_i + u_j\right) \otimes v_k = u_i \otimes v_k + u_j \otimes v_k$ where i, j, k are arbitrary dummy indices and for three vectors, $u_i\, v_j$ and v_k, we have $u_i \otimes \left(v_j + v_k\right) = u_t \otimes v_j + u_i \otimes v_k$.

$\otimes$ **is associative with respect to the multiplication by a scalar:**

For all real numbers α, $\alpha\left(u_i \otimes v_j\right) = u_i \otimes \left(\alpha v_j\right) = \left(\alpha u_i\right) \otimes v_j$.

Every element of the tensor product space $\otimes$ can be written once as a linear combination of the elements of the basis $\left(u_i \otimes v_j\right)$, that is, the tensor T can be written as follows:

$$T = t^{ij} u_i \otimes v_j = t^{ij} e_{ij}, \text{ where } e_{ij} = \left(u_i \otimes v_j\right).$$

The coefficients t^{ij} are the entries or the components of tensor T. Since e_{ij} is a basis, $T = 0$ if and only if all the components t^{ij} are zero. The concept of the tensor can be generalized using several vector spaces. Thus, the multiple tensor product space $E_{n_1} \otimes E_{n_2} \otimes E_{n_3} \otimes E_{n_4} \ldots \otimes E_{n_p}$ with p an integer, can be constructed.

DOI: 10.1201/9781003478812-5

Example 1

Assume that we have two vector spaces, E_2 with the basis $\left(\vec{i},\vec{j}\right)$ and E_3 with the basis $\left(\vec{e}_1,\vec{e}_2,\vec{e}_3\right)$. The basis of the cross product of the two spaces, E_2 and E_3, represented by $E_2 \otimes E_3$, will be a set of $2 \times 3 = 6$ elements $\left(\vec{i}\otimes\vec{e}_1,\vec{i}\otimes\vec{e}_2,\vec{i}\otimes\vec{e}_3,\vec{j}\otimes\vec{e}_1,\vec{j}\otimes\vec{e}_2,\vec{j}\otimes\vec{e}_3\right)$.

5.1.1 EINSTEIN NOTATION

Einstein summation convention can be used to simplify tensor notation. For example, the product of two tensors, $T = \sum_i^3\sum_j^3 A_{ij}\widehat{e_i}\widehat{ej}$, where $\widehat{e_i},\widehat{e_j}$ are unit vectors, will be written with the omission of the sum sign, $T = A_{ij}\widehat{e_i}\widehat{e_j}$, in Einstein notation.

5.2 COORDINATES AND TENSORS

Consider the space of real numbers of dimension n, $\mathbb{R}^n$, and a single real time, t. Continuum properties in this space can be described by arrays of different dimensions, m, such as scalars ($m = 0$), vectors ($m = 1$), matrices ($m = 2$) and general multi-dimensional arrays. In this space, we shall introduce a coordinate system $\{x^i\}_{i=1\times n}$, as a way of assigning n real numbers for every point of space. There can be a variety of possible coordinate systems. A general transformation rule between the coordinate systems is

$$\tilde{x}^i = \tilde{x}^i\left(x^1...x^n\right) \tag{5.1}$$

where the symbol $\sim$ is called the tilde symbol.

The small displacement dx^j can be transformed from the coordinate system (E_{n_j}) to a new coordinate system ($\tilde{x}^i$) using the partial differentiation rules applied to equation (5.1):

$$d\tilde{x}^i = \sum_j \frac{\partial \tilde{x}^i}{\partial x^j} dx^j$$

or in Einstein notation

$$d\tilde{x}^i = \frac{\partial \tilde{x}^i}{\partial x^j} dx^j \tag{5.2}$$

This transformation rule can be generalized using a tensor A to a set of vectors that we shall call contravariant vectors:

$$A\tilde{x}^i = \frac{\partial \tilde{x}^i}{\partial x^j} Ax^j \tag{5.3}$$

Overall, a contravariant vector is defined as a vector which transforms to a new coordinate system according to (5.3). We can also introduce the transformation matrix as follows:

$$a^i_j = \frac{\partial \tilde{x}^i}{\partial x^j} \tag{5.4}$$

with which equation (5.3) can be written as

$$A^i = a^i_j A^j \tag{5.5}$$

Transformation rule (5.3) does not apply to all the vectors in the space. For example, a partial derivative $\partial / \partial x_i$ will transform as follows:

$$\frac{\partial}{\partial \tilde{x}^i} = \frac{\partial}{\partial \tilde{x}^i} \frac{\partial x^j}{\partial x^j} = \frac{\partial x^j}{\partial \tilde{x}^i} \frac{\partial}{\partial x^j} \tag{5.6}$$

that is, the transformation coefficients are the other way up compared to (5.2). Now, we can generalize this transformation rule, so that each vector that transforms according to (5.6) will be called a covariant vector:

$$\tilde{A}_i = \frac{\partial x^j}{\partial \tilde{x}^i} A_j \tag{5.7}$$

This provides the reason for using lower and upper indices in a general tensor notation.

5.3   TENSOR RANK

A tensor of order m in a n-dimensional space is a set of n^m elements identified by m integer indices. For example, a third-order tensor A can be written as A_{ijk} and an m-order tensor can be written as $A_{i_1 \ldots i_m}$. Each index of a tensor changes between 1 and n. For example, in a three-dimensional space ($n = 3$) a second-order tensor will be represented by $3^2 = 9$ components or elements. The tensor index should also comply with one of the two transformation rules: (5.3) or (5.7). An index that complies with the rule (5.7) is called a covariant index and is denoted as a sub-index, and an index complying with the transformation rule (5.3) is called a contravariant index and is denoted as a super-index. Each index of a tensor can be either covariant or contravariant. For example, A^k_{ij} is a third-order tensor of second-order covariant and first-order contravariant. Tensors are usually functions of space and time. As an example, $A_{i_1, i_m} = A_{i_1, i_m} \left(x^1 \ldots x^n, t \right)$ defines a tensor field, i.e., for every point x^i and time t there are a set of m^n numbers, $A_{i_1 \ldots i_m}$.

Remark
Note that the coordinates x^i are not tensors, since generally, they are not transformed as (5.5). The transformation law for the coordinates is actually given by (5.1). Nevertheless, we shall use the upper (contravariant) indices for the coordinates.

5.4   DEFINITION OF KRONECKER DELTA TENSOR

The second-order delta tensor, δ_{ij} is defined as

$$\begin{array}{llll} i = j & \text{implies} & \delta_{ij} = 1 \\ i \neq j & \text{implies} & \delta_{ij} = 0 \end{array} \tag{5.8}$$

From the definition and since coordinates x^i are independent of each other, it follows that

$$\frac{\partial x^i}{\partial x^j} = \delta_{ij} \tag{5.9}$$

5.4.1 Delta Product

From equation (5.3) and the summation convention (5.2), it follows that

$$\delta_{ij}A_j = A_i \qquad (5.10)$$

Assume that there exists the transformation inverse to equation (5.5), which we call b_j^i:

$$dx^i = b_j^i d\tilde{x}^j \qquad (5.11)$$

Then by analogy to equation (5.4) b_j^i can be defined as

$$b_j^i = \frac{dx^i}{d\tilde{x}^j} \qquad (5.12)$$

From this relation and the independence of coordinates in (5.9), it follows that $a_j^i b_k^j = b_j^i a_k^j = \delta_{ik}$, namely:

$$a_j^i b_k^j = \frac{d\tilde{x}^i}{dx^j}\frac{dx^j}{d\tilde{x}^k} = \frac{\partial x^j}{\partial x^j}\frac{d\tilde{x}^i}{d\tilde{x}^k} = \frac{\partial \tilde{x}^i}{\partial \tilde{x}^k} = \delta_{ik} \qquad (5.13)$$

5.5 CARTESIAN TENSORS

Cartesian tensors are a subset of general tensors for which the transformation matrix (5.4) satisfies the following relation:

$$a_i^k b_j^k = \frac{\partial \tilde{x}^k}{\partial x^i}\frac{\partial \tilde{x}^k}{\partial x^j} = \delta_{ij} \qquad (5.14)$$

For Cartesian tensors, we have

$$\frac{\partial \tilde{x}^i}{\partial x^k} = \frac{\partial x^k}{\partial \tilde{x}^i} \qquad (5.15)$$

which means that both equations (5.5) and (5.6) are transformed with the matrix a_k^i. This, in turn, means that the difference between the covariant and contravariant indices vanishes for the Cartesian tensors. Considering this we shall only use the sub-indices whenever we deal with Cartesian tensors.

5.6 TENSOR RULES

Tensor rules guarantee that if an expression follows these rules, it represents a tensor according to Section 5.1. Thus, following tensor rules, one can build tensor expressions that will preserve tensor properties of coordinate transformations (5.1) and coordinate invariance. Tensor rules are based on the following definitions and propositions.

5.6.1 Tensor Terms

A tensor term is a product of tensors. For example:

$$A_{ijk}B_{jk}C_{pq}E_qF_p \qquad (5.16)$$

5.6.2 TENSOR EXPRESSION

A tensor expression is a sum of tensor terms. For example:

$$A_{ijk}B_{jk} + C_i D_{pq}E_q F_p \tag{5.17}$$

Generally, the terms in the expression may come with the + or the × sign.

5.6.3 ALLOWED OPERATIONS

The only allowed algebraic operations in tensor expressions are addition, subtraction and multiplication. Divisions are only allowed for constants, like $1/C$. If a tensor index appears in a denominator, such term should be redefined, so as not to have tensor indices in a denominator. For example, $1/A_i$ should be redefined as $B_i \equiv 1/A_i$.

5.6.4 TENSOR EQUALITY

Tensor equality is an equality of two tensor expressions. For example:

$$A_{ij}B_j = C_{ikp}D_k E_p + E_j C_{jki}B_k \tag{5.18}$$

5.7 FREE AND DUMMY INDICES

A free index is any index that occurs only once in a tensor term. For example, index i is a free index in the term (5.16). Every term in tensor equality should have the same set of free indices. If index i is a free in any term of tensor equality, such as (5.18), it should be the free index in all other terms. For example, $A_{ij}B_j = C_j D_j$. It is not a valid tensor equality since index i is a free index in the term on the right side but not on the left side. Any free index in a tensor expression can be named by any symbol if this symbol does not already occur in the tensor expression. For example, the equality

$$A_{ij}B_j = C_i D_j E_j \tag{5.19}$$

is equivalent to

$$A_{kj}B_j = C_k D_j E_j \tag{5.20}$$

Here, we replaced the free index i with k.

A dummy index is any index that occurs twice in a tensor term. For example, indices j, k, p and q in (5.16) are dummy indices. No index can occur more than twice in any tensor term. Any dummy index in a tensor term can be renamed to any symbol if this symbol does not already occur in this term. For example, the term $A_i B_i$ is equivalent to $A_j B_j$, and so are the terms $A_{ijk}B_j C_k$ and $A_{ipq}B_p C_q$.

5.8 RANK OF A TERM

A rank of a tensor term is equal to the number of its free indices. For example, the rank of the term $A_{ijk}B_j C_k$ is equal to 1. It follows from (5.7) that the ranks of all the terms in a valid tensor expression should be the same. Note that the difference between the order and the rank is that the order is equal to the number of indices of a tensor, and the rank is equal to the number of free indices in a tensor term.

5.9 SUMMATION RULE AND EXCEPTION

Any dummy index implies summation, i.e.,

$$A_i B_i = \sum_i^n A_i B_i \tag{5.21}$$

If there should be no summation over the repeated indices, it can be indicated by enclosing such indices in parentheses. For example, the expression $C_{(i)} A_{(i)} B_j = D_{ij}$ does not imply summation over i.

5.10 SCALAR PRODUCT

5.10.1 DEFINITION

A scalar product operation notation from vector algebra: $\vec{A} \cdot \vec{B}$ is expressed in the tensor notation $A_i B_i$. The scalar product operation is also called a contraction of indices.

5.10.2 REPEATED INDICES

In case an index occurs more than twice in a term, this term should be redefined so as not to contain more than two occurrences of the same index. For example, the term $A_{ik} B_{jk} C_k$ should be rewritten as $A_{ik} D_{jk}$, where D_{jk} is defined as $D_{jk} = B_{j(k)} C_{(k)}$ with no summation over k in the last term.

5.10.3 RENAMING RULES

Note that while the dummy index renaming rule (5.16) is applied to each tensor's term separately, the free index renaming rule (5.19) should apply to the whole tensor expression. For example, the equality (5.19) above, $A_{ij} B_j = C_i D_j E_j$ can also be rewritten as

$$A_{kp} B_p = C_k D_j E_j \tag{5.22}$$

without changing its meaning.

5.11 PERMUTATION TENSOR

The components of a third-order permutation tensor ε_{ijk} are defined to be equal to 0 when any two of i, j and k are equal, 1 when these indices are all different and come in successive order or obtained by cyclic permutation of 123 and $\times 1$ when they are all different and come in the order 132. In a mathematical language, this can be expressed as

$$i = j \cup i = k \cup j = k \Rightarrow \varepsilon_{ijk} = 0$$

$$ijk \in PG(123) \Rightarrow \varepsilon_{ijk} = 1$$

$$ijk \in PG(132) \Rightarrow \varepsilon_{ijk} = -1 \tag{5.23}$$

where $PG(abc)$ is a permutation group of a triple of indices abc, i.e., $PG(abc) = \{abc, bca, cab\}$. For example, the permutation group of 123 will consist of three combinations: 123, 231 and 312, and the permutation group of 123 consist of 132, 321 and 213.

5.12   PERMUTATION OF THE TENSOR INDICES

From the definition of the permutation tensor, it follows that the permutation of any of its two indices changes its sign:

$$\varepsilon_{ijk} = -\varepsilon_{ikj} \tag{5.24}$$

A tensor with this property is called skew-symmetric.

5.13   VECTOR PRODUCT

A vector product (cross-product) of two vectors in vector notation is expressed as

$$\vec{A} = \vec{B} \times \vec{C} \tag{5.25}$$

which in tensor notation can be expressed as

$$A_i = \varepsilon_{ijk} B_j C_k \tag{5.26}$$

5.14   CROSS PRODUCT

Tensor expression (5.26) is more accurate than its vector counterpart (5.25), since it explicitly shows how to compute each component of a vector product.

5.15   SYMMETRIC IDENTITY

If A_{ij} is a symmetric tensor, then the following identity is true:

$$\varepsilon_{ijk} A_{jk} = 0 \tag{5.27}$$

Proof
From the symmetry of A_{ij} we have

$$\varepsilon_{ijk} A_{jk} = \varepsilon_{ijk} A_{kj} \tag{5.28}$$

Let's rename index j into k and k into j in the right-hand side (RHS) of this expression.

$$\varepsilon_{ijk} A_{kj} = \varepsilon_{ijk} A_{jk}$$

Using (5.24), we obtain:

$$\varepsilon_{ijk} A_{kj} = -\varepsilon_{ijk} A_{jk}$$

Comparing the right-hand side (RHS) of this expression to the left-hand side (LHS) of (5.28) we have

$$\varepsilon_{ijk} A_{kj} = \varepsilon_{ijk} A_{jk}$$

from which we conclude that (5.27) is true.

5.16   TENSOR IDENTITY

The following tensor identity is true:

$$\varepsilon_{ijk}\varepsilon_{ipq} = \delta_{jp}\delta_{kq} - \delta_{jq}\delta_{kp} \tag{5.29}$$

Proof

The above identity can be proved by examining the components of equality (5.29) component-by-component.

5.17   VECTOR IDENTITY

Using the tensor identity (5.29), it is possible to prove the following important vector identity:

$$\vec{A} = \vec{B}\times\vec{C} = \vec{B}\left(\vec{A}\cdot\vec{C}\right) - \vec{C}\left(\vec{A}\cdot\vec{B}\right) \tag{5.30}$$

5.18   TENSOR DERIVATIVES

There are two important types of derivatives in tensor analysis, that is, the time derivatives and the spatial derivatives.

5.18.1   TIME DERIVATIVE OF A TENSOR

A partial derivative of a tensor over time is designated as

$$\dot{A} = \frac{\partial A}{\partial t}$$

5.18.2   SPATIAL DERIVATIVE OF A TENSOR

A partial derivative of a tensor A over one or its special components is denoted as $A_{,i}$:

$$A_{,i} \equiv \frac{\partial A}{\partial x_i} \tag{5.31}$$

that is, the index of the spatial component that the derivation is done over is delimited by a comma (',') from other indices. For example, $A_{ij,k}$ is a derivative of a second-order tensor, A_{ij}.

5.18.3   THE DEL OR NABLA OPERATOR

Nabla operator acting on a tensor A is defined as

$$\nabla_i A \equiv A_{,i} \tag{5.32}$$

Even though the notation in (5.31) is sufficient to define the derivative, in some instances, it is convenient to introduce the del or nabla operator.

5.18.4   RANK OF A TENSOR DERIVATIVE

The derivative of a zero-order tensor (scalar) as given by (5.31) forms a first-order tensor (vector). Generally, the derivative of an m-order tensor forms an $m+1$ order tensor. However, if the derivation index is a dummy index, then the rank of the derivative will be lower than that of the original tensor. For example, the rank of the derivative $A_{ij,j}$ is one, since there is only one free index in this term.

5.18.5   GRADIENT

Expression (5.31) represents a gradient, which in a vector notation is ∇A:

$$\nabla A \to A_{,i}$$

5.18.6   DERIVATIVE OF A COORDINATE

From (5.9) it follows that

$$x_{i,j} = \delta_{ij} \tag{5.33}$$

In particular, the following identity is true:

$$x_{i,i} = x_{1,1} + x_{2,2} + x_{3,3} = 1 + 1 + 1 = 3 \tag{5.34}$$

5.18.7   DIVERGENCE OPERATOR

A divergence operator in a vector notation is represented in a tensor notation as $A_{i,i}$:

$$\left(\nabla \cdot \vec{A}\right) \to A_{i,i}$$

5.18.8   LAPLACE OPERATOR

The Laplace operator in vector notation is represented in tensor notation as $A_{,ii}$:

$$\Delta A \to A_{,ii}$$

5.18.9   TENSOR NOTATION

Equations (5.30), (5.32) and (5.33) clearly show that tensor notation is more concise and accurate than vector notation, since it explicitly shows how each component should be computed. It is also more general since it covers cases that don't have representation in vector notation, for example: $A_{ik,kj}$.

5.19   CURVILINEAR COORDINATES

In this section we introduce the idea of tensor invariance and the invariant forms.

5.19.1   TENSOR INVARIANCE

The distance between the material points in a Cartesian coordinate system is computed as $dl^2 = dx_i$ $_\times dx_i$. The metric tensor, g_{ij}, is introduced to generalize the notation of distance (5.35) to curvilinear coordinates.

5.19.2   METRIC TENSOR

The distance element in a curvilinear coordinate system is computed as

$$dl^2 = g_{ij}dx^i dx^j \qquad (5.35)$$

where g_{ij} is called the metric tensor.

Thus, if we know the metric tensor in each curvilinear coordinate system the distance element is computed by (5.35). The metric tensor is defined as a tensor since we need to preserve the invariance of distance in different coordinate systems, that is, the distance should be independent of the coordinate system, thus:

$$dl^2 = g_{ij}dx^i dx^j = \tilde{g}_{ij}d\tilde{x}^i d\tilde{x}^j \qquad (5.36)$$

The metric tensor is symmetric, which can be shown by rewriting (5.35) as follows:

$$g_{ij}dx^i dx^j = g_{ij}dx^j dx^i = g_{ji}dx^j dx^i$$

where we first swapped places for dx^i and dx^j and then renamed index i in j and j in i. We can rewrite the equality above as

$$g_{ij}dx^i dx^j - g_{ji}dx^i dx^j = \left(g_{ij} - g_{ji}\right)dx^i dx^j = 0$$

Since the equality above should hold for any $dx^i dx^j$, we get:

$$g_{ij} = g_{ji} \qquad (5.37)$$

The metric tensor is also called the fundamental tensor. The inverse of the metric tensor is also called the conjugate metric tensor, g^{ij}, which satisfies the relation:

$$g^{ik}g_{kj} = \delta^i_j \qquad (5.38)$$

Let x^i be a Cartesian coordinate system and $\tilde{x}^j$ the new curvilinear coordinate system. Both systems are related by transformation rules (5.5) and (5.11). Then from (5.36) we get

$$dl^2 = dx^i dx^i = \frac{\partial x^i}{\partial \tilde{x}^j}d\tilde{x}^j \frac{\partial x^i}{\partial \tilde{x}^k}d\tilde{x}^k = \frac{\partial x^i}{\partial \tilde{x}^j}\frac{\partial x^i}{\partial \tilde{x}^k}d\tilde{x}^j d\tilde{x}^k \qquad (5.39)$$

When we transform from a Cartesian coordinate to curvilinear coordinates, the metric tensor in a curvilinear coordinate system, $\tilde{g}_{ij}$, can be determined by comparing relations (5.39) and (5.35):

$$\tilde{g}_{ij} = \frac{\partial x^k}{\partial \tilde{x}^i}\frac{\partial x^k}{\partial \tilde{x}^j} \qquad (5.40)$$

Using (5.38) we can also find its inverse as

$$\tilde{g}^{ij} = \frac{\partial \tilde{x}^i}{\partial x^k} \frac{\partial \tilde{x}^j}{\partial x^k} \tag{5.41}$$

From these expressions one can compute g_{ij} and g^{ij} in various curvilinear coordinate systems.

5.20   CONJUGATE TENSORS

For each index of a tensor, we introduce the conjugate tensor where this index is transferred to its counterpart (covariant/contravariant) using the relations:

$$A^i = g^{ij} A_j \tag{5.42}$$

$$A_i = g_{ij} A^j \tag{5.43}$$

A conjugate tensor is also called an associate tensor. Relations (5.42) and (5.43) are also called the operations of raising/lowering of indices.

5.21   TENSOR INVARIANCE

Since the transformation rules defined by expression (5.1) have a simple multiplicative character, any tensor expression should retain its original form under transformation into a new coordinate system. Thus, if an expression is given in a tensor form it will be invariant under coordinate transformations.

Not all the expressions constructed from tensor terms in curvilinear coordinates will be tensors themselves. For example, if vectors A_i and B_i are tensors, then $A_i B_i$ is not generally a tensor. However, if we consider the same operation on a contravariant tensor A_i and a covariant tensor B_i, then the product will form an invariant:

$$\tilde{A}^i \tilde{B}_i = A^i B_i \tag{5.44}$$

Thus, in curvilinear coordinates we have to refine the definition of the scalar product or the index contraction operation to make it invariant.

5.21.1   INVARIANT SCALAR PRODUCT

The invariant form of the scalar product between two covariant vectors A_i and B_i is $g^{ij} A_i B_j$. Similarly, the invariant form of a scalar product between two contravariant vectors A^i and B^i is $g_{ij} A^i B^i$, where g_{ij} is the metric tensor (5.40) and g^{ij} is its conjugate (5.38).

5.21.2   TWO FORMS OF A SCALAR PRODUCT

According to (5.42) and (5.43), the scalar product can be represented by two invariant forms: $A^i B_i$ and $A_i B^i$. It can be shown that these two forms have the same values.

5.21.3   RULES OF INVARIANT EXPRESSIONS

To build invariant tensor expressions we add two more rules to the Cartesian tensor rules previously outlined:

Each free index should keep its vertical position in every term, i.e., if the index is covariant in one term, it should be covariant in every other term, and vice versa. Every pair of dummy indices should be complementary, that is one should be covariant, and the other contravariant.

For example, a Cartesian formulation of a momentum equation for an incompressible viscous fluid is $\dot{u}_i + u_k u_{i,k} = -\dfrac{P_i}{\rho} + v\tau_{ik,k}$.

The invariant form of this equation is as follows:

$$\dot{u}_i + u^k u_{i,k} = -\frac{P_{,i}}{\rho} + v\tau^k_{i,k} \tag{5.45}$$

where the rising of indices was done using relation (5.42): $u^k = g^{kj}u_j$ and $\tau^k_i = g^{kj}\tau_{ij}$.

5.22   COVARIANT DIFFERENTIATION

A simple scalar value, S, is invariant under coordinate transformations. A partial derivative of an invariant is a first-order covariant tensor (vector):

$$A^i = S_{,i} = \frac{\partial S}{\partial x^i}$$

However, a partial derivative of a tensor of the order one and greater is not generally an invariant under coordinate transformations of type (5.7) and (5.3). In a curvilinear coordinate system we should use more complex differentiation rules to preserve the invariance of the derivative. These rules are called the rules of covariant differentiation and they guarantee that the derivative itself is a tensor. According to these rules the derivatives for covariant and contravariant indices will be slightly different. They are expressed as follows:

$$A_{i,j} \equiv \frac{\partial A_i}{\partial x^j} - \begin{Bmatrix} k \\ ij \end{Bmatrix} A_k \tag{5.46}$$

$$A^i_{,j} \equiv \frac{\partial A^i}{\partial x^j} + \begin{Bmatrix} i \\ kj \end{Bmatrix} A^k \tag{5.47}$$

where the construct, $\begin{Bmatrix} k \\ ij \end{Bmatrix}$ is defined as

$$\begin{Bmatrix} k \\ ij \end{Bmatrix} = \frac{1}{2} g^{kl} \left(\frac{\partial g_{il}}{\partial x^j} + \frac{\partial g_{jl}}{\partial x^i} - \frac{\partial g_{ij}}{\partial x^k} \right)$$

and is also known as Christoffel's symbol of the second kind. Tensor g^{ij} represents the inverse of the metric tensor g_{ij} see (5.38). As it could be seen, the differentiation of a single component of a vector will involve all other components of this vector. By differentiating higher-order tensors each index should be treated independently. Thus, differentiating a second-order tensor, A^{ij}, should be performed as

$$A_{ij,\,k} = \frac{\partial A_{ij}}{\partial x^k} + \begin{Bmatrix} m \\ ik \end{Bmatrix} A_{mj} + \begin{Bmatrix} m \\ jk \end{Bmatrix} A_{im}$$

And, as it can be seen, it also involves all the components of this tensor. Likewise, for the contravariant second-order tensor A^{ij}, we have

$$A^{ij}_{,k} = \frac{\partial A^{ij}}{\partial x^k} + \left\{ \begin{matrix} i \\ mk \end{matrix} \right\} A^{mj} + \left\{ \begin{matrix} j \\ mk \end{matrix} \right\} A^{im} \tag{5.48}$$

For a general n-covariant, m-contravariant tensor, we have

$$A^{j_1 \ldots j_m}_{i_1 \ldots i_n, p} = \frac{\partial}{\partial x^p} A^{j_1 \ldots j_m}_{i_1 \ldots i_n, k} + \left\{ \begin{matrix} j_1 \\ qp \end{matrix} \right\} A^{q j_2 \ldots j_m}_{i_1 \ldots i_n} + \ldots + \left\{ \begin{matrix} j_m \\ qp \end{matrix} \right\} A^{j_1 \ldots j_m - 1 q}_{i_1 \ldots i_n}$$
$$+ \left\{ \begin{matrix} q \\ i_1 p \end{matrix} \right\} A^{j_1 \ldots j_m}_{q i_1 \ldots i_n} + \ldots + \left\{ \begin{matrix} j_m \\ qp \end{matrix} \right\} A^{j_1 \ldots j_m}_{i_1 \ldots i_n - 1 q} \tag{5.49}$$

Despite their seeming complexity, the relations of covariant differentiation can be implemented algorithmically and used in numerical solutions on arbitrary curved computational grids.

5.23   RULES OF INVARIANT EXPRESSIONS

As it was pointed out, the rules to build invariant expressions involve raising or lowering indices (5.42) and (5.43). However, since we did not introduce the notation for covariant derivatives, the only way to raise the index of a covariant derivative, say $A_{,p}$, is to use the relation (5.42) directly, that is, $g^{ij} A_{,j}$.

For example, we can re-formulate the momentum equation (5.45) in terms of contravariant free index i as follows:

$$\dot{u}^i + u^k u^i_{,k} = -\frac{g^{ik} P_{,k}}{\rho} + v \tau^{ik}_k \tag{5.50}$$

where the index of the pressure term was raised by means of (5.42).

Using the invariance of the scalar product one can construct two important differentiation operators in curvilinear coordinates: divergence of a vector $divA \equiv A^i_{,i}$ (5.51) and Laplacian, $\Delta A \equiv g^{ik} A_{,ki}$.

5.23.1   DIVERGENCE

Divergence of a vector is defined as A^i_j:

$$divA \equiv A^i_{,i} \tag{5.51}$$

From this definition and the rule of covariant differentiation (5.47) we have

$$A^i_{,i} = \frac{\partial A^i}{\partial x^i} + \left\{ \begin{matrix} i \\ ki \end{matrix} \right\} A^k \tag{5.52}$$

This can be shown to be equal to

$$A^i_{,i} = \frac{\partial A^i}{\partial x^i} + \left(\frac{1}{\sqrt{g}} \frac{\partial}{\partial x^i} \sqrt{g} \right) A^i = \frac{1}{\sqrt{g}} \frac{\partial}{\partial x^i} \left(\sqrt{g} A^i \right) \tag{5.53}$$

where g is the determinant of the metric tensor g_{ij}.

The divergence of a covariant vector A_i is defined as a divergence of its conjugate contravariant tensor (5.42):

$$A(1) = A_1 / h_1 \qquad (5.54)$$

5.23.2 Laplacian

A Laplace operator or a Laplacian of a scalar A is defined as

$$\Delta A \equiv g^{ik} A_{,ki} \qquad (5.55)$$

The definitions (5.54) and (5.55) of differential operators are invariant under coordinate transformations. They can be programmed using a symbolic manipulation package. They can also be used to derive expressions in different curvilinear coordinate systems.

5.24 ORTHOGONAL COORDINATES

5.24.1 Unit Vectors and Stretching Factors

The coordinate system is orthogonal if the tangential vectors to coordinate lines are orthogonal at each point. Consider three-unit vectors, a^i, b^i, c^i, each directed along one of the coordinate axis (tangential unit vectors), that is:

$$a^i = \left\{ a^1, 0, 0 \right\} \qquad (5.56)$$

$$b^i = \left\{ 0, b^2, 0 \right\} \qquad (5.57)$$

$$c^i = \left\{ 0, 0, c^3 \right\} \qquad (5.58)$$

The condition of orthogonality means that the scalar product between any two of these unit vectors should be zero. According to the definition of a scalar product, it should be written in form (5.44), that is, a scalar product between vectors a_i and b_i can be written as $a^i b_i$ and $a_i b^i$. By using the first form and applying the operation of rising indices (5.42), we can express the scalar product in contravariant components only:

$$0 = a^i b_i = g_{ij} a^i b^j$$

$$g_{11} a^1 0 + g_{12} a^1 b^2 + g_{13} 0 0$$

$$g_{21} a^2 b^1 + g_{22} 0 b^2 + g_{23} 0 0$$

$$g_{31} a^3 0 + g_{32} 0 b^2 + g_{33} 0 0$$

$$= \left(g_{12} + g_{21} \right) a^1 b^2 = 2 g_{12} a^1 b^2 = 0 \qquad (5.59)$$

where we used the symmetry of g_{ij} (5.37). Since vectors a^1 and b^2 were chosen to be non-zero, we have $g_{12} = 0$. Applying the same reasoning for scalar products of other vectors, we conclude that the metric tensor has only diagonal components non-zero:

$$g_{ij} = \delta_{ij} g_{(ii)} \tag{5.60}$$

Let us introduce stretching factors, h_i, as the square roots of these diagonal components of g_{ij}:

$$h_1 \equiv \left(g_{11}\right)^{1/2}; \; h_2 \equiv \left(g_{22}\right)^{1/2}; \; h_3 \equiv \left(g_{33}\right)^{1/2} \tag{5.61}$$

Now, consider the scalar product of each of the unit vectors (5.56)–(5.58) with itself. Since all vectors are unit, the scalar product of each with itself should be one:

$$a^i a_i = b^i b_i = c^i c_i = 1$$

otherwise, expressed in contravariant components only the condition of unity is

$$g_{ij} a^i a^j = g_{ij} b^i b^j = g_{ij} c^i c^j = 1$$

Now, consider the first term above and substitute the components of a form (5.56). The only non-zero term will be

$$g_{ij} a^1 a^1 = \left(h_1\right)^2 \left(a^1\right)^2 = 1$$

and consequently:

$$a^1 = \pm \frac{1}{h_1} \tag{5.62}$$

where the negative solution identifies a vector directed into the opposite direction, and we can neglect it for definiteness. Applying the same reasoning for each of the three unit vectors a_i, b_i, c_i, we can rewrite (5.56), (5.57) and (5.58) as follows:

$$a^i = \left\{\frac{1}{h_1}, 0, 0\right\} \tag{5.63}$$

$$b^i = \left\{0, \frac{1}{h_2}, 0\right\} \tag{5.64}$$

$$c^i = \left\{0, 0, \frac{1}{h_3}\right\} \tag{5.65}$$

which means that the components of unit vectors in a curved space should be scaled with coefficients h_i. It follows that the expression for the element of length in curvilinear coordinates (5.35) can be written as follows:

$$dl^2 = g_{ij} d\tilde{x}^i d\tilde{x}^j = h_i^2 \left(d\tilde{x}^i\right)^2 g_{ij} = \delta_{ij} \left(h^{(i)}\right)^2 \tag{5.66}$$

Similarly, we introduce the $h^{(i)}$ coefficients for the conjugate metric tensor (5.38):

$$g_{ij} = \delta_{ij} \left(h^{(i)} \right)^2 \tag{5.67}$$

Combining the latter with (5.38), we obtain: $\delta_{ij} h_{(i)} h^{(i)} = \delta_{ij}$, from which it follows that

$$h_{(i)} = 1 / h^{(i)} \tag{5.68}$$

5.25   PHYSICAL COMPONENTS OF TENSORS

Consider a direction in space determined by a unit vector e_i. Then the physical component of a vector A_i in the direction e_i is given by a scalar product between A_i and e_i, namely:

$$A(e) = g^{ij} A_i e_j$$

The above can also be rewritten as

$$A(e) = A_i e^i = A^i e_j \tag{5.69}$$

Suppose the unit vector is directed along one of the axes: $A^i_{,i} = \dfrac{1}{H} \dfrac{\partial}{\partial x_i} \left(\dfrac{H}{h_{(i)}} A_i \right) i$. From (5.63), it

follows that

$$e^1 = 1/h_1$$

where h_1 is defined by (5.61). Thus, according to (5.69), the physical component of vector A_i in direction 1 in the orthogonal coordinate system is equal to

$$A(1) = A_1 / h_1$$

or, repeating the argument for other components, we have for the physical components of a covariant vector:

$$A_1/h_1, \; A_2/h_2, \; A_3/h_3 \tag{5.70}$$

Following the same reasoning, for the contravariant vector A^i, we have

$$A_1 h^1, \; A_2 h^2, \; A_3 h^3$$

General rules of covariant differentiation previously defined simplify considerably in orthogonal coordinate systems. We can define the nabla operator by the physical components of a covariant vector composed of partial differentials:

$$\nabla_i = \frac{1}{h_{(i)}} \frac{\partial}{\partial x^i} \tag{5.71}$$

where the parentheses indicate that there's no summation with respect to index $A^i_{,i} = \dfrac{1}{H} \dfrac{\partial}{\partial x_i} \left(\dfrac{H}{h_{(i)}} A_i \right) i$.

In the orthogonal coordinate system, the general expressions for divergence (5.53) and Laplacian (5.55) operators can be expressed in terms of stretching factors only:

$$\Delta A_{,i}^{i} = \frac{1}{H} \frac{\partial}{\partial x_i} \left(\frac{H}{h_{(i)}} \frac{\partial A_i}{\partial x_i} \right)$$

$$H = \prod_{i=1}^{n} h_i \tag{5.72}$$

Important examples of orthogonal coordinate systems are spherical and cylindrical coordinate systems. Consider the example of a cylindrical coordinate system: $x_i = \{x_1, x_2, x_3\}$ and $\tilde{x}_i = \{r, \theta, l\}$:

$$\begin{cases} x_1 = r\cos(\theta) \\ x_2 = r\sin(\theta) \\ \quad x_3 = l \end{cases} \tag{5.73}$$

According to (5.40) only a few components of the metric tensor will survive. Then, we can compute the nabla, the divergence and the Laplacian operators according to (5.71), (5.52) and (5.55), or using simplified relations (5.72):

$$\alpha, \alpha \left(u_i \otimes v_j \right) = u_i \otimes \left(\alpha v_j \right) = \left(\alpha u_i \right) \otimes v_j$$

$$divA = \frac{\partial A_1}{\partial \tilde{x}^1} + \frac{1}{\tilde{x}^1} \frac{\partial A_2}{\partial \tilde{x}^2} + \frac{\partial A_3}{\partial \tilde{x}^3} + \frac{1}{\tilde{x}^1} A_1 \tag{5.74}$$

$$divA = \frac{\partial A_r}{\partial r} + \frac{1}{r} \frac{\partial A_\theta}{\partial \theta} + \frac{\partial A_z}{\partial z} + \frac{1}{r} A_r$$

Note that, instead of using the contravariant components as implied by the general definition of the divergence operator (5.51), we are using the covariant components as dictated by relation (5.70). The expression of the Laplacian becomes

$$\vec{\nabla} = \left(\frac{\partial}{\partial r}, \frac{1}{r} \frac{\partial}{\partial \theta}, \frac{\partial}{\partial z} \right)$$

$$\Delta A = \frac{\partial^2 A}{(\partial \tilde{x}_1)^2} + \frac{1}{\tilde{x}_1^2} \frac{\partial^2 A}{(\partial \tilde{x}_2)^2} + \frac{\partial^2 A}{(\partial \tilde{x}_3)^2} + \frac{1}{\tilde{x}_1} \frac{\partial A}{\partial \tilde{x}_1} = \frac{\partial^2 A}{(\partial r)^2} + \frac{1}{r^2} \frac{\partial^2 A}{(\partial \theta)^2} + \frac{\partial^2 A}{(\partial z)^2} + \frac{1}{r} \frac{\partial A}{\partial r} \tag{5.75}$$

The advantage of the tensor approach is that it can be used for any type of curvilinear coordinate transformations, not necessarily analytically defined, like cylindrical or spherical.

5.26   CHRISTOFFEL SYMBOL

Christoffel symbol Γ_{ij}^{m} is defined as follows:

$$\Gamma_{ij}^{m} = \varepsilon^{m} \frac{\partial \varepsilon_i}{\partial q^j} \tag{5.76}$$

Also, by inverting, one can have

$$\frac{\partial \varepsilon_i}{\partial q^j} = \Gamma_{ij}^k \varepsilon^k \tag{5.77}$$

From these two previous expressions, we can prove the symmetry of the Christoffel symbol in the two lower indices. Indeed, if we define, $\vec{\varepsilon}_i = \dfrac{\partial \vec{r}}{\partial q^i}$, then we can write

$$\Gamma_{ij}^m = \varepsilon^m \frac{\partial}{\partial q^j}\left(\frac{\partial \vec{r}}{\partial q^i}\right) = \varepsilon^m \frac{\partial}{\partial q^j}\left(\frac{\partial \vec{r}}{\partial q^j}\right) = \varepsilon^m \frac{\partial \vec{\varepsilon}_j}{\partial q^i} = \Gamma_{ji}^m \tag{5.78}$$

which proves the symmetry of the Christoffel symbol in i and j.

Using the notation:

$$\left[ij,k\right] = g_{mk}\Gamma_{ij}^{\ k} \tag{5.79}$$

let us prove the symmetry in this notation with respect to i, j.

$$\left[ji,k\right] = g_{mk}\Gamma_{ji}^{\ k} = g_{mk}\Gamma_{ij}^{\ k} = \left[ij,k\right] \tag{5.80}$$

We can also write $[i, j, k]$ as a derivative of ε_i:

$$\left[ij,k\right] = g_{mk}\Gamma_{ij}^{\ m} = \varepsilon_m \varepsilon_k \Gamma_{ij}^{\ m} = \varepsilon_m \varepsilon_k \varepsilon^m \frac{\partial \varepsilon_i}{\partial q^j} = \varepsilon_k \delta_m^m \frac{\partial \varepsilon_k}{\partial q^j} = \varepsilon_k \frac{\partial \varepsilon_i}{\partial q^j}$$

Let's now differentiate $g_{mk} = \varepsilon_i \varepsilon_j$ with respect to q^k. We have

$$\frac{\partial g_{ij}}{\partial q^k} = \frac{\partial \varepsilon_i}{\partial q^k}\varepsilon_j + \varepsilon_i \frac{\partial \varepsilon_j}{\partial q^k}$$

$$= [ik, j] + [jk, i] \tag{5.81}$$

From the previous relation, we can deduce successively:

$$\frac{\partial g_{ik}}{\partial q^j} = \left[ij,k\right] + \left[kj,i\right] = \left[ij,k\right] + \left[jk,i\right] \tag{5.82}$$

$$\frac{\partial g_{jk}}{\partial q^i} = \left[ij,k\right] + \left[ik,j\right] \tag{5.83}$$

Combining equations 5.81, 5.82 and 5.83 we have:

$$\left[ij,k\right] = \frac{1}{2}\left[\frac{\partial g_{ik}}{\partial q^j} + \frac{\partial g_{jk}}{\partial q^i} - \frac{\partial g_{ij}}{\partial q^k}\right] \tag{5.84}$$

Meanwhile, we can transform:

$$
\begin{aligned}
g^{ks}\left[ij,k\right] &= \varepsilon^k\varepsilon^s\left[ij,k\right]\\
&= \varepsilon^k\varepsilon^s g_{mk}\Gamma_{ij}{}^m\\
&= \varepsilon^k\varepsilon^s\varepsilon_m\varepsilon_k\Gamma_{ij}{}^m\\
&= \varepsilon^s\varepsilon_m\Gamma_{ij}{}^m\\
&= \Gamma_{ij}{}^s
\end{aligned}
\tag{5.85}
$$

As a result, the Christoffel symbol becomes

$$
\Gamma_{ij}{}^s = g^{ks}\left[ij,k\right] = \frac{1}{2}g^{ks}\left[\frac{\partial g_{ik}}{\partial q^j}+\frac{\partial g_{jk}}{\partial q^i}-\frac{\partial g_{ij}}{\partial q^k}\right]
\tag{5.86}
$$

5.27   ANTISYMMETRIC TENSOR AND VECTOR

An antisymmetric tensor is a tensor with components $a_{ij} = -a_{ji}$. Therefore for $i = j$, $a_{ii} = 0$, and the trace is zero. We see that the components of the principal diagonal are all zero. The antisymmetric tensor can be then written in the matrix form:

$$
\begin{pmatrix}
0 & a_{12} & a_{13}\\
-a_{12} & 0 & a_{23}\\
-a_{13} & a_{23} & 0
\end{pmatrix}
\tag{5.87}
$$

a_{23}, a_{31} and a_{21} are components of the vector $\overrightarrow{A}$, which we can denote by

$$
a_i = \frac{1}{2}\varepsilon_{ikm}A_{km}
\tag{5.88}
$$

This property is a direct consequence of the fact the number of independent components of an antisymmetric tensor is peculiar to three dimensions. We shall also notice that in n-dimensions the antisymmetric tensor has $2^{\frac{1}{2}n(n-1)}$ independent components, while a vector has n components. The antisymmetric tensor can have the form,

$$
\begin{pmatrix}
0 & a_3 & -a_2\\
-a_3 & 0 & a_1\\
a_2 & -a_1 & 0
\end{pmatrix}
\tag{5.89}
$$

That is $a_{12} = a_3$, $a_{21} = \times a_3$.
   And $a_{ik} = 0$, if $i = k$
$a_{ik} = a_m$, if ikm is in even order.
$a_{ik} = \times a_m$, if ikm is in odd order.
In general, we can write:

$$
a_{ik} = \varepsilon_{ikm}a_m
\tag{5.90}
$$

5.28 DYADS

Assume two fixed vectors $T \cdot \vec{c} = \vec{a}\left(\vec{b} \cdot \vec{c}\right)$ and $\vec{c}$ a varying vector. Consider the vector $T \cdot \vec{c}$ defined by $T \cdot \vec{c} = \vec{a}\left(\vec{b} \cdot \vec{c}\right)$. Since $\vec{a}\left[\vec{b} \cdot \left(\lambda_1 \vec{c_1} + \lambda_2 \vec{c_2}\right)\right] = \lambda_1 \vec{a}\left(\vec{b} \cdot \vec{c_1}\right) + \lambda_2 \vec{a}\left(\vec{b} \cdot \vec{c_2}\right), T \cdot \vec{c}$ is a linear function of the vector $\vec{c}$. The quantity T is called a second-order tensor. Let us now use the symbol $\otimes$ and write $\left(\vec{a} \otimes \vec{b}\right) \cdot \vec{c} = \vec{a}\left(\vec{b} \cdot \vec{c}\right)$.

As we can see, the right side of the equality is a vector quantity. Therefore, the left side also is a vector quantity. Tensor $T . \vec{e_1} = T_{11}\vec{e_1} + T_{21}\vec{e_2} + T_{31}\vec{e_3}$ is called the dyad or tensor product of the two vectors. A dyad operates on a vector. A dyad cannot be defined except in terms of its operation vector.

5.28.1 PROPERTIES OF DYAD

For three vectors, $\vec{a}$, $\vec{b}$ and $\vec{c}$, we have

$$\left(\vec{a} \otimes \vec{b}\right) \cdot \vec{c} = \vec{a}\left(\vec{b} \cdot \vec{c}\right) \tag{5.91}$$

Then, for the linear combination, we have

$$\left(\overrightarrow{\alpha_1 a_1} \otimes \vec{b_1} + \overrightarrow{\alpha_2 a_2} \otimes \vec{b_2}\right) \cdot \vec{c} = \overrightarrow{\alpha_1 a_1}\left(\vec{b_1} \cdot \vec{c}\right) + \overrightarrow{\alpha_2 a_2}\left(\vec{b_2} \cdot \vec{c}\right) \tag{5.92}$$

The previous property shows that a dyad is a second-order tensor. In particular, a second-order tensor can be written uniquely as the sum of nine dyads.

5.28.2 COMPONENTS OF A DYAD

The component of a dyad $\left(\vec{a} \otimes \vec{b}\right)$ can be written as follows:

$$(\vec{a} \otimes \vec{b})_{ik} = e_i \cdot \left[\left(\vec{a} \otimes \vec{b}\right) \cdot e_k = a_i b_k \tag{5.93}$$

In matrix notation, we can have

$$\vec{a} \otimes \vec{b} = \begin{bmatrix} a_1 b_1 & a_1 b_2 & a_1 b_3 \\ a_2 b_1 & a_2 b_2 & a_2 b_3 \\ a_3 b_1 & a_3 b_2 & a_3 b_3 \end{bmatrix} \tag{5.94}$$

5.29 OTHER PROPERTIES OF A DYAD

Consider $\left(\vec{e_1}, \vec{e_2}, \vec{e_3}\right)$ a given basis. The product $T . \vec{v}$ is a vector and so is $T . \vec{e_1}$, which can be expressed as a linear combination of the vectors $\vec{e_1}, \vec{e_2}, \vec{e_3}$ of the basis. Hence, for some numbers λ, μ and ν, we have

$$T . \vec{e_1} = \lambda \vec{e_1} + \nu \vec{e_2} + \nu \vec{e_3}. \tag{5.95}$$

Note that the numbers λ, μ and ν can also be renamed T_{11}, ∇A; then we may rewrite:

$$T . \vec{e_1} = T_{11}\vec{e_1} + T_{21}\vec{e_2} + T_{31}\vec{e_3}. \tag{5.96}$$

i.e.,

$$T.e_i = T_{ki}\, e_k \tag{5.97}$$

As a result, the tensor T may be written in the form of a matrix:

$$T = \begin{bmatrix} T_{11} & T_{12} & T_{13} \\ T_{21} & T_{22} & T_{23} \\ T_{31} & T_{32} & T_{33} \end{bmatrix} \tag{5.98}$$

In the previous matrix (5.98) the entries T_{ik} represent the elements in the ith row and the jth column. However, rows and columns are not, in general, interchangeable. More importantly, in $\mathbb{R}^3$, the scalar has $3^0 = 1$ component; a vector has $3^1 = 3$ components and a second-order tensor has $3^2 = 9$ components.

Suppose now that T can be written as the sum of 9 dyads. Thus,

$$T = T_{ki}\, \vec{e}_k \otimes \vec{e}_i \tag{5.99}$$

$$T\,\vec{e}_m = T_{ki} \left(\vec{e}_k \otimes \vec{e}_i \right).\vec{e}_m$$

$$T\,\vec{e}_m = T_{ki}\, \vec{e}_k \left(\vec{e}_i.\vec{e}_m \right)$$

$$T\,\vec{e}_m = T_{km}\, \vec{e}_k$$

$$\text{Similarly, } T\,\vec{e}_i = T_{ki}\, \vec{e}_k \tag{5.100}$$

Thus, the supposition that T can be written as the sum of 9 dyads is consistent. Also, the expression represented by equation (5.99) is unique. To show this, it is important to find explicitly the expressions of the numbers T_{ki}.

We know that $T_{ki}\, \vec{e}_k.\vec{e}_m = \vec{e}_m.\left(T\,\vec{e}_i \right)$, that is, $T_{mi} = \vec{e}_m.\left(T\,\vec{e}_i \right)$.

The numbers T_{mi} are called the components of the tensor T in the basis $\vec{e}_i$.

5.30   MORE PROPERTIES OF THE SECOND-ORDER TENSORS

5.30.1   Pre-multiplication and Post-multiplication of a Tensor and a Vector

Post-multiplication is the product in which the vector follows the tensor. Pre-multiplication is the process in which the tensor follows the vector. To define pre-multiplication, let us use this relation:

$\left(\vec{a}.T \right)\vec{c} = \vec{a}.\left(T\,\vec{c} \right)$ for all $\vec{c}$

$\vec{a}.T$ is a vector whose components may be obtained by setting $\vec{c} = \vec{e}_i$, that is:

$$\left(\vec{a}.T \right)\vec{e}_i, = \vec{a}.\left(T\,\vec{e}_i \right) \tag{5.101}$$

By using the relation $T\vec{e_i} = T_{ki}\vec{e_k}$, we have

$$(\vec{a}.T)\vec{e_i}, = \vec{a}.(T_{ki}\vec{e_k}) = a_k.T_{ki} \tag{5.102}$$

Example 2

$$(a_1, a_2, a_3) \begin{bmatrix} T_{11} & T_{12} & T_{13} \\ T_{21} & T_{22} & T_{23} \\ T_{31} & T_{32} & T_{33} \end{bmatrix}$$

$$= (a_1T_{11} + a_2T_{21} + a_3T_{31}, \ a_1T_{12} + a_2T_{22} + a_3T_{32}, \ a_1T_{13} + a_2T_{23} + a_3T_{33})$$

5.31   THE TRANSPOSE TENSOR

Consider the tensor T expressed by the sum of dyads:

$$T = \vec{a} \otimes \vec{b} + \vec{c} \otimes \vec{d} + \dots \tag{5.103}$$

The transpose of T is expressed by

$$T' = \vec{b} \otimes \vec{a} + \vec{d} \otimes \vec{c} + \dots \tag{5.104}$$

It can be shown that the above definition is independent of the sum of the dyads chosen to represent T. For example, consider a vector $\vec{v}$ and let us form the dot product $T'.\vec{v}$. We have

$$T'.\vec{v} = (\vec{b} \otimes \vec{a} + \vec{d} \otimes \vec{c} + \cdots).\vec{v}$$

$$T'.\vec{v} = \vec{b} \otimes \vec{a}).\vec{v} + (\vec{d} \otimes \vec{c}).\vec{v} + \cdots$$

$$T'.\vec{v} = \vec{b}.(\vec{a}.\vec{v}) + (\vec{d}.(\vec{c}.\vec{v}) + \cdots$$

$$T'.\vec{v} = (\vec{a}.\vec{v}).\vec{b} + (\vec{c}.\vec{v}).\vec{d} + \cdots$$

$$T'.\vec{v} = (\vec{v}.\vec{a}).\vec{b} + (\vec{v}.\vec{c}).\vec{d} + \cdots$$

$$T'.\vec{v} = \vec{v}.T \tag{5.105}$$

Equation (5.105) is true for all $\vec{v}$. It is therefore an alternative for the transpose of a tensor. The Cartesian components of $T\times$ can be found by using $\vec{a} = a_i\vec{e_i}$. So,

$$T' = (b_i.a_k + d_i.c_k + \dots \vec{e_i} \otimes \vec{e_k}$$
$$T'_{ik} = b_i.a_k + d_i.c_k + \dots$$

$$T'_{ik} = T_{ki} \tag{5.106}$$

Example 3
Consider the tensor T with the matrix representation:

$$T = \begin{bmatrix} T_{11} & T_{12} & T_{13} \\ T_{21} & T_{22} & T_{23} \\ T_{31} & T_{32} & T_{33} \end{bmatrix} \text{ in a given basis.}$$

The matrix representation of the transpose of T will be

$$T' = \begin{bmatrix} T_{11} & T_{21} & T_{31} \\ T_{12} & T_{22} & T_{32} \\ T_{13} & T_{23} & T_{33} \end{bmatrix}$$ with respect to the same basis.

Note: the transpose of the transpose of a tensor T is itself:

$$\left(T'\right)' = T \tag{5.107}$$

5.32   SYMMETRIC AND ANTISYMMETRIC TENSOR

5.32.1   SYMMETRIC TENSOR

A symmetric tensor is a tensor which is equal to its transpose, that is $T' = T$.

Most of the tensors encountered in physics are symmetric. As examples, the stress tensor, the strain tensor and the rate of the strain tensor are all symmetric. Symmetric tensors have specific types of properties that most general tensors don't have. For a symmetric tensor, $T_{ik} = T_{ki}$ in all frames of reference. The matrix representation of a symmetric tensor will be

$$T = \begin{bmatrix} T_{11} & T_{12} & T_{13} \\ T_{12} & T_{22} & T_{23} \\ T_{13} & T_{23} & T_{33} \end{bmatrix} \tag{5.108}$$

So, as we can see, symmetric tensors have only six independent components rather than nine.

5.32.2   ANTISYMMETRIC OR SKEW-SYMMETRIC TENSOR

A tensor is said to be antisymmetric or skew-symmetric if it is equal to the opposite of its transpose. It is worthy of note to mention that some phenomena in physics represent antisymmetric tensors. These phenomena involve rotation such as the spin tensor.

For an antisymmetric tensor, $T = -T'$, that is, $T_{ki} = -T_{ik}$. The matrix representation of an antisymmetric tensor is

$$T = \begin{bmatrix} 0 & T_{12} & T_{13} \\ -T_{12} & 0 & T_{23} \\ -T_{13} & -T_{23} & 0 \end{bmatrix} \tag{5.109}$$

So, as we can see, for the antisymmetric tensor, all the components of the leading diagonal are identically zero. Apart from the zero, the antisymmetric tensor has only three independent components. Also, in general, the classification of tensors in symmetric and antisymmetric is of great importance in physics and engineering.

Theorem

Every second-order tensor may be expressed uniquely as the sum of a symmetric and an antisymmetric tensor.

Proof

Suppose S is a symmetric tensor and A is an antisymmetric tensor. Their transpose can be written as

$$S' = S \text{ and } A' = -A.$$

Assume $T = S + A$; the transpose of T will be $T' = S' + A' = S - A$; therefore, $T + T' = 2S$ and $T - T' = 2A$.

$S = \dfrac{1}{2}(T + T')$ and $A = \dfrac{1}{2}(T - T')$. As a result, the tensor T can be written as

$$T = \frac{1}{2}(T + T') + \frac{1}{2}(T - T').$$

In this form, we can say that a tensor has two parts: the symmetric part, $\dfrac{1}{2}(T + T')$, and the anti-symmetric part, $\dfrac{1}{2}(T - T')$.

PROBLEM SET 5

5.1 This problem consists of writing Christoffel symbols $\Gamma_{ij}{}^{m}$ as the sum of the partial derivatives of the metric tensor. Knowing that $\Gamma_{ij}^{m} = \varepsilon^{m}\dfrac{\partial \varepsilon_i}{\partial q^j}$.

   1. Prove that the Christoffel symbol $\Gamma_{ij}{}^{m}$ is symmetric.

   2. Considering the notation: $[ij,k] = g_{mk}\Gamma_{ij}{}^{k}$.

      a. Prove the symmetry with respect to this notation in the indices, i, j.

      b. Show that $[ij,k] = \varepsilon_k \dfrac{\partial \varepsilon_i}{\partial q^j}$

      c. Prove that $[ij,k] = \dfrac{1}{2}\left[\dfrac{\partial g_{ik}}{\partial q^j} + \dfrac{\partial g_{jk}}{\partial q^i} - \dfrac{\partial g_{ij}}{\partial q^k}\right]$.

      d. Symmetric identity: Prove that if A_{ij} is a symmetric tensor, then the following identity is true: $\varepsilon_{ijk}A_{jk} = 0$.

   3. Write the metric tensor in cylindrical coordinates.

5.2 Using the tensor identity: $\varepsilon^{ijk}\varepsilon_{ipq} = \delta_{jp}\delta_{kq} - \delta_{jq}\delta_{kp}$

   a. prove the vector identity

   $$\vec{A} \times (\vec{B} \times \vec{C}) = \vec{B}(\vec{A} \cdot \vec{C}) - \vec{C}(\vec{A} \cdot \vec{B}).$$

   b. Prove that $A^i B_i = A_i B^i$ and both are invariant while $A_i B_i$ is not.

5.3 In Classical mechanics angular momentum is given by $\vec{L} = \vec{r} \times \vec{p}$. In quantum mechanics the linear momentum $\vec{P}$ is replaced by $-i\vec{\nabla}$.

   a. Write down the Cartesian components, L_x, L_y, L_z, of the angular momentum operator L as functions of partial derivatives of x, y and z.

   b. Prove that the commutator bracket notation, $[L_x, L_y] = iL_z$.

   c. Prove that $\vec{L} \times \vec{L} = i\vec{L}$.

   d. Two other vectors $\vec{a}$ and $\vec{b}$ commute with each other and with L, that is $[\vec{a}, \vec{b}] = [\vec{a}, \vec{L}] = [\vec{a}, \vec{L}] = 0$. Show that $[\vec{a} \cdot \vec{L}, \vec{b} \cdot \vec{L}] = i\vec{a} \times \vec{b} \cdot \vec{L}$.

5.4   Find all the eigenvalues and eigenvectors of the tensor

$$T = \begin{bmatrix} 2 & 0 & 3i \\ 0 & 1 & 2 \\ 0 & 0 & -1 \end{bmatrix}$$ and of the Hermitian transpose tensor T^H. Verify that the eigenvalues

of the Hermitian transpose tensor T^H are the complex conjugates of the eigenvalues of T.

5.5   Prove that for an antisymmetric tensor, $a_{ii} = 0$ and $a_{ij} = -a_{ji}$.

5.6   If $A_{ik} = 0$ is a tensor, prove that

   a. $$\begin{vmatrix} A_{22} & A_{23} \\ A_{32} & A_{33} \end{vmatrix} + \begin{vmatrix} A_{33} & A_{31} \\ A_{13} & A_{11} \end{vmatrix} + \begin{vmatrix} A_{11} & A_{12} \\ A_{21} & A_{22} \end{vmatrix} = \frac{1}{2} A_{ii} A_{kk} - \frac{1}{2} A_{ik} A_{ki}$$

   b.   Prove that the previous expression is a scalar. Relate it to an invariant of the roots of the equation $\|A_{ik} - \lambda \delta_{ik} = 0\|$.

5.7   Use Einstein notation to show that the tensor $A + A^T$ is symmetric and that $A \times A^T$ is antisymmetric.

5.8   The magnitude of a tensor A is defined by $|A| = abs\left(\sqrt{\dfrac{A:A}{2}}\right)$, with $A:A$ the tensor scalar product.

5.9   What is the Laplacian of a tensor T in Einstein notation? What is the order of this quantity?

5.10 Using Einstein notation to show that $(A + B)^T = A^T + B^T$.

5.11 Prove the following equality:

$$\frac{\partial f}{\partial t} + \frac{\partial f}{\partial t_i} v_i = \frac{\partial f}{\partial t} + \vec{v} \cdot \vec{\nabla} f$$

5.12 Show that for orthogonal coordinate systems,

$$\left\{ \begin{matrix} i \\ i\, i \end{matrix} \right\} = \frac{1}{2} \frac{\partial}{\partial x^i} \left(\ln(g_{ii}) \right); \quad \left\{ \begin{matrix} i \\ ij \end{matrix} \right\} = \frac{1}{2} \frac{\partial}{\partial x^j} \left(\ln(g_{ii}) \right);$$

$\Delta A \to A_{,ii}$ for i,j,k distinct,
and that for $i \neq j$,

$$\left\{ \begin{matrix} i \\ i\, i \end{matrix} \right\} = \frac{1}{2} \frac{\partial}{\partial x^i} \left(\ln(g_{ii}) \right); \quad \left\{ \begin{matrix} i \\ ij \end{matrix} \right\} = \frac{1}{2} \frac{\partial}{\partial x^j} \left(\ln(g_{ii}) \right);$$

$$\left\{ \begin{matrix} i \\ j\quad j \end{matrix} \right\} = -\frac{1}{2g_{ii}} \frac{\partial g_{jj}}{\partial x^i} c \text{ (no sum).}$$

5.13 The Laplacian of a scalar ϕ is defined in elementary vector analysis by $\left(\dfrac{\partial^2}{\partial x^2} + \dfrac{\partial^2}{\partial y^2} + \dfrac{\partial^2}{\partial z^2} \right)\phi$, where (x,y,z) constitutes a rectangular Cartesian coordinate system. Show that the correct general tensor expression is $g^{ij}\phi_{,ij}$.

5.14 Consider $\vec{a} = (3,4,1)$; $\vec{b} = (1,0,1)$; $\vec{c} = (2,1,0)$;

$$T = \begin{bmatrix} 3 & 1 & 1 \\ 2 & 4 & -2 \\ 1 & -1 & 0 \end{bmatrix}; \quad S = \begin{bmatrix} 1 & 1 & -1 \\ 2 & 2 & 3 \\ 4 & 0 & 2 \end{bmatrix}.$$

Write down $\vec{a}.T$; $\vec{a}.T$; $T.\vec{b}$; $\vec{a}.\left(T.\vec{b}\right)$; $\left(\vec{a}.T\right).\vec{b}$; $\vec{b}.\left(T.\vec{a}\right)$; $\vec{a}.S.\vec{b}$; $\vec{b}.S.\vec{a}$.

6 Differential Forms

> I do not think the division of the subject into two parts – into applied mathematics and experimental physics a good one, for natural philosophy without experiment is merely mathematical exercise, while experiment without mathematics will neither sufficiently discipline the mind or sufficiently extend our knowledge in a subject like physics.
>
> *–Balfour Stewart*

INTRODUCTION

This chapter introduces the product of 2-basic forms such as $dx_1 \wedge dx_2$, Hamilton's equations and the Maxwell relations, the Legendre transformations and the thermodynamical potentials, D'Alembert's principle and Lagrange's equation. In this chapter, some thermodynamical concepts and their constructions using differential forms are also introduced. For example, the differential of the internal energy, the Helmholtz free energy and other potentials are explained.

6.1 DEFINITION

Consider the vector force $\vec{F}$ whose point of application displaces in 3-D by $d\vec{r}$. Assume $\vec{F} = F_1\vec{i} + F_2\vec{j} + F_3\vec{k}$ and $d\vec{r} = dx_1\vec{i} + dx_2\vec{j} + dx_3\vec{k}$ in an orthonormal basis $(\vec{i},\vec{j},\vec{k})$. The work done by the force $\vec{F}$ is given as follows:

$$w = \int \vec{F} \cdot d\vec{r} = \int \left(F_1 dx_1 + F_2 dx_2 + F_3 dx_3 \right) \tag{6.1}$$

We can generalize this form for the case where $\vec{F}$ has n components $(F_1, F_2, ..., F_n)$ in the vector space V_n of n dimensions:

$$w = \int \left(F_1 dx_1 + F_2 dx_2 + F_3 dx_3 + ... + F_n dx_n \right) \tag{6.2}$$

We can also have a situation where each of these components $F_1, F_2, ..., F_n$ depends on $(x_1, x_2, ..., x_n)$. Then,

$$dw_x = F_1 dx_1 + F_2 dx_2 + ... + F_n dx_n \tag{6.3}$$

is called a differential 1-form or a 1-form.

Example 1
Let $f: \mathbb{R}^2 \to \mathbb{R}$ be a differentiable function in a region D of $\mathbb{R}^3$ for $f = f(x,y)$;

$f = \dfrac{\partial f}{\partial x} dx + \dfrac{\partial f}{\partial y} dy$ is a differential 1-form.

DOI: 10.1201/9781003478812-6

Exact Differential

Suppose $f = F(x,y)dx + G(x,y)dy$ a differential on $\mathbb{R}^2$, with $F(x,y) = \left(\dfrac{\partial f}{\partial y}\right)_x$ and $G(x,y) = \left(\dfrac{\partial f}{\partial y}\right)_x$; df is an exact differential if

$$\left(\frac{\partial F}{\partial y}\right)_x = \left(\frac{\partial G}{\partial x}\right)_y \tag{6.4}$$

$\left(\dfrac{\partial F}{\partial y}\right)_x$ is the partial derivative of F with respect to y as x is maintained constant.

We can also write

$$\frac{\partial^2 f}{\partial x \partial y} = \frac{\partial^2 f}{\partial y \partial x} \tag{6.5}$$

Example 2

Consider the function of three variables (x,y,z):

$$f(x,y,z) = 2xy^2z + x^2y^2z^2.$$

Calculate the partial derivatives: $\left(\dfrac{\partial f}{\partial y}\right)_{xz}$, $\left(\dfrac{\partial f}{\partial x}\right)_{yz}$, $\left(\dfrac{\partial^2 f}{\partial y^2}\right)_{xz}$ and $\left(\dfrac{\partial^2 f}{\partial z \partial y}\right)_x$.

Solution

$$\left(\frac{\partial f}{\partial y}\right)_{xz} = 4xyz + 2x^2yz^2$$

$$\left(\frac{\partial f}{\partial x}\right)_{yz} = 2y^2z + 2xy^2z^2$$

$$\left(\frac{\partial^2 f}{\partial y^2}\right)_{xz} = \frac{\partial\left(4xyz + 2x^2yz^2\right)}{\partial y} = 4xz + 2x^2z^2$$

$$\left(\frac{\partial^2 f}{\partial z \partial y}\right)_x = \frac{\partial}{\partial z}\left(\frac{\partial f}{\partial y}\right)_x = 4xy + 4x^2yz$$

6.2   PRODUCT OF 2-BASIC FORMS $dx_1 \wedge dx_2$

Geometrically, $dx_1 \wedge dx_2(\vec{a},\vec{b})$ is the area of the parallelogram spanned by the projections of vector $\vec{a}$ and vector b into the x_1, x_2 plane.

Assume $\vec{a} = (a_1, a_2, a_3)$ and $\vec{b} = (b_1, b_2, b_3)$, then,

$$dx_1 \wedge dx_2(\vec{a},\vec{b}) = \det\begin{pmatrix} a_1 & b_1 \\ a_2 & b_2 \end{pmatrix} \tag{6.6}$$

In general,

$dx_i \wedge dx_j = -dx_j \wedge dx_i$ (same property as the vector product)

$dx_i \wedge dx_j = 0$.

One can always define a linear combination of the functions $dx_i \wedge dx_j$. For instance, if $\vec{F} = (F_1, F_2, F_3)$ is a vector field in a given region D of $\mathbb{R}^3$, it can be defined for each vector $\vec{x}$ in D, the function of ordered pairs $(\vec{a}, \vec{b})$ of vectors in $\mathbb{R}^3$:

$$\tau_x = F_1(\vec{x}) dx_2 \wedge dx_3 + F_2(\vec{x}) dx_3 \wedge dx_1 + F_3(\vec{x}) dx_1 \wedge dx_2$$

This function τ_x is called a differential 2-form or simply 2-form.

Example 3
For the 2-form

$$\tau_x = 2dx_2 \wedge dx_3 + dx_3 \wedge dx_1 + 5dx_1 \wedge dx_2$$

$$\vec{F} = (2,1,5) \quad \vec{a} = (1,2,3) \quad \vec{b} = (0,1,1)$$

$$\tau(\vec{a}, \vec{b}) = 2\det \begin{pmatrix} 2 & 1 \\ 3 & 1 \end{pmatrix} + \det \begin{pmatrix} 3 & 1 \\ 1 & 0 \end{pmatrix} + \det \begin{pmatrix} 1 & 0 \\ 2 & 1 \end{pmatrix}$$

Remember that the components of the vector product $\vec{a} \times \vec{b}$ are as follows:

$$\det \begin{pmatrix} 2 & 1 \\ 3 & 1 \end{pmatrix}, \quad \det \begin{pmatrix} 3 & 1 \\ 1 & 0 \end{pmatrix}, \quad \det \begin{pmatrix} 1 & 0 \\ 2 & 1 \end{pmatrix}.$$

For the 3-forms, we can define for $\vec{a} = (a_1, a_2, a_3) \vec{b} = (b_1, b_2, b_3)$ and $\vec{c} = (c_1, c_2, c_3)$.

$$dx_1 \wedge dx_2 \wedge dx_3 (\vec{a}, \vec{b}, \vec{c}) = \det \begin{pmatrix} a_1 & b_1 & c_1 \\ a_2 & b_2 & c_2 \\ a_3 & b_3 & c_3 \end{pmatrix} \tag{6.7}$$

which is the parallelepiped spanned by the vectors $\vec{a}$, $\vec{c} = (c_1, c_2, c_3)$ and $\vec{c}$.

Examples: Deriving the Maxwell Relations
According to the first law of thermodynamics, if we consider only the variables entropy S, and volume V, then the change in the internal energy U of the given system is as follows:

$$dU = TdS - PdV \tag{6.8}$$

T and $-P$ are the intensive parameters associated with the extensive parameters S and V, respectively.

$$T = \left. \frac{\partial U}{\partial S} \right|_V \tag{6.9}$$

and

$$P = -\left. \frac{\partial U}{\partial S} \right|_V \tag{6.10}$$

dU is an exact total differential; therefore,

$$\frac{\partial T}{\partial V}\bigg|_S = -\frac{\partial P}{\partial S}\bigg|_V \tag{6.11}$$

which is the first Maxwell relation and

$$\frac{\partial^2 U}{\partial V \partial S} = \frac{\partial^2 U}{\partial S \partial V} \tag{6.12}$$

The second Maxwell relation is the equation

$$\frac{\partial S}{\partial V}\bigg|_T = -\frac{\partial P}{\partial T}\bigg|_V \tag{6.13}$$

Proof

Consider the free energy function (also called Helmholtz function) denoted A (or F in some literature).

$$A = U - TS \tag{6.14}$$

Let's take the differential of A

$$\begin{aligned} dA &= dU - TdS - SdT \\ &= TdS - PdV - TdS - SdT \\ &= -SdT - PdV \end{aligned} \tag{6.15}$$

We can successively derive

$$S = -\frac{\partial A}{\partial T}\bigg|_V \tag{6.16}$$

and

$$P = -\frac{\partial A}{\partial V}\bigg|_T \tag{6.17}$$

Since dA is an exact total differential, we have

$$\frac{\partial^2 A}{\partial V \partial T} = \frac{\partial^2 A}{\partial T \partial V} \tag{6.18}$$

and

$$\frac{\partial S}{\partial V}\bigg|_T = -\frac{\partial P}{\partial T}\bigg|_V \tag{6.19}$$

which is the second Maxwell relation.

6.3   HAMILTON'S EQUATIONS AS MAXWELL RELATIONS

Consider some smooth paths S starting at a fixed position and fixed time and ending at the point q, at the time t. $S=S(q,t)$ is also called the Hamilton's principal function. Nature will always choose a path with the least action or at least one that's a stationary point of the action. S being smooth, we have $dS = pdq - Hdt$ for some functions p and H called the momentum and the energy, which obey

$$p = \frac{\partial S}{\partial q}\Big|_{t} \tag{6.20}$$

and

$$H = -\frac{\partial S}{\partial t}\Big|_{q} \tag{6.21}$$

Since dS is an exact differential, we have

$$\frac{\partial p}{\partial t}\Big|_{q} = -\frac{\partial H}{\partial q}\Big|_{t} \tag{6.22}$$

which is the first of Hamilton's equations.

6.4   LEGENDRE TRANSFORMATION AND THERMODYNAMICAL POTENTIALS

Assume a given function f that we are trying to change in g dependent in $\frac{\partial f}{\partial x} = p$. We also assume that f is differentiable. For each point $(x, f(x))$, assume that it maps a given point $(p, g(p))$ and inversely. This means that the function $\frac{\partial f}{\partial x}$ is injective and monotonic; in other words, the graph of f is either concave or convex, that is, $\frac{\partial f}{\partial x}$ does not change the sign. The Legendre transformation is defined by $(p) = \pm(f(x) - xp)$, where $p = \frac{\partial f}{\partial x}$. As property, Legendre transformation is an involution, that is, we can have the correspondence, $f \to g \to f$ through the transformation.

6.5   D'ALEMBERT'S PRINCIPLE AND LAGRANGE'S EQUATION

Consider a system of particles which undergoes an infinitesimal displacement $\delta\vec{r}$, under a force $\vec{F}$; $\delta\vec{r}$ can be seen as the sum of all the displacements of all individual points or infinitesimal particles forming the system. As a result, we can write $\delta\vec{r} = \delta\vec{r_1} + \delta\vec{r_2} + \ldots + \delta\vec{r_n}$, and the individual force exerted on each of these particles will be written as $\vec{F}$; $\delta\vec{r}$. For a system in equilibrium, $\vec{F_i} = 0$, and the work done by each of the infinitesimal force is also zero: $\sum_i \vec{F_i}.\delta\vec{r_i} = 0$.

The force $\vec{F_i}$ can be seen as the summation of two major forces. One of them is the applied force, $\vec{F_i}^{\,a}$, and the other is the force called friction force. If there's no friction or the friction force is zero, the work due to the friction force is zero. As a result, the only force that works is the applied force. We assume that this applied force is not perpendicular to the displacement at each instant t, and we can write the principle of virtual work q_j.

If the system moves with a given acceleration and then a given momentum $\vec{p}_i$, Newton's second law of motion leads to the equation: $\vec{F}_i = \dot{\vec{p}}_i$, which yields:

$$\sum\left(\vec{F}_i - \dot{\vec{p}}_i\right) = \vec{0} \tag{6.23}$$

which is called D'Alembert's principle.

We now assume that the vector position $\vec{r}_i = \vec{r}_i\left(q_1, q_2, \ldots, q_n, t\right)$ is the function of the generalized coordinates q_j and time t. The differential of $\vec{r}_i$ is given by $\vec{r}_i = \sum_i \frac{\partial \vec{r}_i}{\partial q_j} dq_j + \frac{\partial \vec{r}_i}{\partial t} dt$. The velocity of each particle of the system is $Q_j = \sum_i \vec{F}_i \cdot \frac{\partial \vec{r}_i}{\partial q_j}$. The equation of motion of the system can be written as follows:

$$\frac{d}{dt}\left(\frac{\partial T}{\partial \dot{q}_j}\right) - \left(\frac{\partial T}{\partial q_j}\right) = Q_j \tag{6.24}$$

where $Q_j = \sum_i \vec{F}_i \cdot \frac{\partial \vec{r}_i}{\partial q_j}$ represents the components of the generalized force.

Remember that L is the Lagrangian of the system. In fact, the Lagrangian is the difference between the kinetic energy and the potential energy: $L=T-V$.

The kinetic energy can be written as $= \sum_i \frac{1}{2} m_i v_i^2$, where m_i is the mass of each individual particle of the system. More importantly,

$$\frac{d}{dt}\left(\frac{\partial L}{\partial \dot{q}_j}\right) - \left(\frac{\partial L}{\partial q_j}\right) = 0 \tag{6.25}$$

which is called the Lagrange equation of motion.

Example 4

$f(x) = x^2$. $p = \frac{\partial f}{\partial x} = 2x$ $x = \frac{p}{2}$, so $g(p) = \left(x^2 - xp\right) = \left(\frac{p}{2}\right)^2 - \frac{p^2}{2} = \frac{p^2}{4}$.

We can notice that $G(P,T) = U - S\frac{\partial U}{\partial S} - V\frac{\partial U}{\partial V} = U - TS + PV$ is convex and so is $g(p)$.

Example 5

Considering the Lagrangian, $L(q, \dot{q})$.

$p = \frac{\partial L}{\partial \dot{q}}$ is the impulsion. The Legendre transformation of L with respect to $\dot{q}$, $H = \dot{q}p - L(q, \dot{q})$ is the Hamiltonian.

Example 6

Consider the internal energy $U(S, V, N)$. We perform the Legendre transformation on S to introduce F:

$$F = U - S\frac{\partial U}{\partial S} = U - ST = F(T, V, N)$$

We can differentiate F

$$dF = dU - SdT - TdS = TdS - PdV + \mu dN - SdT - TdS$$
$$= -PdV + \mu dN - SdT. \qquad (6.26)$$

where μ is the chemical potential, N is the number of particles in the thermodynamical system and F represents the maximum work exchanged by a system undergoing an isothermal process.

Legendre transformation of $U(S, V, N)$ with respect to V will introduce the function enthalpy.

$$H = U - V\frac{\partial U}{\partial V} = U + PV, \text{ because } -P = \frac{\partial U}{\partial V}$$

Legendre transformation of $U(S, V, N)$ with respect to V and N will introduce the grand potential energy or Landau free energy or Landau potential.

$$\Phi_G = U - S\frac{\partial U}{\partial S} - N\frac{\partial U}{\partial N} = U - TS - \mu N \qquad (6.27)$$

Legendre transformation of $U(S,V,N)$ with respect to S and V will introduce the Gibbs free energy or Gibbs function.

$$G(P,T) = U - S\frac{\partial U}{\partial S} - V\frac{\partial U}{\partial V} = U - TS + PV \qquad (6.28)$$

6.6 THE EXTERIOR DERIVATIVE

The fundamental theorem of calculus states that if $\dfrac{df}{dx}$ is continuous on an interval $[a,b]$, we have

$$\int_l^b \frac{df}{dx}dx = f(b) - f(a) \qquad (6.29)$$

the stokes formula

$$\int_A curlF \cdot dA = \int_{\partial A} F \cdot \vec{t}ds \qquad (6.30)$$

and the Gauss formula

$$\int_V divF \cdot dV = \int_{\partial \sigma} F \cdot \vec{n}ds \qquad (6.31)$$

Since they express a kind of derivative of a function in terms of itself on a set of lower dimensions, we can define the exterior differentiation of the real-valued differentiable function f of several variables x_1, x_2, x_3, x_n by

$$df = \frac{\partial f}{\partial x_1}dx_1 + \frac{\partial f}{\partial x_2}dx_2 + \frac{\partial f}{\partial x_3}dx_3 + \ldots + \frac{\partial f}{\partial x_n}dx_n \qquad (6.32)$$

PROBLEM SET 6

6.1.   Calculate the following differential forms on the indicated vectors.

   a. $dx_1 + 2dx_2$; (1 2)

   b. $dx_1 + 2dx_2$; (−1, 4)

   c. $3dx_2 + 2dx_3$; (2, 3)

   d. $3dx + -dy + 5dz$; (1, −1, 3)

   e. $xdx + ydy - zdz$; (1, −1, 1).

6.2.   Find the value of

   a. $dx \wedge dy + 3dx \wedge dz, \big((1,2,1),(2,1,3)\big)$

   b. $dx \wedge dy; \big((-1,-2,0),(1,0,3)\big)$

   c. $\big(3\sin zdx + 2\cos xdy\big) \wedge \big(dx + dz\big)$

6.3.   Compute $\displaystyle\int_\gamma ydx + xdy$, where $d\varphi$ is given by $g(t) = (\cos t, \sin t), 0 \le t \le \pi/2$

6.4.   Compute $dx_1 + 2dx_2$ where γ is given by

   a. $g(t) = \big(-t^2, t^2, t^2\big)$ for $-1 \le t \le 1$

   b. $h(t) = \big(t^2, t^4, 2t\big)$ for $0 \le t \le 1$

6.5.   Consider the internal Energy U as a function of the entropy S, the volume V and the number of particles N. Prove and write all Maxwell equations.

6.6.   A soap film is stretched between two coaxial circular rings of radius R. The distance between the rings is d. You may ignore gravity. Find the shape of the soap.

6.7.   Given the electromagnetic wave equations,

$$\frac{\partial^2 \varphi}{\partial x^2} + \frac{\partial^2 \varphi}{\partial y^2} + \frac{\partial^2 \varphi}{\partial z^2} = \frac{1}{c^2}\frac{\partial^2 \varphi}{\partial t^2},$$

   a.   Write Galileo coordinate transformations from the inertial reference frame $S(x,y,z,t)$ to the inertial reference frame $S-(x-,y-,z-,t-)$. Consider that $S-$ travels with a constant velocity v with respect to S along their common axis $x-x-$.

   a.   Express $d\varphi$ as a function of x, y and t.

   b.   Prove that the wave equation is not invariant under Galileo transformations.

6.8.   Given the wave equations $\dfrac{\partial^2 \varphi}{\partial x^2} + \dfrac{\partial^2 \varphi}{\partial y^2} + \dfrac{\partial^2 \varphi}{\partial z^2} = \dfrac{1}{c^2}\dfrac{\partial^2 \varphi}{\partial t^2}$,

   a.   Write Lorentz' coordinate transformations from the inertial reference frame $S(x,y,z,t)$ to the inertial reference frame $S-(x-,y-,z-,t-)$. Consider that $S-$ travels with a constant velocity v with respect to S along their common axis $x-x-$.

   a.   Express $d\varphi$ as a function of x, y, z and t and as a function of $x-$, $y-$, $z-$ and $t-$.

   b.   Prove that the wave equation is invariant Lorentz transformations

6.9.   Show that the differential expression

   $dx^2 + dy^2 + dz^2 - c^2dt^2$ is invariant under Lorentz transformations

6.10.   Prove that the expressions $dx^2 + dy^2 + dz^2 - c^2dt$ and $x^2 + y^2 + z^2 - c^2t^2$ are not invariant under Galileo transformations.

7 Infinite Sequences and Series

The mathematical sciences particularly exhibit order, symmetry, and limitation; and these are the greatest forms of the beautiful.

–Aristotle

INTRODUCTION

This chapter reviews arithmetic and geometric sequences, series and their convergence, harmonic series, integral test theorem, comparison test, ratio test and several other convergence tests. Infinite sequences and series are useful in mathematical physics and related topics. Special functions such as the Bessel functions and the Legendre polynomials can be defined as series

7.1 DEFINITION OF A SEQUENCE

A set of numbers arranged in a definite order according to some definite rule is called a sequence. A sequence is a function whose domain is the set $\mathbb{N}$ of natural numbers. Sequences of objects are most commonly denoted using braces. A sequence is called finite if the number of terms is finite. A finite sequence has always a last term.

7.1.1 NOTATION

The sequence $\{u_1, u_2, u_3, \ldots, \{u_n\}_{n=1}^{\infty} \ldots\}$ can be denoted by $\{u_n\}$ or $\{u_n\}_{n=1}^{\infty}$.

An infinite sequence is an unending string of numbers, such as $u_1, u_2, u_3, \ldots, u_n \ldots$ with $n = 1, 2, 3, \ldots$.

Example 1
$(-1), (+1), (-1), (+1), (-1), (+1), (-1), (+1), (-1), (+1), \ldots$ is a sequence the general term of which is $u_n = (-1)^n$.

Example 2
Write the four first terms of the sequence $u_n = \left\{ \dfrac{n}{n+1} \right\}$.

Solution
$u_1 = 1/2, u_2 = 2/3, u_3 = 3/4, u_4 = \dfrac{4}{5}$ are the first four terms.

7.2 ARITHMETIC SEQUENCES

An arithmetic sequence is defined by the recursive formula:

$$u_1 = u$$

$$u_n - n_{n-1} = d \tag{7.1}$$

DOI: 10.1201/9781003478812-7

d is a constant called the common difference of the arithmetic sequence. It can be proved that

$$u_n = u_1 + (n-1)d \tag{7.2}$$

7.3 GEOMETRIC SEQUENCES

A sequence of numbers u_n is called a geometric sequence if the quotient of successive terms is a constant called a common ratio. So, if r is the common ratio of the sequence $(u_n)_n$, then

$$\forall n \in N^*, u_n = r u_{n-1} \tag{7.3}$$

It can be proved (see Rubric 6.7, example 7) that

$$u_n = u_1 r^{n-1} \tag{7.4}$$

7.4 CONVERGENCE OF A SEQUENCE

A sequence $\{S_n\}$ is said to converge to the limit L if for any $\varepsilon > 0$, there is a positive integer N such that $\{S_n\}$ for $n \geq N$. In this case, we can write $\lim_{n\to\infty} S_n = L$.

Theorem 1
For a real number c, $\{S_n\}$

Proof
$S_n = c$ for all n. For any $\varepsilon > 0$, there is a positive integer N such that for $\{u_n\}$, $|c - c| = 0 < \varepsilon$ is true.

Suppose now that the sequences $\{u_n\}$ and $\{v_n\}$ converge to limits l_1 and l_2, respectively, and that c is a constant. Then

$$\lim_{n\to+\infty} (cu_n) = c \lim_{n\to+\infty} (u_n) = cl_1$$

$$\lim_{n\to+\infty} (a_n \pm b_n) = l_1 \pm l_2$$

$$\lim_{n\to+\infty} (a_n b_n) = \lim_{n\to+\infty} (a_n) \lim_{n\to+\infty} (b_n) = l_1 l_2$$

$$\lim_{n\to+\infty} \left(\frac{a_n}{b_n}\right) = \frac{\lim_{n\to+\infty} (a_n)}{\lim_{n\to+\infty} (b_n)} = \frac{l_1}{l_2} (if l_2 \neq 0) \tag{7.5}$$

Some sequences do not converge; they diverge. A sequence that doesn't converge to a finite limit, diverges.

Example 3.a. Consider the sequences.

$$u_n = \frac{n(n+1)}{2} \text{ and } a_n = 2n$$

1. Find $u_1, u_2, u_3, u_2 - u_1, x\, u_2 - u_2,$

$$u_n - u_{n-1}, ^- u_{2n} - u_n$$

2. Find $a_1 - a_2 - a_3$

$$a_2 - a_1 \ a_3 - a_3 -$$

$$a_n - a_3 -, \ a_{2n} - a_n$$

3. Express $\dfrac{u_n}{na_n}$ and calculate $\lim\limits_{n \to \infty} \left(\dfrac{u_n}{na_n} \right)$.

Solution

1. $u_1 = \dfrac{1(1+1)}{2} = 1$

$u_2 = \dfrac{2(2+1)}{2} = 3$

$u_3 = \dfrac{3(3+1)}{2} = 6$

$u_2 - u_1 = 3 - 1 = 2$

$u_3 - u_2 = 6 - 3 = 3$

$u_n - u_{n-1} = \dfrac{n(n+1)}{2} - \dfrac{(n-1)(n-1+1)}{2} = \dfrac{n(n+1)}{2} - \dfrac{n(n-1)}{2} = n$

$u_{2n} - u_n = \dfrac{2n(2n+1)}{2} - \dfrac{(n)(n+1)}{2} = \dfrac{(n)(3n+1)}{2}$

2. $a_1 = 2.1 = 2, \ a_2 = 2.2 = 4, \ a_3 = 2.3 = 6$

$a_2 - a_1 = 2, \ a_3 - a_2 = 2$

$a_n - a_{n-1} = 2n - 2(n-1) = 2$

$a_{2n} - a_n = 2(2n) - 2n = 2n$

3. $\dfrac{u_n}{na_n} = \dfrac{n(n+1)}{2n(2n)} = \dfrac{n^2 + n}{4n^2} = \dfrac{1}{4} + \dfrac{1}{4n}$

$\lim\limits_{n \to \infty} \left(\dfrac{u_n}{na_n} \right) = \lim\limits_{n \to \infty} \left(\dfrac{1}{4} + \dfrac{1}{4n} \right) = \dfrac{1}{4}$

Example 3.b.

Study the convergence or divergence of the sequence a. $S_n = (-2)^n$, b. $S_n = (\ln n)/n$, c. $S_n = (\ln n)/n$.

Solution

The sequence $S_n = (\ln n)/n$ can be expanded by: $-2, 4, -8, 16, \ldots$, which has no limit because the absolute value of the terms grows large without bound. This means that the sequence diverges.

The sequence $S_n = (\ln n)/n$ can be expanded by: 1, 1/4, 1/9, 1/16, $\ldots$, and $S_n = (\ln n)/n$. Hence, the sequence converges and the limit is $L = 0$.

For the sequence $S_n = (\ln n)/n$, let us consider the function $f(x) = (\ln x)/x$, which is defined for all $x > 0$. Using the L'Hopital rule, we obtain

$$\lim_{x \to \infty} \frac{\ln x}{x} = \lim_{x \to \infty} \frac{(\ln x)'}{(x)'} = \lim_{x \to \infty} \frac{1/x}{1} = 0 \ .$$

Hence, the sequence converges to 0.

7.5   BOUNDED SEQUENCE

7.5.1   DEFINITION 1

A sequence $\{S_n\}_{n=1}^{\infty}$ is bounded above if there is some number b that is greater than each term of the sequence, so that for every positive number n, $S_n \le b$.

A sequence $\{S_n\}_{n=1}^{\infty}$ is bounded below if there is some number a, so that for every positive number n, $a \le S_n$.

A sequence that is bounded above and below is bounded.

Note that the lower bound a and the upper bound b are not unique.

7.5.2   MONOTONIC SEQUENCE

A sequence $\{S_n\}_{n=1}^{\infty}$ is said to be increasing if $S_1 \le S_2 \le S_3 \le \cdots \le S_n \le S_{n+1} \le \cdots$.

A sequence $\{S_n\}_{n=1}^{\infty}$ is said to be decreasing if $S_1 \ge S_2 \ge S_3 \ge \cdots \ge S_n \ge S_{n+1} \ge \cdots$.

A monotonic (or monotone) sequence is a sequence that is either increasing or decreasing.

Theorem 2

An increasing sequence that is bounded above converges.

A decreasing sequence that is bounded below converges.

Example 4

The terms $S_n = \sqrt{2} + \sin(\pi/n)$ satisfy the inequalities:

$\sqrt{2} \le \sqrt{2} + \sin(\pi/n) \le \sqrt{2} + 1$ for $n = 1,2,3$.

We can conclude that the sequence is bounded below by $\sqrt{2}$ and above by $\sqrt{2}+1$.

7.6   SERIES REVIEW

7.6.1   DEFINITION 2

A series is the result of the sum of all the terms of a sequence. For example, if $a_1, a_2, a_3, \ldots, a_n$ are the n terms of a sequence, then, the associated series is given by

$$\sum_{k=1}^{n} a_k = a_1 + a_2 + a_3 + \ldots + a_n \tag{7.6}$$

7.6.2   ARITHMETIC SERIES

An arithmetic series is the sum of the terms of an infinite arithmetic sequence. The sum of the n terms of an arithmetic sequence is given by

$$S_n = \Sigma_{k=1}^n a_k = \frac{n}{2}(a_1 + a_n) = \frac{n}{2}\left[2a_1 + (n-1)d\right] \tag{7.7}$$

with a_1 being the first term.

7.6.3　Geometric Series

A geometric series is the sum of the terms of a geometric sequence. In other words, the series $\sum_{n=1}^{\infty} a_n$ is called geometric if there is a fixed constant number r such that for every $n > 0$, $a_{n+1} = ra_n$.

Theorem 3
If $a \neq 0$ and if $-1 < r < 1$, the geometric series converges and its sum is given by

$$\sum_{k=0}^{\infty} ar^k = \frac{a}{1-r} \tag{7.8}$$

If $a \neq 0$ and $|r| \geq 1$, the series diverges. If $a = 0$, the series converges to the sum 0 for all r.

7.7　SERIES IDENTITIES

7.7.1　Mathematical Induction

7.7.1.1　Principle of Mathematical Induction
Assume that we want to prove that for all positive integers $|r| \geq 1$, the statement $P(n)$ holds. Mathematical induction requires two steps:

(a) **First step or Base case**

The first step consists of verifying that the statement, $|r| \geq 1$, holds at first order, $n = 0$, that is, show that $P(n+1)$ holds.

(b) **Second step or induction step**

In the second step, we assume that $P(n)$ holds and show then that $P(n+1)$ holds. From there, we can conclude. In this step, the assumption that $P(n)$ holds is the induction hypothesis and the implication $P(n+1)$ is the induction conclusion.

Conclusion
We can now conclude that the statement $P(n)$ holds for all $n > 0$.

Example 5
Prove the relation in equation (7.4), that is, if $(u_n)_n$ is a geometric sequence, then

$$\forall c \in \mathbf{N}^*, u_n = u_1 r^{n-1}.$$

Solution
Let us prove by mathematical induction that for all non-zero integers n, $u_n = r^{n-1}u_1$ (1).
　　First, $(u_n)_n$ is a geometric series; therefore, the successive terms are as follows:
　　$u_1 = u_1$, property (1) is true for the order $n = 1$
　　$u_2 = ru_1$, property (1) is true for the order $n = 2$

$u_3 = r^2 u_1$, property (1) is true for the order $n = 3$

$u_4 = r u_3 = r^3 u_1$, property (1) is true for the order $n = 4$

Suppose that property (1) is true for all nonzero natural numbers, that is $u_n = r^{n-1} u_1$ and let us prove by induction that $u_{n+1} = r^n u_1$.

$u_{n+1} = r u_n = r r^{n-1} u_1 = r^n u_1$, which proves that for all non-zero integers, $u_n = r^{n-1} u_1$.

Example 6

Consider $S_n = 1 + 3 + 5 + 7 + ... + (2n - 1)$

Prove by induction the statement: $S_n = n^2$ for $n \geq 1$.

Solution

We need to proceed with the two steps before we conclude:

First Step

For $n = 1$

$S_1 = 1$ and $1^2 = 1$, so $S_1 = 1^2$. Hence, the statement $P(1)$ holds.

Second Step

Assume $P(k)$ holds, that is:

$S_k = 1 + 3 + 5 + 7 + ... + (2k - 1) = k^2$.

Now evaluate S_{k+1}.

$$\begin{aligned}
S_{k+1} &= 1 + 3 + 5 + 7 + ... + [2(k+1) - 1)] \\
&= 1 + 3 + 5 + 7 + ... + (2k - 1) + [2(k+1) - 1)] \\
&= k^2 + 2k + 1 \\
&= k^2 + [2(k+1) - 1)] \\
&= k^2 + 2k + 1 \\
&= (k+1)^2
\end{aligned}$$

This proves that $P(k+1)$ holds.

7.7.2 CONCLUSION

Combining the results from the first and the second steps, we can conclude that the statement $P(n)$ holds for $n \geq 1$.

7.7.2.1 A Few Induction Identities

The following identities can be proved by mathematical induction:

$$\sum_{k=1}^{n}(a_k \pm b_k) = \sum_{k=1}^{n} a_k \pm \sum_{k=1}^{n} b_k \tag{7.9}$$

$$\forall c \in R, \sum_{k=1}^{n}(a a_k) = \pm \left(\sum_{k=1}^{n} a_k \right) \tag{7.10}$$

$$\forall c \in R, \sum_{k=1}^{n} c = nc \qquad (7.11)$$

$$\sum_{k=1}^{n}(k) = \frac{n(n+1)}{2} \qquad (7.12)$$

$$\sum_{k=1}^{n}(k^2) = \frac{n(n+1)(2n+1)}{2} \qquad (7.13)$$

$$\sum_{k=1}^{n} k^3 = \frac{n^2(n+1)^2}{4} \qquad (7.14)$$

$$\sum_{k=1}^{n}(k^4) = \frac{n(n+1)(2n+1)(3n^2+3n-1)}{30} \qquad (7.15)$$

7.8   CONVERGENT SERIES

7.8.1   DEFINITION 3

The series S_{n_i} is said to be convergent and to have the value S as convergence value if

$$\lim_{n \to \infty} S_n = S.$$

The condition for the existence of a limit is that:
   for all $\varepsilon > 0$, there exists $N = N(\varepsilon)$ such that $N = N(\varepsilon)$ for $i > N$.

7.8.2   CAUCHY CRITERION

Cauchy's criterion for convergence expresses a necessary and sufficient condition for a sequence to be convergent. Cauchy's criterion is satisfied when, for all $\varepsilon > 0$, there exists $N = N(\varepsilon)$ such that

$$|S_j - S_i| < \varepsilon \quad \text{for all } i, j > N \qquad .(7.16)$$

Proof
To prove one implication, suppose that the sequence S_n converges to a given S, then using the definition of the convergent series,
   for all $\varepsilon > 0$, there exists $N = N(\varepsilon)$ such that $N = N(\varepsilon)$ for $i > N$

   However, $|S_j - S_i| = |S_j + S - S - S_i| < |S - S_i| + |S - S_j| < \varepsilon/2 + \varepsilon/2 = \varepsilon.$

   For $i, j > N$.
   To prove the other implication, suppose the Cauchy criterion holds, then there exists a subsequence S_n which converges to S. Every sequence S_i converges to S. Then it is true that, for all $\varepsilon > 0$, there exists N_1 such that $|S - S_m| < \varepsilon$ for $i \geq N_1$. It is also true that if we are given $\varepsilon > 0$, we can find an N_2 such that $|S_n - S_m| < \varepsilon$ for $n, m \geq N_2$. Let us now choose N_1 such that $|S - S_{n_i}| < \varepsilon/2$ for $i \geq N_1$. We also need to choose N_2 such that $\varepsilon > 0$, for $n, m \geq N_2$. Suppose $n \geq N_2$, and choose S_{n_i} with $n_i \geq N_2$ and $i \geq N_1$. Therefore,

$$|S_n - S| = |S_n + S_{ni} - S_{ni} - S| < |S_n - S_{ni}| + |S_{ni} - S| < \varepsilon/2 + \varepsilon/2 = \varepsilon.$$

7.9   THE HARMONIC SERIES

The series, $\sum_{n=1}^{\infty}\frac{1}{n}=1+\frac{1}{2}+...+\frac{1}{n}...$, is called the Harmonic series. The Harmonic series diverges.

To prove this, let us arrange the series in partial sums such as

$$\sum_{n=1}^{\infty}\frac{1}{n}=1+\frac{1}{2}+\left(\frac{1}{3}+\frac{1}{4}\right)+\left(\frac{1}{5}+\frac{1}{6}+\frac{1}{7}+\frac{1}{8}\right)+\left(\frac{1}{9}+\frac{1}{10}+\frac{1}{11}+\cdots\frac{1}{16}\right)+...+ \tag{7.17}$$

Each term in parenthesis is greater than $\frac{1}{2}$; so, we can see that for $n=2^k$, the partial sum s_n is greater than $\frac{k}{2}$. The harmonic series is not bounded from above; it diverges. Here is an important theorem on these types of series.

Theorem 4

A series $\sum_{n=1}^{\infty}a_n$ of nonnegative terms converges if and only if its partial sums are bounded from the above.

Example 7

Examine the convergence of the series $\sum_{n=1}^{\infty}\frac{1}{n^2}$.

Solution

Consider the function $f(x)=\frac{1}{x^2}$, which is continuous and positive. We can compare the series to the integral of the function f. The partial sum of the sequence is less than the graph under the curve of $y=\frac{1}{x^2}$. Therefore, this series is bounded from above and is convergent. In fact,

$$\begin{aligned}S_n &= 1+\frac{1}{2^2}+...+\frac{1}{n^2}+...\\&=f(1)+f(2)+....+f(n)+...\\&<1+\int_1^{\infty}f(x)dx\\&<1+1\\&<2\end{aligned} \tag{7.18}$$

The integral appearing in formula (6.18) can be calculated:

$$\int_1^{\infty}f(x)dx=\int_1^{\infty}\frac{1}{x^2}dx=\left[-\frac{1}{x}\right]_1^{\infty}=1.$$

7.10   COMPARISON TEST

The comparison test is embedded in the following theorem:

Theorem 5

Assume three nonnegative term – series, $\sum_{n=1}^{\infty}a_n$, $\sum_{n=1}^{\infty}c_n$ and $\sum_{n=1}^{\infty}c_n$.

Suppose $a_n \le b_n \le c_n$ for all $n>N$ where N is an integer; then,

If $\sum c_n$ converges, then $\sum b_n$ converges.

If $\sum a_n$ diverges, then $\sum b_n$ diverges.

Example 8

Find out if the series $\sum_{n=0}^{\infty} \frac{1}{n!}$ converges and diverges.

Solution

It is interesting to find out that the series

$$\sum_{n=o}^{\infty} \frac{1}{n!} = 1 + \frac{1}{1!} + \frac{1}{2!} + \frac{1}{3!} \ldots \frac{1}{n!} + \ldots$$

converges and to observe that

$$\frac{1}{3!} = \frac{1}{6} < \frac{1}{2^2}$$

$$\frac{1}{4!} = \frac{1}{24} < \frac{1}{2^3}$$

So, the terms of the series $\sum_{n=o}^{\infty} \frac{1}{n!}$ are successively less than the terms of the series $1 + \sum_{n=o}^{\infty} \frac{1}{2^n}$

which converges to $1 + \frac{1}{1-1/2} = 3$. Therefore, the series $\sum_{n=0}^{\infty} \frac{1}{n!}$ converges. It is interesting to point

out that the series $\sum_{n=o}^{\infty} \frac{1}{n!}$ does not converge to 3. This technique helps us to only prove that it is a convergent series.

7.11   LIMIT COMPARISON TEST

Theorem 6

Suppose two series of terms strictly positive, a_n, b_n for all $n > N$, with N an integer.

If $\lim_{n \to \infty} \frac{a_n}{b_n} = c > 0$, then both $\sum a_n$ and $\sum b_n$ converge or both diverge.

If $\lim_{n \to \infty} \frac{a_n}{b_n} = 0$ and $\sum b_n$ converges then $\sum a_n$ converges.

If $\lim_{n \to \infty} \frac{a_n}{b_n} = \infty$ and $\sum b_n$ diverges then $\sum a_n$ diverges.

Example 9

Find if the sequence $\frac{a_n}{b_n} = \frac{2^n}{2^n - 1}$ converges or diverges.

Solution

Consider $a_n = \dfrac{1}{2^n - 1}$ and $b_n = \dfrac{1}{2^n}$; $\dfrac{a_n}{b_n} = \dfrac{2^n}{2^n - 1}$, and $\lim_{n\to\infty} \dfrac{a_n}{b_n} = \lim_{n\to\infty} \dfrac{2^n}{2^n - 1} = 1 > 0$; but $\sum_{n=1}^{\infty} b_n$

$= \sum_{n=1}^{\infty} \dfrac{1}{2^n}$ is a geometric series that converges with common ratio $r = \dfrac{1}{2}$ to $\dfrac{1/2}{1-1/2} = 1$. Therefore,

$\sum_{n=1}^{\infty} \dfrac{1}{2^n - 1}$ converges according to the theorem of limit comparison test in condition a.

7.12   THE RATIO TEST

Theorem 7

Consider a series of positive terms, a_n with $\lim_{n\to\infty} \dfrac{a_{n+1}}{a_n} = \alpha$; then

   If $\alpha < 1$, the series converges.
   If $\alpha > 1$ or $n = 1,2,3$ is infinite, the series diverges.
   If $\alpha = 1$, then there's no conclusion or the test is inconclusive.

Example 10

Test the convergence of the series $\sum_{n=1}^{\infty} a_n = \sum_{n=1}^{\infty} \dfrac{1}{2^n}$ using the ratio test.

Solution

$\lim_{n\to\infty} \dfrac{a_{n+1}}{a_n} = \lim_{n\to\infty} \dfrac{1}{2^{n+1}} \dfrac{2^n}{1} = \dfrac{1}{2} = \alpha < 1$. Therefore, the series $\displaystyle\sum_{n=1}^{\infty} a_n = \sum_{n=1}^{\infty} \dfrac{1}{2^n}$ is convergent.

7.13   THE ROOT TEST

Theorem 8

Consider a series of terms $a_n \geq 0$ and $a_n \geq 0$. Then,
   If $\alpha < 1$, the series converges.
   If $\alpha > 1$, or α is infinite, the series diverges.
   If $\alpha = 1$, then there's no conclusion or the test is inconclusive.

Example 11

Study the convergence of the series $\sqrt[n]{a_n} = 1/2 = \alpha < 1$, using the root test.

Solution

$\sqrt[n]{a_n} = 1/2 = \alpha < 1$. Hence, the convergence of the series.

Example 12

Does the series $\displaystyle\sum_{n=1}^{\infty} n^2 \left(\dfrac{3}{4}\right)^n$ converges or diverges?

Solution

Let us apply the ratio test:

$$\lim_{n\to\infty} \dfrac{(n+1)^2 \left(\dfrac{3}{4}\right)^{n+1}}{n^2 \left(\dfrac{3}{4}\right)^n} = \lim_{n\to\infty} \dfrac{(n+1)^2}{n^2} \dfrac{3}{4} = \dfrac{3}{4} < 1$$

Hence, by Theorem 7, the series converges.

7.14   DEFINITION OF THE ABSOLUTE CONVERGENCE TEST

The absolute convergence test helps study the convergence of $\sum_{n=1}^{\infty}|a_n|$ and is defined as follows:

If $\sum_{n=1}^{\infty}|a_n|$ converges, then the series $\sum_{n=1}^{\infty}a_n$ absolutely converges.

Theorem 9 (the absolute convergence test)
An absolutely convergent series converges.

Example 13

Let us consider the alternating harmonic series $\sum_{n=1}^{\infty}a_n = \sum_{n=1}^{\infty}(-1)^{n+1}\dfrac{1}{2^n}$.

The series $\sum_{n=1}^{\infty}|a_n| = \sum_{n=1}^{\infty}\left|\dfrac{1}{2^n}\right|$, converges; therefore, the series $\sum_{n=1}^{\infty}a_n$ converges.

Theorem 10 (the Cauchy integral test)
Suppose $f(x)$ is a continuous function and has the following properties:

$f(n) = a_n$ for each integer $n > 0$

$\sum_{n=1}^{\infty}a_n$ is a decreasing function for $x \geq 1$

$f(x) > 0$ for $\sum_{n=1}^{\infty}a_n$,

then the series $\sum_{n=1}^{\infty}a_n$ and the integral $\displaystyle\int_{1}^{\infty} f(x)\,dx$ converge or diverge together.

When the integral converges, the sum can be estimated by

$$\int_{1}^{\infty}f(x)\,dx \leq \sum_{1}^{\infty}a_n \leq a_1 + \int_{1}^{\infty}f(x)\,dx.$$

An application of the Cauchy integral test can be summarized in the following examples.

Example 14

Study the convergence of the series $\displaystyle\sum_{n=3}^{\infty}\dfrac{n}{(4+n^2)^{3/4}}$

Solution
Using the Cauchy integral test in the previous theorem, with $x \geq 3$ we have

$$I = \int_{3}^{M}\frac{x}{(4+x^2)^{3/4}}\,dx = \frac{1}{2}\frac{(4+x^2)^{1/4}}{1/4}\Bigg|_{3}^{M} = 2\sqrt[4]{4+M^2} - \sqrt[4]{13}$$

$\lim\limits_{M\to\infty} I_M = \infty$ which means that the series diverges.

Example 15

Test the series $\displaystyle\sum_{n=1}^{\infty}\dfrac{1}{n^2+1}$ for convergence or divergence.

Solution

The function $f(x) = \dfrac{1}{x^2 + 1}$ is continuous, positive and decreasing in the interval $[1, \infty)$. We can therefore use the integral test by finding the improper integral:

$$\int_1^\infty \frac{1}{x^2 + 1}\,dx = \lim_{t\to\infty}\int_1^t \frac{1}{x^2+1}\,dx = \lim_{t\to\infty}\tan^{-1}(x)\Big|_1^t = \lim_{t\to\infty}\left(\tan^{-1}(t) - \frac{\pi}{4}\right) = \frac{\pi}{4}$$

As a result, the integral $\displaystyle\int_1^\infty \frac{1}{x^2+1}\,dx$ is convergent; thus by the integral test the series $\displaystyle\sum_{n=1}^\infty \frac{1}{n^2+1}$ is a convergent series.

Corollary

The p-series $\displaystyle\sum_{n=1}^\infty \frac{1}{n^p} = 1 + \frac{1}{2^p} + \frac{1}{3^p} + ... + \frac{1}{n^p} + ...$ converges if $p > 1$ and diverges if $p \leq 1$.

Example 16
Study the convergence or divergence of the series u_n.

Solution

The series $\displaystyle\sum_{n=1}^\infty \frac{1}{n^3} = 1 + \frac{1}{2^3} + \frac{1}{3^3} + ... + \frac{1}{n^3} + ...$ is a p-series with $p = 3 > 1$; therefore, it converges.

7.15   SERIES OF FUNCTIONS

We know that the terms of the infinite series u_n can be regarded as functions of a variable x. $u_n = u_n(x)$. The series sum can be written as u_n. We can now define the uniform convergence.

7.15.1   UNIFORM CONVERGENCE

Consider x an element of the closed interval $[a,b]$. If for a sufficiently small positive number $\varepsilon > 0$, there exists a number $N \leq n$, such that $|s(x) - s_n(x)| < \varepsilon$, then the series is said to be uniformly convergent in the interval $[a,b]$.

7.15.2   WEIERSTRASS M (MAJORANT) TEST

The Weierstrass M (Majorant) Test is the most common test used in uniform convergence. If one can construct a series of number $\displaystyle\sum_{i=1}^\infty M_i$ so that $M_i \geq |u_i(x)|$ for all x in the interval $[a,b]$ and $\displaystyle\sum_{i=1}^\infty M_i$ is convergent, then the series $u_i(x)$ is uniformly convergent.

Proof

If we consider $\sum M_i$ a convergent series, some number N exists such that for $n+1 \geq N$, $\displaystyle\sum_{i=1}^\infty M_i < \varepsilon$.

Therefore, $\displaystyle\sum_{i=n+1}^\infty |u_i(x)| < \varepsilon$. Hence, $|s(x) - s_n(x)| = \left|\displaystyle\sum_{i=n+1}^\infty u_i(x)\right| < \varepsilon$, and the series $\displaystyle\sum_{i=1}^\infty u_i(x)$ is uniformly convergent.

7.16   ABEL'S TEST

Abel's test is also used to prove uniform convergence of a series of functions. Consider $f_n(x)$; if $\sum a_n = A$ is convergent and the function $f_n(x)$ is monotonic and bounded for all x element of $[a,b]$, then the series of functions $\sum_{i=1}^{\infty} u_i(x)$ is uniformly convergent in the interval $[a, b]$.

Note that Abel's test is useful when dealing with the power series which will be our next topic to be discussed.

7.17   TAYLOR'S EXPANSION

7.17.1   POWER SERIES

Definition 4

Power series is an infinite series of the form $\sum_{n=0}^{\infty} a_n(x-a)^n = a_0 + a_1(x-a) + a_2(x-a)^2 + ...$ where $a_1, a_2, a_3, ...$ are constants and x a variable.

Example 17

$\sum_{n=0}^{\infty} 3^n(x-2)^n, \sum_{n=0}^{\infty} 2^n \frac{(x+1)^n}{n!}, -\sum_{n=0}^{\infty} (-1)^n x^n$ are power series.

The question which arises is one of the intervals of convergence. The next theorem will address this question.

Theorem 11

For each power series $\sum_{n=0}^{\infty} a_n(x-a)^n$, there exists a number $R \geq 0$ called the radius of convergence of the series, such that if $|x-a| < R$, then the series converges absolutely; and if $|x-a| > R$, the series diverges.

This means that if the series converges, it converges inside the interval $a - R < x < a + R$, and if it diverges it happens in the closed interval $a - R \leq x \leq a + R$.

7.17.2   TAYLOR AND MACLAURIN SERIES

Definition 5

If the derivatives of the function f exists at $x = a$, then the series given by the equality

$$\sum_{n=0}^{\infty} \frac{f^{(n)}(a)}{n!}(x-a)^n = f(a) + f'(a)(x-a)$$
$$+ \frac{f''(a)}{2!}(x-a)^2 + ... + \frac{f^{(n)}(a)}{n!}(x-a)^n + ...$$

(7.19)

is called Taylor's series for f about a.

For $a = 0$, the series becomes $\sum_{n=0}^{\infty} \frac{f^{(n)}(0)}{n!}x^n$ and is called the Maclaurin series for f. So, the Maclaurin series can be seen as Taylor's series about $a = 0$.

Example 18

Find the Maclaurin series expansion of $f(x) = e^x$.

Solution

$$f'(0) = f''(0) = \ldots = f^{(n)}(0) = 1$$

All the derivatives are the same.

$$f'(0) = f''(0) = \ldots = f^{(n)}(0) = 1$$

$$e^x = \sum_{n=0}^{\infty} \frac{f^{(n)}(0)}{n!} x^n$$

$$= f(0) + \frac{f'(0)}{1!} x + \frac{f''(0)}{2!} x^2 + \ldots + \frac{f^{(n)}(0)}{n!} x^n$$

$$= 1 + \frac{1}{1!} + \frac{1}{2!} x^2 + \ldots + \frac{1}{n!} x^n$$

Let us now study the convergence of $\sum_{n=0}^{\infty} \frac{x^n}{n!}$. We will apply the ratio test by evaluating the limit:

$$\lim_{n \to \infty} \left| \frac{a_{n+1}}{a_n} \right| = \lim_{n \to \infty} \left\{ \frac{|x|^{n+1}}{(n+1)!} \times \frac{n!}{|x|^n} \right\} = \lim_{n \to \infty} \frac{|x|}{(n+1)} = 0; \text{ hence the series } \sum_{n=0}^{\infty} \frac{x^n}{n!} \text{ converges.}$$

Example 19

Find the Taylor series for $f(x) = \ln x$ about $x = 1$.

Solution

We can see the pattern $f^{(n)}(1) = (-1)^{n+1}(n-1)!$ for $n = 1, 2, 3, \ldots$ and Taylor expansion of $\ln x$ is

$$\ln x = \sum_{n=0}^{\infty} \frac{(-1)^{n+1}}{n} (x-1)^n = (x-1) - \frac{(x-1)^2}{2} + \frac{(x-1)^3}{3} - \frac{(x-1)^4}{4} + \ldots .$$

7.17.3 TAYLOR'S THEOREM WITH REMAINDER

Consider Taylor's expansion of a function $f(x)$

$$f(x) = f(a) + \frac{f'(a)}{1!}(x-a) + \frac{f''(a)}{2!}(x-a)^2 + \ldots + \frac{f^{(n)}(a)}{n!}(x-a)^n + \ldots$$

We can observe that the first $(n + 1)$ terms in Taylor's series of $f(x)$ is a polynomial $P_n(x)$, which can be written as follows:

$$P_n(x) = f(a) + \frac{f'(a)}{1!}(x-a) + \frac{f''(a)}{2!}(x-a)^2 + \ldots + \frac{f^{(n)}(a)}{n!}(x-a)^n .$$

We can also notice that

$$P_n(a) = f(a)$$

$$P_n'(a) = f'(a)$$

$$-$$

$$P_n^{(n)}(a) = f^{(n)}(a).$$

We now need to think about measuring how close the polynomial is to the Taylor expansion of $f(x)$. We then introduce the quantity $R_n(x) \equiv f(x) - P_n(x)$ called the remainder. We can therefore write $f(x)$ as follows:

$$f(x) = f(a) + \frac{f'(a)}{1!}(x-a) + \frac{f''(a)}{2!}(x-a)^2 + ... + \frac{f^{(n)}(a)}{n!}(x-a)^n + R_n(x)$$

In order to prove the convergence of the series $\displaystyle\sum_{n=0}^{\infty} \frac{f^{(n)}(a)}{n!}(x-a)^n$, it is sufficient to prove that $R_n(x)$ tends to 0 as n tends to ∞.

Theorem 12
The remainder in Taylor's expansion is generally given by the integral expression:

$$R_n(x) = \frac{1}{n!} \int_a^x f^{(n+1)}(t)(x-t)^n dt$$

with $n = 0,1,2,3,....$.

Corollary (Taylor's Inequality)
Let M be the maximum of $\left| f^{(n+1)}(t) \right|$, for $t \in [a,x]$; then $\left| R_n(x) \right| \le \dfrac{M}{(n+1)!} \left| x-a \right|^{n+1}$.

Proof
It is interesting to prove the corollary at the order $n = 1$. Thus, assume $\left| f''(x) \right| \le M$ for $-d \le x - a \le d$

(a) First, let us consider the case where $f''(x) \ge 0$, which means that we have:

$$f''(x) \le M .$$

By integrating sidewise, we have:

$$\int_a^x f''(t)dt \le \int_a^x Mdt$$

$$f'(t)\Big|_a^x \le Mt\Big|_a^x$$

$$f'(x) - f'(a) \le M(x-a)$$

$$f'(x) \le f'(a) + M(x-a)$$

Let us integrate a second time; we will then have:

$$\int_a^x f'(t)\,dt \le \int_a^x f'(a)\,dt + \int_a^x M(t-a)$$

$$f(x)-f(a)\le f'(a)(x-a)+M\left[\frac{t^2}{2}-at\right]_a^x$$

$$f(x)-f(a)\le f'(a)(x-a)+M\left[\frac{x^2}{2}-ax-\frac{a^2}{2}+a^2\right]$$

$$f(x)-f(a)\le f'(a)(x-a)+\frac{M}{2}\left[x^2-2ax+a^2\right]$$

$$f(x)-f(a)\le f'(a)(x-a)+\frac{M}{2}(x-a)^2$$

$$f(x)-f(a)-f'(a)(x-a)\le \frac{M}{2}(x-a)^2$$

$$R_1(x)\le \frac{M}{2}(x-a)^2.$$

(b) Second, let us consider the case where $f''(x)\le 0$, which means that we have $f''(x)\ge -M$. We will have a similar remainder $R_1(x)\ge -\frac{M}{2}(x-a)^2$; therefore, we have $|R_1(x)|\le \frac{M}{2}|x-a|^2$, which proves Taylor's inequality for $n = 1$. In general, the result for all n can be proved in a similar way by integrating $n + 1$ times.

Example 20
Prove that e^x is equal to the sum of its Maclaurin series.

Solution
$f(x)=e^x$; all the derivatives are equal to e^x, that is $f^{(n+1)}(x)=e^x$ for all n.

If d is any positive number and $|x|\le d$, then $\left|f^{(n+1)}(x)\right|=e^x<e^d$. Taylor inequality with $a = 0$ and $M = e^d$ says that $\left|R_n(x)\right|\le \frac{e^d}{(n+1)!}|x|^{n+1}$ for $|x|\le d$.

$\displaystyle\lim_{n\to\infty}\frac{e^d}{(n+1)!}|x|^{n+1}=0$. Therefore, the limit of $\left|R_n(x)\right|$ as n tends to infinity is zero: $\displaystyle\lim_{n\to\infty}\left|R_n(x)\right|=0$.

Hence, e^x is equal to the sum of its Maclaurin series, that is, $e^x=\displaystyle\sum_{n=0}^{\infty}\frac{x^n}{n!}$ for all x.

Example 21

In Einstein theory of special relativity, the mass of a particle moving with a velocity v is given by $m = \dfrac{m_0}{\sqrt{1 - \dfrac{v^2}{c^2}}}$, where m_0 is the rest mass of the particle and c is the speed of light in a vacuum. The kinetic energy K of the particle is given by the difference between its energy $E = mc^2$ and its rest energy $E_0 = m_0c^2$. $K = E - E_0 = mc^2 - m_0c^2$.

a. Show that when the speed of the particle is very small compared to the speed of light in free space, the kinetic energy of the particle also called the relativistic kinetic energy agrees with the classical kinetic energy.

b. Use Taylor's inequality to estimate the difference in the expressions of the kinetic energy K when $|v| \le 100 m / s$.

Solution

a.

$$K = E - E_0 = mc^2 - m_0c^2$$

$$= c^2 \left(\frac{m_0}{\sqrt{1 - \dfrac{v^2}{c^2}}} - m_0 \right)$$

$$= m_0 c^2 \left[\left(1 - \frac{v^2}{c^2} \right)^{-1/2} - 1 \right].$$

If we consider $v \ll c$, then $x = \dfrac{v^2}{c^2}$ is very small and we can use Maclaurin expansion to the first order of the expression $f(v) = \left(1 - \dfrac{v^2}{c^2} \right)^{-1/2}$ or $f(x) = \left(1 - x^2 \right)^{-1/2}$

$$f'(x) = \frac{1}{2} \left(1 - x^2 \right)^{-3/2}$$

$$f'(0) = \frac{1}{2}.$$

Using Maclaurin expansion and stopping at the first order, we have

$$f(x) = 1 + \frac{1}{2} x + \dots$$

and $K = m_0 c^2 \left[1 + \dfrac{1}{2} x + \dots - 1 \right]$

Replacing x by $\dfrac{v^2}{c^2}$ gives $K = m_0 c^2 \left[\dfrac{1}{2} \dfrac{v^2}{c^2} \right] = \dfrac{1}{2} m_0 v^2 = K_{classical}$.

b. If $x = \dfrac{v^2}{c^2}$, then $f(x) = \left(1 - x^2\right)^{-1/2}$, and M is a number such that, $\left|f(x)\right| \le M$, then Taylor's inequality can give us:

$$R_1(x) \le \frac{M}{2!} x^2.$$

We have $\left|f''(x)\right| = \dfrac{3 m_0 c^2}{4\left(1 - v^2 / c^2\right)^{5/2}} \le \dfrac{3 m_0 c^2}{4\left(1 - 100^2 / c^2\right)^{5/2}}$; using the value of $c = 3.10^8$ m / s, we have

$$R_1(x) \le \frac{1}{2} \cdot \frac{3 m_0 c^2}{4\left(1 - 100^2 / c^2\right)^{5/2}} \cdot \frac{100^4}{c^4} < 4.17 \times 10^{-10} m_0.$$

So, when $|v| \le 100$ m / s, the magnitude of the error in using the classical expression of the kinetic energy is about $4.17 \times 10^{-10} m_0$.

7.17.4   DIFFERENTIATION AND INTEGRATION OF A POWER SERIES

An important application of the power series is that when a power series has a positive radius of convergence, it can be differentiated or integrated term by term to give the derivative or the integral of the original series. It is embedded in the following theorem:

Theorem 13

If the series $\displaystyle\sum_{n=0}^{\infty} a_n (x-a)^n$ converges to $f(x)$ in the interval given by $|x-a| < R$, where $R > 0$,

then $f'(x)$ exists for $\displaystyle\sum_{n=1}^{\infty} n a_n (x-a)^{n-1}$ and the series $\displaystyle\sum_{n=1}^{\infty} n a_n (x-a)^{n-1}$ converges to $f'(x)$ for $|x-a| < R$.

Furthermore, the series

$$\sum_{n=0}^{\infty} \frac{a_n}{n+1} (x-a)^{n+1} \text{ converges to the definite integral } \int_a^x f(t)\,dt \text{ for } |x-a| < R.$$

Furthermore,

All the derivatives $f^{(n)}(x)$ exists for $F(x) = \displaystyle\int_0^x \sin t^2 dt$, and $a_n = \dfrac{f^{(n)}(a)}{n!}$.

Example 22

Evaluate the expression of the integral $\int \sin x^2 dx$.

Solution

$I = \int \sin x^2 dx$ is an indefinite integral and cannot be expressed with a finite number of combinations of elementary functions, but an infinite series for it can be found. In fact, from the series,

$\sin x = \displaystyle\sum_{n=0}^{\infty} \frac{(-1)^n x^{2n+1}}{(2n+1)!}$, we can have by substitution, $\sin t^2 = \displaystyle\sum_{n=0}^{\infty} \frac{(-1)^n t^{4n+2}}{(2n+1)!}$, for all t.

From the previous theorem, $F(x) = \int_0^x \sin t^2 dt$ can be written as the infinite series:

$$F(x) = \sum_{n=0}^{\infty} \int_0^x \frac{(-1)^n t^{4n+2}}{(2n+1)!} dt = \sum_{n=0}^{\infty} \frac{(-1)^n x^{4n+3}}{(2n+1)!(4n+3)},$$

which is the antiderivative of $\sin x^2$.

Overall, $\int \sin x^2 dx = C + \sum_{n=0}^{\infty} \frac{(-1)^n x^{4n+3}}{(2n+1)!(4n+3)}.$

7.18   TELESCOPING SERIES

The series $\lim_{n \to \infty} \left[\frac{1}{n^2 + c^2} - \frac{1}{c^2} \right] = -\frac{1}{c^2}$ has the partial sum

$$B_n = \sum_{k=1}^{n} (b_k - b_{k-1}) \tag{7.20}$$

which can be expanded as follows:

$$B_n = (b_1 - b_0) + (b_2 - b_1) + \ldots + (b_n - b_{n-1})$$

$$B_n = b_n - b_0 \tag{7.21}$$

Equation (7.21) can also be obtained from equation (7.20) by using a shift of index. From (7.20) we have

$$B_n = \sum_{k=1}^{n} b_k - \sum_{k=1}^{n} b_{k-1} = \sum_{k=1}^{n} b_k - \sum_{k=1}^{n-1} b_k = b_n - b_0$$

Because of the way the partial sum in (7.20) collapses into the form represented in (7.21), the series (7.20) is called the telescoping series. When such a series arises naturally, the telescoping of its partial sum can be useful.

Example 23

Find the sum of the series $\sum_{k=1}^{n} \left[\frac{1}{n^2 + c^2} - \frac{1}{(n-1)^2 + c^2} \right]$, with $c \neq 0$

Solution

For this series, the partial sum is given by $S_n = \frac{1}{n^2 + c^2} - \frac{1}{c^2}$. Since

$\lim_{n \to \infty} \left[\frac{1}{n^2 + c^2} - \frac{1}{c^2} \right] = -\frac{1}{c^2}$, the series converges to $-\frac{1}{c^2}$.

Example 24 (the Telescoping Sum)

Show that the series $\sum_{n=1}^{\infty} \dfrac{1}{n(n+1)}$ is convergent and find its sum.

Solution

The series $s_n = \sum_{n=1}^{\infty} \dfrac{1}{n(n+1)}$ is not a geometric series, so we will use the definition of a convergent

series and compute the partial sums.

$$\sum_{n=1}^{\infty} \frac{1}{n(n+1)} = \frac{1}{1.2} + \frac{1}{2.3} + \frac{1}{3.4} + \dots + \frac{1}{n(n+1)}$$

We can simplify this expression by using the partial fraction decomposition:

$$\frac{1}{i(i+1)} = \frac{1}{i} - \frac{1}{i+1}$$

$$\sum_{i=1}^{n} \frac{1}{i(i+1)} = \sum_{i=1}^{n} \left(\frac{1}{i} - \frac{1}{i+1} \right)$$

$$= \left(1 - \frac{1}{2} \right) + \left(\frac{1}{2} - \frac{1}{3} \right) + \left(\frac{1}{3} - \frac{1}{4} \right) + \dots + \left(\frac{1}{n} - \frac{1}{n+1} \right)$$

$$= 1 - \frac{1}{n+1}$$

Now, the limit, $\lim_{n \to \infty} s_n = \lim_{n \to \infty} \left(1 - \dfrac{1}{n+1} \right) = 1 - 0 = 1.$

Therefore, the series s_n is convergent and converges to 1.

In the previous example, note that the terms cancel two by two and the sum collapses like a pirate's collapsing telescope into two terms.

7.19 BINOMIAL SERIES

Consider a real number k and a variable x such that $|x| < 1$; then,

$$\frac{1}{i(i+1)} = \frac{1}{i} - \frac{1}{i+1}$$

$$= 1 + kx + \frac{k(k-1)}{2!} x^2 + \frac{k(k-1)(k-2)}{3!} x^3 + \dots$$

is called binomial series.

The notation $\dbinom{k}{n} = \dfrac{k!}{n!(k-n)!} = \dfrac{k(k-1)(k-2)\dots(k-n+1)}{n!}$ represents the notation of the binomial

coefficients.

Example 25

In modern physics, the total relativistic energy E is given by

$$E = m_0 c^2 \left[\left(1 - \frac{v^2}{c^2} \right)^{-1/2} \right].$$

The binomial theorem can be used to find the expression of the expansion of E. Therefore, let us consider $x = -\frac{v^2}{c^2}$ and $k = -1/2$; then we have

$$E = m_0 c^2 \left[\begin{array}{l} 1 - \frac{1}{2}\left(-\frac{v^2}{c^2} \right) + \frac{(-1/2)(-3/2)}{2!}\left(-\frac{v^2}{c^2} \right)^2 \\ \\ + \frac{(-1/2)(-3/2)(-5.2)}{3!}\left(-\frac{v^2}{c^2} \right)^3 + \cdots \end{array} \right]$$

$$E = m_0 c^2 + \frac{1}{2} m_0 v^2 + \frac{3}{8} m_0 \frac{v^4}{c^2} + \frac{5}{16} m_0 v^2 \frac{v^4}{c^4} + \cdots$$

$E_0 = m_0 c^2$ is the rest mass energy and the relativistic energy is $E_0 = m_0 c^2$.

$$K = \frac{1}{2} m_0 v^2 + \frac{3}{8} m_0 \frac{v^4}{c^2} + \frac{5}{16} m_0 \frac{v^6}{c^4} + \cdots$$

$$K = \frac{1}{2} m_0 v^2 \left(1 + \frac{3}{4} \frac{v^2}{c^2} + \frac{5}{8} \frac{v^4}{c^4} + \cdots \right)$$

When v is very small, that is $v \ll c$, the term $\left(1 + \frac{3}{4} \frac{v^2}{c^2} + \frac{5}{8} \frac{v^4}{c^4} + \cdots \right)$ reduces to 1 and the kinetic energy reduces to $K = \frac{1}{2} m_0 v^2$, which agrees with the classical result.

PROBLEM SET 7

7.1   Prove the following statement:

   a.   The series $\sum_{n=1}^{\infty} \frac{1}{n^2}$ is convergent.

   b.   The Harmonic series $\sum_{n=1}^{\infty} \frac{1}{n}$ diverges.

7.2   Show that if $s > 1$, then $\sum_{v=1}^{s} \left(\frac{s}{v} - \frac{1}{v} \right) - s\log(n)$ as $n \to \infty$, s remaining fixed and that if the limit is $\phi(s)$, then $0 < \phi(s) + \frac{1}{s-1} \leq s - 1$.

7.3   Prove that the series

$$\sum_{1}^{\infty} \sum_{1}^{\infty} \frac{l+m}{\left(l^2 + m^2 \right)^2} \text{ converges or not accordingly as } s > 3/2 \text{ or } s \leq 3/2.$$

7.4   Use mathematical induction to prove the following identities:

a. $\sum_{k=1}^{n}\left(a_k \pm b_k\right) = \sum_{k=1}^{n} a_k \pm \sum_{k=1}^{n} b_k$

b. $\forall \alpha \in R,\ \sum_{k=1}^{n}\left(\alpha a_k\right) = \alpha \left(\sum_{k=1}^{n} a_k\right)$

c. $\forall c \in R,\ \sum_{k=1}^{n} c = nc$

d. $\sum_{k=1}^{n}\left(k^2\right) = \dfrac{n\left(n+1\right)\left(2n+1\right)}{2}$

e. $\sum_{k=1}^{n}\left(k^2\right) = \dfrac{n\left(n+1\right)\left(2n+1\right)}{2}$

f. $\sum_{k=1}^{n} k^3 = \dfrac{n^2\left(n+1\right)^2}{4}$

g. $\sum_{k=1}^{n}\left(k^4\right) = \dfrac{n\left(n+1\right)\left(2n+1\right)\left(3n^2 + 3n - 1\right)}{30}$

7.5   Prove that the p–series $p_{\text{series}}^{\infty}\ \dfrac{1}{n^p} = \dfrac{1}{1^p} + \dfrac{1}{2^p} + \ldots + \dfrac{1}{n^p} + \ldots$ for p a constant real, converges

for $p > 1$ and diverges for $p \le 1$

7.6   Determine if the sequence converges or diverges.

a. $\sum_{n=1}^{\infty}\dfrac{1}{10^n}$; b. $\sum_{n=1}^{\infty}\dfrac{1}{2^n}$; c. $\sum_{n=1}^{\infty}\dfrac{2^n}{3^n}$; d. $\sum_{n=1}^{\infty}\dfrac{1}{n^4+1}$; e.$\sum_{n=1}^{\infty}\dfrac{1}{1+e^n}$; f.$\sum_{n=1}^{\infty}\dfrac{1}{n\ln\left(n\right)}$.

7.7   Use the limit comparison test to determine if the series converges or diverges.

a. $\sum_{n=1}^{\infty}\dfrac{n+1}{n^2-5n+6}$; b. $\sum_{n=1}^{\infty}\ln\left(1+\dfrac{1}{n^2}\right)$

7.8   Determine the convergence or the divergence.

a. $\sum_{n=1}^{\infty}\dfrac{1+\cos n}{n^2}$, b. $\sum_{n=1}^{\infty}\dfrac{1+\sin n}{n^2}$, c. $\sum_{n=1}^{\infty}\dfrac{1}{5^{n-1}+1}$.

7.9   Use the ratio test to determine the convergence or the divergence of the series.

a. $\sum_{n=1}^{\infty}\dfrac{1}{10^n}$, b. $\sum_{n=1}^{\infty}\dfrac{1}{2^n}$, c.$\sum_{n=1}^{\infty}\dfrac{2^n}{3^n}$, d. $\sum_{n=1}^{\infty}\dfrac{n!}{\left(2n\right)!}$

7.10 Use the root test to determine if the sequence is convergent or divergent.

a. $\sum_{n=1}^{\infty}\dfrac{\left(n!\right)^2}{2^n}$, b. $\sum_{n=1}^{\infty}\dfrac{1+n}{3^n}$, $\sum_{n=1}^{\infty}\dfrac{1}{10^n}$, c. $\sum_{n=1}^{\infty} n^2 e^{-n}$, d. $\sum_{n=1}^{\infty}\dfrac{\left(2n\right)}{n!\left(n+1\right)!\left(n+2\right)!}$.

7.11 In the following question, the nth term of the sequence is given. In each case write the first five terms of the sequence and state whether the sequence is convergent or divergent. If it is convergent give its limit.

a. $1/\sqrt{n}$, b. $\sum_{n=1}^{\infty}\dfrac{\left(2n\right)}{n!\left(n+1\right)!\left(n+2\right)!}$, c. $\ln(n)/100$, d. $\sum_{n=1}^{\infty}\dfrac{\left(2n\right)}{n!\left(n+1\right)!\left(n+2\right)!}$, e. $\ln\left(\dfrac{1+2n}{n}\right)$

7.12 Prove that the sequence $S_n = r^n$ diverges if $\left|r\right| > 1$.

7.13 Determine whether the following sequences are increasing or decreasing. You can study the difference $S_{n+1} - S_n$.

a. $\dfrac{\left(n+1\right)!}{1.3.5.7\ldots\left(2n+1\right)}$, b. $\dfrac{\left(n\right)!}{2^n}$.

7.14   (Taylor's inequality) Prove Taylor's inequality for $n = 2$, that is prove that if $\left| f''(x) \right| \le M$

for $|x - a| \le d$, then $\left| R_n(x) \right| \le \dfrac{M}{6} |x - a|^3$ for $|x - a| \le d$.

7.15   (Maclaurin series) Find the Maclaurin series for $\cos(x)$, $\sin(x)$, $\dfrac{1}{1-x}$, $\tan^{-1}(x)$.

7.16   Prove that for an arithmetic sequence $(a_n)_n$, the sum of the n terms is given by

$$S_n = \sum_{k=1}^{n} a_k = \frac{n}{2}(a_1 + a_n) = \frac{n}{2}\left[2a_1 + (n-1)d \right]$$

7.17   Find the sum of the first 20 terms of the sequence
4, 9, … .

7.18   Prove that for a geometric sequence, $(a_n)_n$, the sum S_n of the n terms is given by

$$S_n = \sum_n a_n = a_1 \frac{1 - r^n}{1 - r}, \text{ with } r \ne 1.$$

7.19   Write the first six terms of the geometric sequence: 4, 6, …

7.20   Find the sum of the first eight terms of the previous sequence.

7.21   Show that

a. $\sin x = \displaystyle\sum_{n=0}^{\infty} (-1)^n \frac{x^{2n+1}}{(2n+1)!}$

b. $\cos x = \displaystyle\sum_{n=0}^{\infty} (-1)^n \frac{x^{2n}}{(2n)!}$

7.22

a.   Expand by the binomial theorem and integrate term by term to obtain the Gregory series for $y = \tan^{-1}(x)$. Note that $\tan(x) = y$

$$y = \tan^{-1}(x) = \int_0^x \frac{dt}{1 + t^2} = \int_0^x 1 - t^2 + t^4 - t^6 + \ldots]dt.$$

$$y = \sum_{n=0}^{\infty} (-1)^n \frac{x^{2n+1}}{(2n+1)!}, \quad -1 \le x \le 1$$

b.   By comparing the series expansions shows that

$$\tan^{-1}(x) = \frac{i}{2} \ln\left(\frac{1 - ix}{1 + ix} \right)$$

7.23   Find the first three nonzero terms in the Maclaurin series for the function

$$f(x) = e^{-x^2} \tan^{-1}(x)$$

7.24   Find the general term of the series and use the ratio test to show that the series converges

$$1 + \frac{1.2}{1.3} + \frac{1.2.3}{1.3.5} + \frac{1.2.3.4}{1.3.5.7} + \ldots$$

7.25   Use any method to determine whether the series converges.

$$\sum_{k=1}^{\infty} \frac{lnk}{3^k}$$

7.26 Use the ratio test to test the convergence of the series

$$\sum_{n=1}^{\infty} a_n$$

7.27 Use any method to determine whether the series converges.

$$\sum_{n=1}^{\infty} \frac{\ln(n+1)}{10^k}$$

8 Functions of a Complex Variable

All that I've learned, I've forgotten. The little I still know, I've guessed.

—Sebastien-Roch Nicolas de Chamfort

INTRODUCTION

Complex numbers are the extensions of real numbers that help to solve the quadratic equation $x^2 + 1 = 0$ and permit the uniform treatment of linear differential equations and the development of some interesting mathematical physics theories such as two-dimensional fluid flow and the potential theory. The equation $x^2 + 1=0$ has no real solution. To deal with this problem, mathematicians of the eighteenth century introduced the imaginary number $i = \sqrt{-1}$, which gives $i^2 = -1$. Therefore, the equation $x^2 + 1=0$ becomes $x^2 - i^2 = 0$ and can be solved in the set of the complex numbers C. By the beginning of the nineteenth century, it was determined that a complex number has the form $x + iy$, where x and y are real numbers. It could be regarded as an alternative symbol for the ordered pair (x, y).

8.1 DEFINITION OF A COMPLEX NUMBER

A complex number is an ordered pair of real numbers, denoted by (x, y) or $x + iy$.

Example 1
The ordered pair $(1, 1)$ has the alternate form $1 + i$.
   The ordered pair $(-1, 2)$ has the alternate form $-1 + 2i$.
   Sometimes, it is convenient to use a letter such as z to represent a complex number. We can write $z = x + iy$, where x is the real part of z and y the imaginary part, that is, $x = \text{Re}(z)$ and $y = \text{Im}(z)$.

8.2 POLAR COORDINATES AND GEOMETRIC FORM OF A COMPLEX NUMBER

Consider a point M of components (x,y) as represented in Figure 8.1.
   Looking at the geometry of the above figure, we can express $\cos(\theta) = x/r$ and $\sin(\theta) = x/r$. In polar coordinates, (r,θ), the Cartesian components x and y are therefore written as follows:

$$\begin{cases} x = r\cos(\theta) \\ y = r\sin(\theta) \end{cases} \tag{8.1}$$

Moreover, complex number z can be written $z = r\left[\cos(\theta) + i\sin(\theta)\right]$, which is called the polar representation of a complex number and noted $z = \left[r,\theta\right]$ and read z a complex number of modulus $r = |z|$ with $r > 0$ and argument θ with $r = |z|$.

DOI: 10.1201/9781003478812-8

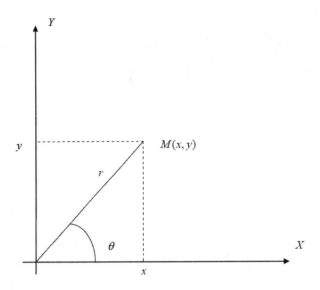

FIGURE 8.1   Geometric interpretation of polar coordinates.

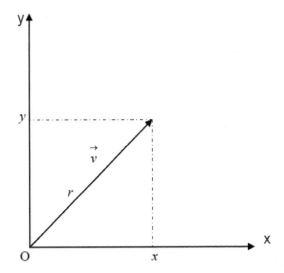

FIGURE 8.2   Geometric interpretation of a complex number.

8.3   GEOMETRICAL REPRESENTATION OF A COMPLEX NUMBER

In the Cartesian plane $\left(\overrightarrow{OX}, \overrightarrow{OY}\right)$, the complex number z can be represented by a vector $\vec{v} = x\vec{i} + y\vec{j}$, where $\vec{i}$ and $\vec{j}$ are, respectively, the units' vectors on the x-axis and on the y-axis. The real number $r = |z|$ is the length of vector $\vec{v}$ in SI unit. It is also the norm of vector $z_1 = 2 + 5i$. Figure 8.2 gives the representation of the vector $\vec{v}$ in the Cartesian coordinates.

8.4   PROPERTIES OF COMPLEX NUMBERS

8.4.1   ADDITION

For two complex numbers $z_1 = a_1 + ib_1$ and $z_2 = a_2 + ib_2$, their sum is defined by the complex number:

$$z_1 + z_2 = (a_1 + a_2) + i(b_1 + b_2). \tag{8.2}$$

Example:

If $z_1 = 2 + 5i$ and $z_2 = 1 - i$, then $z_1 + z_2 = 3 + 4i$.

8.4.2 SUBTRACTION

For two complex numbers $z_1 = a_1 + ib_1$ and $z_2 = a_2 + ib_2$, the subtraction of the two complex numbers is defined as the complex number: $z_1 - z_2 = (a_1 - a_2) + i(b_1 - b_2)$.

Example:

If $z_1 = 2 + 5i$ and $z_2 = 1 - i$, then $z_1 - z_2 = 1 + 6i$.

8.4.3 MULTIPLICATION

For two complex numbers $z_1 = a_1 + ib_1$ and $z_2 = a_2 + ib_2$, the multiplication of the two complex numbers is defined as the complex number:

$$z_1 \cdot z_2 = (a_1 a_2 - b_1 b_2) + i(a_1 b_2 + b_1 a_2) \tag{8.3}$$

Example:

If $z_1 = 2 + 5i$ and $z_2 = 1 - i$ then $z_1.z_2 = 7 + 3i$.

Example: Definition of the Modulus of z

Consider the complex number $z = x + iy$; $z^* = x - iy$ is defined as the complex conjugate of the complex number, $z_1 \cdot z_2 = (a_1 a_2 - b_1 b_2) + i(a_1 b_2 + b_1 a_2)$. The product of z^* and z^* will be given by

$$zz^* = x^2 + y^2 = |z|^2 \tag{8.4}$$

where $|z|$ is called the modulus of z or absolute value of z and is always a positive real number.

Example:

If $z = 1 + i$, $|z| = \sqrt{1^2 + 1^2} = \sqrt{2}$.

8.4.4 DIVISION

For two complex numbers $z_1 = a_1 + ib_1$ and $z_2 = a_2 + ib_2$, the ratio $\dfrac{z_1}{z_2}$ with $z_2 \neq 0$ is given by

$$\frac{z_1}{z_2} = \frac{a_1 + ib_1}{a_2 + ib_2} \tag{8.5}$$

By multiplying the numerator and the denominator by the conjugate of the denominator we rationalize the ratio $\dfrac{1+i}{1-2i}$, we then obtain

$$\frac{z_1}{z_2} = \frac{a_1 + ib_1}{a_2 + ib_2} = \frac{(a_1 + ib_1)(a_2 - ib_2)}{(a_2 + ib_2)(a_2 - ib_2)} = \frac{(a_1 a_2 + b_1 b_2) + i(b_1 a_2 - a_1 b_2)}{a_2^2 + b_2^2}$$

$$= \frac{(a_1 a_2 + b_1 b_2)}{a_2^2 + b_2^2} + i\frac{(b_1 a_2 - a_1 b_2)}{a_2^2 + b_2^2}$$

Example:

Rationalize the denominator of the complex function $\dfrac{1+i}{1-2i}$.

Solution:

$$\frac{1+i}{1-2i} = \frac{(1+i)(1+2i)}{(1-2i)(1+2i)} = \frac{1+3i-2}{1^2+2^2} = \frac{-1+3i}{5}$$

8.4.5 MULTIPLICATION USING GEOMETRIC FORM

Assume two complex numbers z_1 and z_2 with modulus or absolute values and arguments, r_1, θ_1, and r_2, θ_2, respectively. We can use the notation $z_1 = [r_1, \theta_1]$ and $z_2 = [r_2, \theta_2]$. The product of the two complex numbers can be given by

$$z_1 z_2 = [r_1, \theta_1]^-[r_2, \theta_2]$$

$$[r_2, \theta_2] = r_1 r_2 \left[\cos(\theta_1 + \theta_2) + i\sin(\theta_1 + \theta_2) \right]$$

$$z_1 z_2 = \left[r_1 r_2, \theta_1 + \theta_2 \right] \tag{8.6}$$

Proof

Starting from

$z_1 z_2 = [r_1, \theta_1]^-[r_2, \theta_2]$, we have

$$z_1 z_2 = r_1 \left[\cos(\theta_1) + i\sin(\theta_1) \right] r_2 \left[\cos(\theta_2) + i\sin(\theta_2) \right]$$

$$z_1 z_2 = r_1 r_2 \left\{ \left[\cos(\theta_1)\cos(\theta_2) - \sin(\theta_1)\sin(\theta_2) \right] + i\left[\cos(\theta_1)\sin(\theta_2) + \sin(\theta_1)\cos(\theta_2) \right] \right\}$$

$$z_1 z_2 = r_1 r_2 \left[\cos(\theta_1 + \theta_2) + i\sin(\theta_1 + \theta_2) \right]$$

So, $z_1 z_2 = z_1 z_2$.

8.4.5.1 Conclusion

The modulus of the product of two complex numbers is the product of the modulus of each complex number. Similarly, the argument of the product of the two complex numbers is the sum of the arguments of each complex number.

$$|z_1 \cdot z_2| = |z_1| \cdot |z_2|$$

$$Arg(z_1 \cdot z_2) = Arg(z_1) + Arg(z_2)$$

8.4.6 DIVISION USING GEOMETRIC FORM

The complex number z_1 divided by the complex number z_2 is defined by

$$\frac{z_1}{z_2} = \frac{[r_1, \theta_1]}{[r_2, \theta_2]}$$

$$= \frac{r_1 \left[\cos(\theta_1) + i\sin(\theta_1) \right]}{r_2 \left[\cos(\theta_2) + i\sin(\theta_2) \right]} = \frac{r_1}{r_2} \frac{\left[\cos(\theta_1) + i\sin(\theta_1) \right] \cdot \left[\cos(\theta_2) - i\sin(\theta_2) \right]}{\left[\cos(\theta_2) + i\sin(\theta_2) \right] \cdot \left[\cos(\theta_1) - i\sin(\theta_1) \right]}$$

$$= \left(\frac{r_1}{r_2}\right) \frac{\begin{bmatrix} \cos(\theta_1)\cdot\cos(\theta_2)+\sin(\theta_1)\cdot\sin(\theta_2) \\ +i\left[\sin(\theta_1)\cdot\cos(\theta_2)-\cos(\theta_1)\cdot\sin(\theta_2)\right] \end{bmatrix}}{\left[\cos^2(\theta_2)+\sin^2(\theta_2)\right]}$$

$$= \left(\frac{r_1}{r_2}\right)\left[\cos(\theta_1-\theta_2)+\sin(\theta_1-\theta_2)\right]$$

So,

$$|z_1| = \sqrt{\left(\sqrt{3}\right)^2 + 1^2} = \sqrt{3+1} = \sqrt{4} = 2. \qquad (8.7)$$

8.4.6.1   Conclusion

The modulus of the ratio of two complex numbers is the ratio of the modulus of each complex number. Similarly, the argument of the ratio of the two complex numbers is the difference of the arguments of each complex number.

$$\left|\frac{z_1}{z_2}\right| = \frac{|z_1|}{|z_2|} \text{ with } |z_2| \neq 0$$

$$Arg\left(\frac{z_1}{z_2}\right) = Arg(z_1) - Arg(z_2)$$

Example 2

Two complex numbers Z_1 and Z_2 are given as follows:

$z_1 = \sqrt{3} + i$; $z_2 = 1 - i$. Evaluate the length and the argument of $z_1 \cdot z_2$; write down your answer in geometric form.

Solution:

We first need to find the modulus and the argument of each complex number z_1 and z_2, then we can write the geometric form of the each of them and of the product $z_1 \cdot z_2$.

$$|z_1| = \sqrt{\left(\sqrt{3}\right)^2 + 1^2} = \sqrt{3+1} = \sqrt{4} = 2$$

We can write z_1 as follows:

$$z_1 = 2\left(\frac{\sqrt{3}}{2} + \frac{1}{2}i\right) = 2\left[\cos\left(\frac{\pi}{6}\right) + i\sin\left(\frac{\pi}{6}\right)\right]$$

We see that the modulus of z_1 is 2 and its argument appears to be $\frac{\pi}{6}$. Therefore, the geometric form of z_1 can be written: $z_1 = \left[2, \frac{\pi}{6}\right]$. Similarly, the modulus of z_2 is $|z_2| = \sqrt{1^2 + (-1)^2} = \sqrt{1+1} = \sqrt{2}$. The complex number z_2 can also be written as follows: $z_2 = \sqrt{2}\left(\frac{1}{\sqrt{2}} - \frac{1}{\sqrt{2}}i\right) = \sqrt{2}\left[\cos\left(-\frac{\pi}{4}\right) + i\sin\left(-\frac{\pi}{4}\right)\right]$

or in the geometric form, $z_2 = \left[\sqrt{2}, -\frac{\pi}{4}\right]$.

Finally, we can calculate the modulus and the argument of the product $z_1 \cdot z_2$ as follows:

$$|z_1 \cdot z_2| = |z_1| \cdot |z_2| = 2\sqrt{2}$$

$$Arg(z_1 \cdot z_2) = Arg(z_1) + Arg(z_2) = \frac{\pi}{6} - \frac{\pi}{4} = -\frac{\pi}{12}$$

The geometric form of $z_1 \cdot z_2$ is then $z_1 \cdot z_2 = \left[2\sqrt{2}, -\frac{\pi}{12} \right]$.

We can finally write the product

$$z_1 \cdot z_2 = 2\sqrt{2} \left[\cos\left(-\frac{\pi}{12} \right) + i\sin\left(-\frac{\pi}{12} \right) \right].$$

8.4.7 TAYLOR SERIES OF A COMPLEX VARIABLE

For a given complex variable z, Taylor's expansion of $f(z) = e^z$ about $z = 0$ gives:

$$f(z) = 1 + z + \frac{z}{2!} + \cdots + \frac{z^n}{n!} + \cdots = \sum_{n=0}^{+\infty} \frac{z^n}{n!} \tag{8.8}$$

In particular, if z is a pure imaginary complex number, $z = iy$; then

$$e^z = e^{iy} = \sum_{n=0}^{+\infty} \frac{(iy)^n}{n!} = \sum_{j=0}^{+\infty} (-i)^j \frac{(y)^{2j}}{(2j)!} + i\sum_{j=0}^{+\infty} (-i)^j \frac{(y)^{2j+1}}{(2j+1)!}$$

As a result, $e^{iy} = \cos y + i\sin y$, which is called the Euler formula. It represents the complex exponential of a purely imaginary variable in terms of the real trigonometric function.

So, $e^{i\pi/2} = \cos\left(\frac{\pi}{2} \right) + i\sin\left(\frac{\pi}{2} \right) = i.$

$$e^{i\pi} = \cos(\pi) + i\sin(\pi) = -1$$

It is worthy of note to recall that e^z satisfies the properties of exponentials, that is, $e^{z_1 + z_2} = e^{z_1} e^{z_2}$, $e^{z_1 - z_2} = \frac{e^{z_1}}{e^{z_2}}$, $\left(e^z \right)^n = e^{nz}$ and $\frac{d}{dz}\left[e^z \right] = e^z$.

8.4.8 EULER FORMULA

$\cos(\theta) + i\sin(\theta)$ can be represented by an exponential function as expressed by the following Euler's formula:

$$e^{i\theta} = \left[\cos(\theta) + i\sin(\theta) \right] \tag{8.9}$$

Proof

Euler formula can be proved using the exponential of $e^{i\theta}$:

$$e^{i\theta} = \sum_{n=0}^{+\infty} \frac{(i\theta)^n}{n!}$$

$$= \sum_{n=0}^{+\infty} \frac{(-1)^n \theta^{2n}}{(2n)!} + i\sum_{n=1}^{+\infty} \frac{(-1)^{n-1}\theta^{2n-1}}{(2n-1)!}$$

$$= \cos\theta + i\sin\theta$$

The special case of $\theta = \pi$ gives $e^{i\pi} + 1 = 0$.

8.4.9 DE MOIVRE THEOREM

If the complex variable is $z = [r,\theta] = r\left[\cos(\theta) + i\sin(\theta)\right]$, then

$$z^n = \left[r^n, n\theta\right] = r^n\left[\cos(n\theta) + i\sin(n\theta)\right] \qquad (8.10)$$

Proof

We will start by using the Euler formula:

$$e^{i\theta} = \cos\theta + i\sin\theta$$

$$(e^{i\theta})^n = e^{ni\theta} = (\cos\theta + i\sin\theta)^n$$

However, from Euler formula, we can write

$$e^{ni\theta} = \cos n\theta + i\sin n\theta. \qquad (8.11)$$

Using equations 8.10 and 8.11 one obtains:

$$(\cos\theta + i\sin\theta)^n = \cos n\theta + i\sin n\theta \text{ for all values of } n. \qquad (8.12)$$

Example 3

For the complex number $z = 2\left[\cos\left(\frac{\pi}{3}\right) + i\sin\left(\frac{\pi}{3}\right)\right]$, compute z^9.

Solution

Using De Moivre formula, we have

$$Z^9 = 2^9\left[\cos(3\pi) + i\sin(3\pi)\right] = -512$$

Example 4

Find the real and imaginary parts of $(1+i)^{10}$.

Solution

We use the polar representation of $1+i = \sqrt{2}e^{i\pi/4}$. So,

$$z = (1+i)^{10} = (\sqrt{2}e^{i\pi/4})^{10} = 2^5 e^{i5\pi/2} = 2^5 e^{i\pi/2} = 2^5 i.$$

Therefore, $Re(z) = 0$ and $Im(z) = 2^5$.

8.4.10   TRIANGLE INEQUALITIES

For two complex numbers z_1 and z_2,

$$\left| z_1 \right| - \left| z_2 \right| \le \left| z_1 + z_2 \right| \le \left| z_1 \right| + \left| z_2 \right| \tag{8.13}$$

8.5   COMPLEX FUNCTIONS OF COMPLEX VARIABLES

8.5.1   DEFINITION 1

For a complex variable $z = x + iy$, we can associate a complex function,

$$f(z) = f(x, y) = f_1(x, y) + i f_2(x, y) \tag{8.14}$$

where f_1 and f_2 are real functions of the real variables x and y. $f_1(x,\ y)$ is the real part of $f(x,y)$ and $f_2(x,y)$ is the imaginary part of $f(x,y)$.

8.5.2   LOGARITHM OF A COMPLEX VARIABLE

For a non-zero complex number $z = re^{i\theta}, \theta$, θ being the argument of z such that $-\pi < \theta \le \pi$ and $r = \left| z \right| > 0$,

$$\ln(z) = \ln r + i\theta \tag{8.15}$$

In particular $\ln(-1) = \ln(e^{\pi i}) = \ln(1) + i\pi = i\pi$

Example 5

Show that $\sin^{-1}(z) = -i\ln\left(iz \pm \sqrt{1-z^2} \right)$

Solution

To solve this question, let us assume $\sin^{-1}(z) = \theta$. Therefore, $z = \sin(\theta) = \dfrac{e^{i\theta} - e^{-i\theta}}{2i}$. From this expression, we can calculate z as a function of z. Then, we have

$$e^{i\theta} - e^{-i\theta} = 2iz \text{ or } e^{2i\theta} - 1 = 2ize^{i\theta} \text{ or } e^{2i\theta} - 2ize^{i\theta} - 1 = 0.$$

Assume now that $u = e^{i\theta}$, the previous equation becomes

$u^2 - 2izu - 1 = 0$. The reduced discriminant is $\Delta = -z^2 + 1$ and the solutions are $u = \dfrac{iz \pm \sqrt{1-z^2}}{1}$.

Now, we will return to the exponential form. We obtained: $e^{i\theta} = iz \pm \sqrt{1-z^2}$ from which we can have $i\theta = \ln\left(iz \pm \sqrt{1-z^2} \right)$ or $\theta = -i\ln\left(iz \pm \sqrt{1-z^2} \right)$, which gives the solution to the problem: $\sin^{-1}(z) = -i\ln\left(iz \pm \sqrt{1-z^2} \right)$.

8.6 DERIVATIVE OF A COMPLEX-VALUED FUNCTION

8.6.1 DEFINITION 2

Let $f(z)$ be a complex-valued function of the complex variable z. The derivative of f is defined as

$$f'(z) = \lim_{\delta z \to 0} \frac{f(z+\delta z) - f(z)}{z+\delta z - z}. \text{ This limit can also be written, } f'(z) = \frac{df}{dz}.$$

8.6.2 ANALYTICITY

Whenever the first derivative of f at the point z_0, that is, $f'(z_0)$ exists, f is said to be analytic (or regular or holomorphic) at the point z_0. The function is analytic throughout a region in the complex plane if f' exists for every point in that region. Any point at which f' does not exist is called a singularity or singular point of the function f. If $f(z)$ is analytic everywhere in the complex plane, it is called entire.

Example 6
- $1/z$ is analytic except at $z = 0$ and is singular at that point.
- The functions z^n, n a nonnegative integer, and e^z are entire functions.

8.6.2.1 Definition of a Singular Point (of an Analytic Function)

A point at which an analytic function $f(z)$ is not analytic, i.e., at which $f'(z)$ fails to exist, is called a singular point. There are different types of singular points:

Isolated and non-isolated singular points: A singular point z_0 is called an isolated singular point of an analytic function, $f(z)$, if there exists a deleted ε-spherical neighborhood of z_0 that contains no singularity. If no such neighborhood can be found, z_0 is called a non-isolated singular point. Thus, an isolated singular point is a singular point that stands completely by itself, embedded in regular points.

8.7 THE CAUCHY–RIEMANN CONDITIONS

The Cauchy–Riemann conditions are necessary and sufficient conditions for a function to be analytic at a point. Suppose $f(z)$ is analytic at z_0. Then, $f'(z_0)$ may be obtained by taking $f'(z_0)$ to zero through purely real or through purely imaginary values; for example, if we consider the infinitesimal change in z as a complex number on the form: $\delta z = \delta x + i\delta y$ and δx, with $u = u(x,y)$ and $v = v(x,y)$, then, $f'(z) = \dfrac{\delta f}{\delta z} = \dfrac{\delta u + i\delta v}{\delta x + i\delta y}$ should be the same whenever δx or δy tends to zero. First, let us examine the case $\delta x \to 0$. The first derivative becomes $\dfrac{\delta f}{\delta z} = \dfrac{\delta u + i\delta v}{i\delta y} = \dfrac{\partial v}{\partial y} - i\dfrac{\partial u}{\partial y}$. Second, let us examine the case $\delta y \to 0$. We will have $\dfrac{\delta f}{\delta z} = \dfrac{\delta u + i\delta v}{i\delta x} = \dfrac{\partial u}{\partial x} - i\dfrac{\partial v}{\partial x}$. Since the derivative is independent of how the limit is taken, we can equate these two expressions, meaning that they must have equal real parts and equal imaginary parts. Therefore, we have the identities:

$$\frac{\partial u}{\partial x} = \frac{\partial v}{\partial y} \text{ and } \frac{\partial v}{\partial x} = -\frac{\partial u}{\partial y}. \tag{8.16}$$

These are the Cauchy–Riemann conditions. These conditions are not only necessary, but if the partial derivatives are continuous, they are sufficient to assure analyticity.

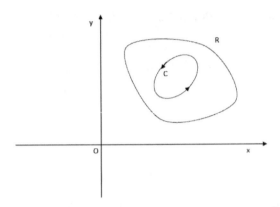

FIGURE 8.3 A closed path C within a region, R.

Example 7

Consider the function defined by the conjugate of z, z^*; that is, if $z = x + iy$, $z^* = x - iy$. The Cauchy–Riemann conditions never hold. In fact, $u(x) = x$ and $v(x) = -y$; so, $\dfrac{\delta f}{\delta z} = \dfrac{\delta u + i\delta v}{i\delta x} = \dfrac{\partial u}{\partial x} - i\dfrac{\partial v}{\partial x}$ and z^* is nowhere an analytic function of z.

Theorem 1a

Cauchy Integral Theorem

Consider a complex plane (x, y) and a simply connected region R, and a closed path C within that region (Figure 8.3).

Assume a function f of a complex variable z. If $f(z)$ is analytic, that is, its partial derivatives continue throughout some simply connected region R for every closed path C in R, and in addition, if f is single valued, then the line integral of $f(z)$ around C is zero, that is:

$$\int_C f(z)\,dz = \oint_C f(z)\,dz = 0 \tag{8.17}$$

Proof

Assume that $f(z) = u(x, y) + iv(x, y)$ is an analytic complex function and $dz = dx + idy$ the differential of z: $\int_C f(z)\,dz = \oint [u(x, y) + iv(x, y)](dx + idy)$. Now, separating the real part and imaginary part gives $\int_C f(z)\,dz = \oint \left[u(x, y)\,dx - v(x, y)\,dy \right] + i\oint \left[u(x, y)\,dy + v(x, y)\,dx \right]$.

Let's use Stokes' theorem, that is, $\oint \vec{V}\,d\vec{\lambda} = \int \vec{\nabla} \times \vec{V}\,d\vec{A}$. Consider $\vec{V} = V_x\vec{i} + V_y\vec{j}c$, then, $\oint V_x\,dx + V_y\,dy$
$= \int \left(\dfrac{\partial V_y}{\partial x} - \dfrac{\partial V_x}{\partial y} \right) dx\,dy$.

Let's consider $u = V_x$ and $-v = V_y$, then

$$\oint v\,dx + u\,dy = \int_s \left(\frac{\partial u}{\partial x} - \frac{\partial v}{\partial y} \right) dx\,dy = -\int_s \left(\frac{\partial u}{\partial x} - \frac{\partial v}{\partial y} \right) dx\,dy$$

Now, if we consider that $v = V_x$ and that $u = V_y$, then we will have

$$\oint v\,dx + u\,dy = \int_s \left(\frac{\partial u}{\partial x} - \frac{\partial v}{\partial y} \right) dx\,dy = -\int_s \left(\frac{\partial u}{\partial x} - \frac{\partial v}{\partial y} \right) dx\,dy$$

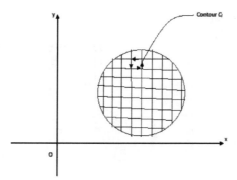

FIGURE 8.4 Cauchy–Goursat contour.

$$\oint f(z)dz = -\int_s \left(\frac{\partial v}{\partial x} + \frac{\partial u}{\partial y}\right)dxdy + \int_s \left(\frac{\partial u}{\partial x} - \frac{\partial v}{\partial y}\right)dxdy.$$

It is worthy of note to notice that each integrand is zero because $f(z) = u + iv$ is analytic, that is $\frac{\partial u}{\partial x} = \frac{\partial v}{\partial y}$ and $\frac{\partial v}{\partial x} = -\frac{\partial u}{\partial y}$. As a result, $\int_C f(z)dz = 0$.

Cauchy–Goursat Proof
In this proof, the interior of the regions will be divided into several small squares. Then, the integral of $f(z)$ is going to be the sum of all the integrals over the small regions. For each small region being closed, the individual integral is zero and their sum is also zero (Figure 8.4).

$$\int_C f(z)dz = \sum_i \oint_{C_i} f(z)dz = 0 \qquad (8.18)$$

Consequence
For analytic functions, the line integral depends only on the initial point and the final point on the integral path, that is:

$$\int_1^{l_2} f(z)dz = F(z_2) - F(z_1) = -\int_1^1 f(z)dz \qquad (8.19)$$

Example 8

Case of Multiple Connected Regions
For the case of multiple connections such as the one shown in Figure 8.5, a circulation around the closed path gives the following relation:

$$\int_{ABCDEFA} = \int_{AB} f(z)dz + \int_{BC} f(z)dz + \int_{CD} f(z)dz$$
$$+ \int_{DE} f(z)dz + \int_{EF} f(z)dz + \int_{FA} f(z)dz. \qquad (8.20)$$

Cauchy Integral Formula
If f is a complex function of the complex variable z and if it is analytic on a closed contour C and within the interior region bounded by C, then

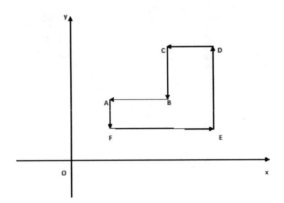

FIGURE 8.5 Multiple connected regions.

$$\frac{1}{2\pi i}\oint \frac{f(z)}{z-z_0}\,dz \int = f(z_0)$$ (8.21)

where z_0 is a point in the interior of the region bounded by the contour C. If z_0 is exterior to the contour, then the Cauchy integral gives zero:

$$\frac{1}{2\pi i}\oint_c \frac{f(z)}{z-z_0}\,dz = 0$$ (8.22)

Derivatives

Cauchy integral can be used to find the successive derivatives of an analytic complex function. In general, if $f(z)$ is analytic, then

$$f^{(n)}(z_0) = \frac{n!}{2\pi i}\oint_c \frac{f(z)}{(z-z_0)^{n+1}}\,dz$$. (8.23)

It is important to note that all the successive derivatives of $f(z)$ are analytic.

Theorem 1b

It is worthy of note to introduce Morera's theorem which states as follows:

If $f(z)$ is continuous in a region D and satisfies $\dfrac{1}{2\pi i}\oint_c \dfrac{f(z)}{z-z_0}\,dz = 0$ for all closed contours " in

D, then $f(z)$ is analytic in D. Morera's theorem does not require simple connectedness. Let D be a region, with $f(z)$ continuous on D, and let its integrals around closed loops be zero. Now, pick any point z_0 in D, and pick a neighborhood of $f(z)$. Evaluate the integral of f as

$$\int_{z_0}^{z} f(z)\,dz = F(z) \text{ and } F'(z) = f(z).$$

Hence F is analytic and has derivatives of all orders, as does f; so f is analytic at any arbitrary point z_0 in D and therefore is analytic in D.

8.8 DEFINITION OF A HOLOMORPHIC FUNCTION

Consider the complex single-valued function $f(z)$ in a given region. $f(z)$ is holomorphic if it possesses a unique derivative at every point in the region, that is, $\lim_{z \to z_0} \dfrac{f(z) - f(z_0)}{z - z_0}$ is unique.

Theorem 2a (Morera's Theorem and the Holomorphy)

If $f(z)$ is a continuous function in a region R and if $\oint f(z)dz = 0$ around all closed contours C lying within R, then $f(z)$ is holomorphic everywhere in R.

Theorem 2b (Theorem on Taylor Series)
Consider z_0 is a point within the interior domain of holomorphy of the complex function, $f(z)$, then, $f(z)$ can be represented by the Taylor series:

$$f(z) = \sum_{n=0}^{\infty} a_n (z - z_0)^n \tag{8.24}$$

where the coefficient

$$a_n = \frac{f^{(n)}(z_0)}{n!} = \frac{1}{2\pi i} \int_{C'} \frac{f(z)dz}{(z - z_0)^{n+1}} \tag{8.25}$$

with $C-$ any closed curve about the point, z_0, enclosing only point of holomorphy of C. Furthermore, this series will converge for all point inside a circle centered about z_0 whose radius is the distance from z_0 to the closest point at which $f(z)$ ceases to be holomorphic.

Theorem 2c
If $f(z)$ is holomorphic in a region R and if $f(z_j) = 0$ for $j = 1, 2, \ldots$ at a set of points $\{z_j\}$ having an accumulation point z_∞ within R, then $f(z) = 0$ everywhere within R.

8.8.1 GENERAL DEFINITION

In general, a function is said to be holomorphic in a region if it is single valued and has a uniquely determined derivative in every neighborhood of R.

The function is said to be analytic if it is holomorphic in the region except for some points that do not interrupt the continuity of R.

The function is meromorphic in a region if it is single valued in R and has at most a finite number of poles but not essential singular points.

Finally, to examine the nature of a function about point $z = \infty$, it is useful to make the substitution $u = 1/z$.

Example 9
For the complex variable function, $z = \pm i$, there exists a pole 2 of order 2. $f(z)$ has two zeros $z = \pm i$, each of order 1. Now, if we assume $u = 1/z$, the function becomes: $f(1/u) = \dfrac{(1/u)^2 + 1}{(1/u - 2)^2} = \dfrac{u^2 + 1}{(1 - 2u)^2}$.

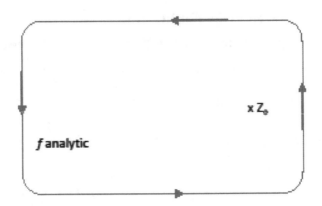

FIGURE 8.6   An analytic function on a closed contour C.

As $z \to \infty$, $u \to 0$ and $z = \infty$. So, there's no singularity at $z = \infty$. Also, $f(z)$ is meromorphic in the entire z plane, including $z = \infty$, and holomorphic beyond the circle of radius $|z| = 2$.

8.9   THEORY OF RESIDUES

In the Cauchy–Goursat theorem, if $f(z)$ is holomorphic within a given region R, enclosed by a curve C also lying within R and enclosing any point, we have, $\oint_c f(z)dz = 0$. In the case where the domain encompasses several singularities, we would need to divide the domain into several contour C_i, each of which encircling one singularity at a time. Considering only one subcontour containing only one singularity, in such region, $f(z)$ can be expanded using the Laurent expansion by

$$(z) = \sum_{n=-\infty}^{n=+\infty} \alpha_n (z - z_0)^n \tag{8.26}$$

Let us take the integral of both side of the Laurent expansion of $f(z)$. We have:

$$\oint f(z)dz = \oint \sum_{n=-\infty}^{n=+\infty} \alpha_n (z - z_0)^n \tag{8.27}$$

Now, we can write $(z - z_0)$ using the trigonometric form

$$z - z_0 = [\rho, \theta] = \rho e^{i\theta} \tag{8.28}$$

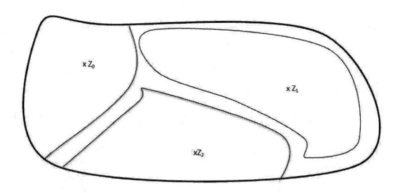

FIGURE 8.7   Contour with singularities z_0, z_1 and z_2.

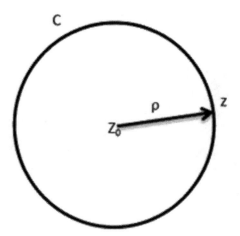

FIGURE 8.8 Contour circle of center O and radius ρ.

$$dz = i\rho e^{i\theta} d\theta = i(z - z_0) d\theta \tag{8.29}$$

We consider that C is a circle of center O and radius $|z - z_0| = \rho$;

$$\oint f(z) dz \oint = \sum_{n=-\infty}^{n=+\infty} i\alpha_n \rho^{n+1} \int^{2\pi} e^{i(n+1)\theta} e^{i(n+1)\theta} d\theta \tag{8.30}$$

Remark

For $n \neq -1$, we have

$$\int^{2\pi} e^{i(n+1)\theta} d\theta = 0 \tag{8.31}$$

and for

$$n = -1, \int^{2\pi} e^{i(n+1)\theta} d\theta = 2\pi \tag{8.32}$$

Therefore,

$$\oint f(z) dz = 2\pi i \alpha_{-1} \tag{8.33}$$

This means that the contour integral is equal to $2\pi i$ times the coefficient of $(z - z_0)^{-1}$, in the Laurent expansion of $f(z)$. This coefficient α_{-1} is called the residue of $f(z)$. As a result, we can write

$$\alpha_{-1} = R_e\left[f(z), z = z_0\right] \equiv R \tag{8.34}$$

In general, if the domain has m isolated singularities enclosed by the contour, C, then the resultant residue is the sum of all the residues and

$$\oint f(z) dz = 2\pi i \sum_{i=1}^{m} R_i \tag{8.35}$$

Example 10

Consider the complex function $f(z) = \dfrac{1}{z^2(1-z)}$.

a. Find the Laurent expansion of $f(z)$ about $z=0$

b. Find the integral $\oint f(z)dz$ around the path of radius $|z| = \dfrac{1}{2}$ about $z=0$

c. Find the integral $\oint f(z)dz$ around the path of radius
   $|z| = 2$ about $z = 0$.

Solution

a. Find the Laurent expansion of $f(z)$ about $z=0$.

   $f(z)$ is not defined for $z = 0$ and $z = 1$. The singularity $z = 0$ is located within the circle of
   center the origin O and radius, $r = |z| = \dfrac{1}{2}$.

   Applying the Laurent expansion, $f(z) = \sum\limits_{n=-\infty}^{n=+\infty} \alpha_n(z - z_0)^n$, with the coefficients

   $\alpha_n = \dfrac{1}{2\pi i} \displaystyle\int_C (t - z_1)^{-n-1} f(t)\,dt$ and C a circle of radius ρ such that $0 < \rho < 1$. Considering

   $t = \rho e^{i\phi}$; $dt = i\rho e^{i\phi}d\phi = itd\phi$.

   $\alpha_n = \dfrac{1}{2\pi i}\displaystyle\int_C t^{-n-1} f(t)\,dt = \dfrac{1}{2\pi i}\int_C \dfrac{t^{-n-3}it}{\left(1 - \rho e^{i\phi}\right)}\,d\phi = \dfrac{\rho^{-n-2}}{2\pi}\sum\limits_{m=0}^{\infty}\rho^m\int_0^{2\pi}e^{-(n-m+2)i\phi}\,d\phi$

   Since $\displaystyle\int_0^{2\pi} e^{-(n-m+2)i\phi}\,d\phi = \begin{cases} 0 & m \neq n+2 \\ 2\pi & m = n+2 \end{cases}$

   then $\alpha_n = \begin{cases} 1 & n = -2,-1,0,+1,+2,... \\ 0 & n < -2 \end{cases}$ and the Laurent expansion about $z = 0$ is given by

   $f(z) = \displaystyle\sum_{n=-2}^{\infty} z^n$ or $f(z) = \dfrac{1}{z^2} + \dfrac{1}{z} + 1 + z + z^2 + ...$

b. Find the integral $\oint f(z)dz$ around the path of radius $|z| = \dfrac{1}{2}$ about $z = 0$; the residue is

   $R_{-1} = 1$ and the integral is $\displaystyle\oint_{\rho=1/2} f(z)dz = 2\pi i$.

   Find the integral $\oint f(z)dz$ around the path of radius
   $|z| = 2$ about $z = 0$.

   Calculations will give $\displaystyle\oint_{\rho=2} f(z)dz = 0$.

PROBLEM SET 8

8.1   For $-1 < p < 1$ prove that the series $\sum_{n=0}^{n=+\infty} p^n \cos(nx)$ and $\sum_{n=0}^{n=+\infty} p^n \sin(nx)$ converge,

respectively, to $\dfrac{1 - p\cos(x)}{1 - 2p\cos(x) + p^2}$ and $\dfrac{p\sin(x)}{1 - 2p\cos(x) + p^2}$.

8.2   Using the same techniques, prove the following equalities:

a. $\cos^{-1}(z) = -i\ln\left(z \pm \sqrt{z^2 - 1}\right)$

b. $\tan^{-1}(z) = \dfrac{i}{2}\ln\left(\dfrac{i+z}{i-z}\right)$

c. $\sinh^{-1}(z) = \ln\left(z + \sqrt{z^2 + 1}\right)$

d. $\cosh^{-1}(z) = \ln\left(z + \sqrt{z^2 - 1}\right)$

e. $\tanh^{-1}(z) = \dfrac{1}{2}\ln\left(\dfrac{1+z}{1-z}\right)$.

8.2   a.   Prove that for a complex variable z, the integral $I = \oint_c e^{1/z}dz = 2\pi i$ knowing that

C is a circle centered about the origin of axis O. Use the expansion $e^{1/z} = \sum \dfrac{1}{n!z^n}$.

b.   Evaluate the integral: $I = \oint_c \dfrac{3z^2 + 2}{z(z+1)}dz$ with C a circle or radius $|z| = 3$. Use the

Cauchy integral and prove that $I = -6\pi i$.

8.3

a.   Compute using Cauchy's integral formula

$$\oint_c \frac{e^z}{z - \pi i / 2}dz$$

where C is the boundary of a square with sides $x \pm 2$ and $y = \pm 2$

b.   Evaluate the real definite integral by contour integration $= \oint_c \dfrac{\log(x)}{b^2 + x^2}dx = \dfrac{\pi \log b}{2b}$.

8.4   Given the system of equations

$$\frac{\partial u}{\partial t} = v\frac{\partial^2 u}{\partial x^2}, 0 < x < d$$
$$u(0,t) = U\cos(\omega t), u(d,t) = 0, t > 0$$

which pertains to an incompressible viscous fluid motion between a pair of (infinite) parallel plate, one of which (at $x = 0$) oscillates in its own plane while the other (at $x = d$) remains fixed. Establish a representation for the solution of this system

$$u(x,t) = U\,\mathrm{Re}\left[\frac{\sinh\lambda(d-x)}{\sinh\lambda d}e^{i\omega t}\right], 0 < x < d, \text{ where } \lambda = \left(\frac{i\omega}{v}\right)^{1/2}$$

where $\mathrm{Re}\big[\Phi\big]$ denotes the real part of the complex valued function Φ. Discuss some approximations forms of the solution in the respective limits, $\dfrac{\omega d^2}{v} \to \infty$ and

$$\frac{\omega d^2}{v} \to 0$$

What is the approximate expression for $u(x,t)$ if $u(x,0) = 0,\, 0 < x < d$?

8.5   Evaluate the integral, $\displaystyle\int^{2\pi} \frac{\sin^2(\varphi)\,d\varphi}{(1+\varepsilon\cos(\varphi))^2},\, 0 < |\varepsilon| < 1.$

8.6   A complex number $z = a + ib$ where a and b are real numbers can be represented by the following $2-2$ matrices:

$$\begin{pmatrix} a & b \\ -b & a \end{pmatrix}$$

a.   Show that this matrix representation holds for addition, subtraction and multiplication.

b.   Find the matrix corresponding to $\dfrac{1}{z} = (a+ib)^{-1}$.

8.7   Verify the Jacobi identity:

$$\big[A,[B,C]\big] = \big[B,[A,C]\big] = \big[C,[A,B]\big].$$

9 Fourier Series

Heat, like gravity, penetrates every substance of the universe, its rays occupy all parts of space. The object of our work is to set forth the mathematical laws which this element obeys. The theory of heat will hereafter form one of the most important branches of general physics.

Joseph Fourier

INTRODUCTION

Fourier series is an important topic in mathematical physics. After a definition, the Euler formula will be given, and its applications will be developed. Fourier series helps expand a periodic function into an infinite sum of harmonic oscillations at given frequencies.

9.1 DEFINITION

A physical parameter or quantity that varies periodically with time, like waves and rotating machines, may be described with a periodic function. We know that a function is said to be periodic if there exists a real number $T > 0$ such that for every real value t, $(t + nT) = f(t)x$, where the multiplicative factor n is an integer. The smallest value T is called the period of the function f. If t is the time variable, then $f(t)$ is a periodic function of period T. The fundamental angular frequency of the function f is known to be $\omega = \dfrac{2\pi}{T}$. The function $f(t)$ can be written as a Fourier series, that is, into an infinite sum of harmonics components at multiples of the fundamental angular frequency, that is:

$$
\begin{aligned}
f(t) &= a_0 + \sum_{n=1}^{\infty}\left[a_n\cos(n\omega t) + b_n\sin(n\omega t)\right] \\
&= a_0 + \sum_{n=1}^{\infty} c_n \cos(n\omega t + \phi_n)
\end{aligned}
\tag{9.1}
$$

$f(t) = a_0 + \sum_{n=1}^{\infty} d_n\sin(n\omega t + \psi_n) = \sum_{n=-\infty}^{n=\infty} g_n\exp(in\omega t)$ could also be written in terms of the sine function of time t and be transformed into the exponential function as follows:

$$
f(t) = a_0 + \sum_{n=1}^{\infty} d_n\sin(n\omega t + \psi_n) = \sum_{n=-\infty}^{n=\infty} g_n\exp(in\omega t)
\tag{9.2}
$$

The coefficients a_n, b_n, c_n, d_n and g_n appearing in the function $f(x)$ are called amplitudes; $\tan\phi_n = \dfrac{-b_n}{a_n}$ and $\tan\psi_n = \dfrac{a_n}{b_n}$ (phase ambiguity) and ψ_n are called the initial phases or the phase at time $t = 0$. The determination of these coefficients is crucial in harmonic analysis.

9.2 EULER FORMULAS

Fourier series can also be described in terms of Euler's formulas for a periodic function f of period T, which can be given as follows:

DOI: 10.1201/9781003478812-9

(a) The zeroth term or the term of rank 0

$$a_0 = \frac{1}{T}\int_0^T f(t)\,dt \tag{9.3}$$

(b) The nth terms of the terms of rank n

$$a_n = \frac{2}{T}\int_0^T f(t)\cos(n\omega t)\,dt \text{ for } n \geq 1 \tag{9.4}$$

$$b_n = \frac{2}{T}\int_0^T f(t)\sin(n\omega t)\,dt \text{ for } n \geq 1 \tag{9.5}$$

All together we have $n+1$ terms.

The coefficient g_n appearing in the exponential form of the Fourier expansion of $f(t)$ given in (9.2) is written as follows:

$$g_n = \frac{1}{T}\int_0^T f(t)e^{-in\omega t}\,dt \tag{9.6}$$

with

$$n = 0,\pm 1,\pm 2,\pm 3,\pm 4,\dots . \tag{9.7}$$

The function f is assumed to be absolutely integrable. The integration sign goes over a full period and can be taken from t_0 to $t_0 + T$, or from $-\frac{T}{2}$ to $\frac{T}{2}$. The range of t chosen appropriately so that the difference between the upper and the lower value of the integration sign is one period, that is T. The following relationships between the coefficients can be proved for $n \geq 1$:

$$a_n = g_n + g_{-n} = c_n\cos\phi_n = d_n\sin\psi_n \tag{9.8}$$

$$b_n = i(g_n - g_{-n}) = -c_n\sin\phi_n = d_n\cos\psi_n \tag{9.9}$$

$$g_n = \frac{1}{2}(a_n - ib_n) \tag{9.10}$$

and

$$g_{-n} = \frac{1}{2}(a_n + ib_n) \tag{9.11}$$

$$c_n{}^2 = d_n{}^2 = a_n{}^2 + b_n{}^2 = 4g_ng_{-n} \; (\text{sign ambiguity})$$

$$\tan\phi_n = \frac{-b_n}{a_n} \text{ and } \tan\psi_n = \frac{a_n}{b_n} \; (\text{phase ambiguity})$$

a_0 can be regarded as the mean value of the function f.

It is important to note that the function $f(t)$ can be a complex function and t can be stretched from $t = -\infty$ to $t = +\infty$. In such a case, Fourier series is called the complex Fourier series and the above equations hold.

Example 1 (using Mathematica Command)

For the third-order Fourier series of $t/2$, we use the commands,

$$\text{FourierSeries}\left[t/2,t,3\right]$$

$$\text{Plot}\left[\%,\{t,-3\text{Pi},3\text{Pi}\},\text{AxesLabel}\rightarrow\{t,\text{FourierSeries}\}\right]$$

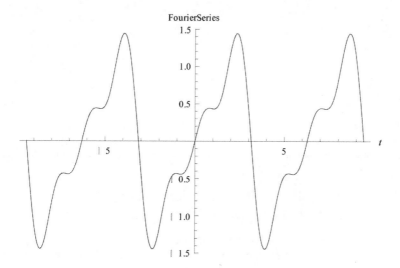

FIGURE 9.1 Schematic of Fourier series of the third order of $t/2$ with the Mathematica Code, FourierSeries$\left[t/2,t,3\right]$.

For the fourth-order Fourier series of t

$$\text{FourierSeries}\left[t,t,4\right]$$

$$\text{Plot}\left[\%,\{t,-4\text{Pi},4\text{Pi}\},\text{AxesLabel}\rightarrow\{t,\text{FourierSeries}\}\right]$$

There are other definitions of the Fourier series where, in the expansion of the function f, a_0 is replaced by $\dfrac{a_0}{2}$. For example, if the function f is defined for all x such that $-\pi \leq x \leq \pi$, then it can be represented as the series:

$$f(x)=\frac{a_0}{2}+\sum_{n=1}^{\infty}\left[a_n\cos(nx)+b_n\sin(nx)\right] \tag{9.12}$$

and Fourier coefficients of $f(x)$ can be given by the Euler formulas:

$$a_n=\frac{1}{\pi}\int_{-\pi}^{\pi}f(x)\cos(nx)dx\,\text{for}\,n\geq1 \tag{9.13}$$

$$a_n=\frac{1}{\pi}\int_{-\pi}^{\pi}f(x)\cos(nx)dx\,\text{for}\,n\geq1 \tag{9.14}$$

The coefficient ½ of a_0 is not a big deal and one should not be confused by that since a_0 is always a constant value.

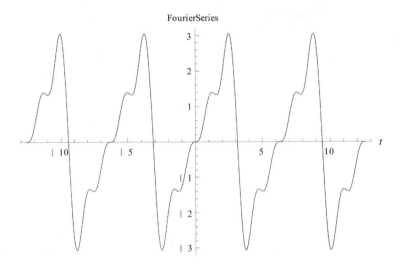

FIGURE 9.2 Schematic of Fourier series of the fourth order of t with the Mathematica Code, FourierSeries$\left[t,t,4\right]$.

9.3 USEFUL INFORMATION RELATING TO BASIC EQUATIONS: THE ALLOWED FORM OF $f(t)$ IN PHYSICS

The function $f(t)$ must stretch from the values of t between $-\infty$ and $+\infty$; it must repeat itself with some characteristic period T and must be single valued. $f(t)$ may be complex as well as the various coefficients involved in the expression of $f(t)$. Moreover, $f(t)$ can present discontinuities (as in a square or a saw tooth waveform) and will be a suitable function if it is everywhere finite and has only a finite real number of finite discontinuities per period. These conditions are known as Dirichlet's conditions. Furthermore, $f(t)$ may contain a finite real number of delta functions per period.

9.3.1 SIMPLIFYING CONDITIONS

If $f(t)$ is an even function, that is $f(-t) = f(t)$, then only the cosine terms are needed in equation (9.1) and if $f(t)$ is an odd function, that is $f(-t) = -f(t)$, then only the sine terms are needed in equation (9.1). So, it is useful to choose functions that are either even or odd to make the calculations easy. Other simplifications may occur when $f(t)$ is real or pure imaginary.

If $f(t)$ is real, then: a_n, b_n, c_n, d_n, φ_n and θ_n will be real, but g_n will be complex.

$$g_n = g_{-n}{}^*, a_n = \mathrm{Re}(2g_n)$$

$$g_n{}^* = g_{-n}, b_n = -Im(2g_n)$$

If $f(t)$ is pure imaginary, then a_n, b_n, c_n and d_n will be pure imaginary. θ_n and φ_n will be real, but g_n will be complex.

$$g_n = -g_{-n}{}^*, a_n = 2iIm(g_n) \text{ for } n \geq 1$$

$$g_n{}^* = -g_{-n}, b_n = 2i\,\mathrm{Re}(g_n)$$

If $f(t)$ is even, then $\varphi_n = -\pi/2$, $\varphi_n = 0$, $\theta_n = \pi/2$, b_n, $n \geq 1$, $a_n = c_n = d_n = 2g_n$ for $n \geq 1$.

If g_n is odd, then $g_n = g_{-n}$, $\varphi_n = -\pi / 2$, $\theta_n = 0$, $a_n = 0$, $b_n = 0$, $b_n = c_n = d_n = 2ig_n$.

Below is a table of Fourier series expansions of some waveform's signals.

Time Domain Signal $x(t)$	Fourier Series Expansion
Positive square wave	$x(t) = \dfrac{A}{2} + \dfrac{2A}{\pi}\left(\sin \omega_0 t + \dfrac{1}{3}\sin 3\omega_0 t + \dfrac{1}{5}\sin 5\omega_0 t + \dfrac{1}{7}\sin 7\omega_0 t + \cdots \right)$
Square wave	$x(t) = \dfrac{4A}{\pi}\left(\cos \omega_0 t - \dfrac{1}{3}\cos 3\omega_0 t + \dfrac{1}{5}\cos 5\omega_0 t - \dfrac{1}{7}\cos 7\omega_0 t + \cdots \right)$
Triangular wave	$x(t) = \dfrac{8A}{\pi^2}\left(\cos \omega_0 t + \dfrac{1}{9}\cos 3\omega_0 t + \dfrac{1}{25}\cos 5\omega_0 t + \dfrac{1}{49}\cos 7\omega_0 t + \cdots \right)$
Sawtooth wave	$x(t) = \dfrac{2A}{\pi}\left(\sin \omega_0 t - \dfrac{1}{2}\sin 2\omega_0 t + \dfrac{1}{3}\sin 3\omega_0 t - \dfrac{1}{4}\sin 4\omega_0 t + \cdots \right)$
Rectangular wave (Pulse train) Duty cycle $= d = \dfrac{\tau}{T_0}$	$x(t) = Ad + 2Ad\left(\dfrac{\sin \pi d}{\pi d} \right)\cos \omega_0 t$ $+ 2Ad\left(\dfrac{\sin 2\pi d}{2\pi d} \right)\cos 2\omega_0 t + 2Ad\left(\dfrac{\sin 3\pi d}{3\pi d} \right)\cos 3\omega_0 t + \cdots$
Ideal impulse train	$x(t) = \dfrac{1}{T_0} + \dfrac{2}{T_0}(\cos \omega_0 t + \cos 2\omega_0 t + \cos 3\omega_0 t + \cos 4\omega_0 t + \cdots)$

FIGURE 9.3   Fourier series expansions for some waveform signals in sine-cosine form.

9.4   APPLICATIONS

In application, let us present some examples of the calculation of the coefficients of the Fourier series.

Example 2

Evaluate the Fourier coefficients of the function $f(x) = x^2 + \pi x$

Solution

$$a_0 = \frac{1}{\pi}\int_{\pi}\left(x^2 + \pi x\right)dx = \frac{2}{3}\pi^2$$

For $n \geq 1$,

$$a_n = \frac{4}{n^2}\cos\left(n\pi\right)$$

$$b_n = \frac{-2\pi}{n}\cos\left(n\pi\right)$$

Example 3

Compute the Fourier coefficients of the function defined as follows:

$$f(x) = \begin{cases} \dfrac{1}{3}(x+\pi) & \text{if} \quad -\pi < x < \dfrac{\pi}{2} \\[3mm] \pi - x & \text{if} \quad \dfrac{\pi}{2} < x < \pi \end{cases}$$

Solution

$$a_0 = \frac{1}{\pi}\int_{-\pi}^{\pi} f(x)\,dx = \frac{\pi}{2}$$

For $n \geq 1$,

$$a_n = \frac{2}{3n^2\pi}\left[\cos(n\pi) - \cos\left(\frac{n\pi}{2}\right) + \frac{n\pi}{2}\sin\left(\frac{n\pi}{2}\right)\right]$$

$$b_n = \frac{2}{3n^2\pi}\left[-3n\pi\cos(n\pi) + \frac{3n\pi}{2}\sin\left(\frac{n\pi}{2}\right) - \sin\left(\frac{n\pi}{2}\right)\right]$$

Example 4

Compute the Fourier coefficients of the function

$$f(x) = \begin{cases} x & 0 \leq x \leq \pi \\ -x & -\pi \leq x \leq 0 \end{cases}$$

Solution

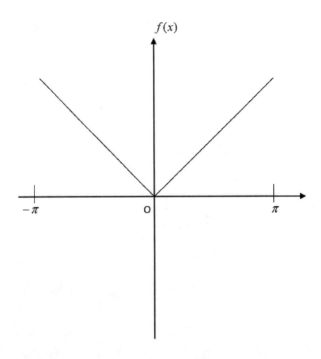

FIGURE 9.4 Function $f(x)$ is even.

Since $f(x)$ is an even function, we have

$$a_0 = \frac{1}{\pi}\int_{-\pi}^{\pi} f(x)\,dx = \frac{2}{\pi}\int_{0}^{\pi} f(x)\,dx = \frac{2}{\pi}\int_{0}^{\pi} x\,dx = \frac{x^2}{\pi}\Big|_0^{\pi} = \pi$$

$$a_n = \frac{1}{\pi}\int_{-\pi}^{\pi} f(x)\cos nx\,dx$$

$$= \frac{2}{\pi}\int_{0}^{\pi} x\cos nx\,dx = \frac{2}{\pi}\left[\frac{1}{n^2}\cos nx + \frac{1}{n}x\sin nx\,dx\right]_0^{\pi} = \pi$$

$$a_n = \frac{2[(-1)^n - 1]}{\pi n^2} = \begin{cases} 0 & n \ \ \text{even} \\ -\dfrac{4}{\pi n^2} & n \ \ \text{odd} \end{cases}$$

$$b_n = \frac{1}{\pi}\int_{-\pi}^{\pi} f(x)\sin nx\,dx = 0$$

so that

$$f(x) = \frac{\pi}{2} - \frac{4}{\pi}\sum_{n=0}^{\infty} \frac{\cos[(2n+1)x]}{(2n+1)^2}.$$

Example 5

Compute the Fourier coefficients of the function $f(x) = \begin{cases} -1 & -\pi < x < 0 \\ +1 & 0 < x < \pi \end{cases}$

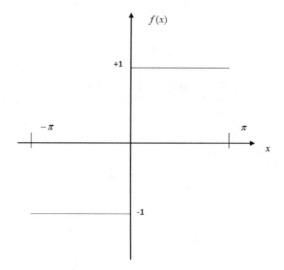

FIGURE 9.5 The function $f(x)$ is odd.

Solution

Since $f(x)$ is an odd function, we have

$$a_0 = 0 = a_n$$

$$b_n = \frac{1}{\pi}\int_{-\pi}^{\pi} f(x)\sin nx\,dx = \frac{2}{\pi}\int_0^{\pi}\sin nx\,dx = -\frac{2}{\pi}\left[\frac{\cos nx}{n}\right]_0^{\pi} = \frac{-2}{n\pi}\left[(-1)^n - 1\right]$$

So that $f(x) = \dfrac{4}{\pi}\sum_{n=0}^{\infty}\dfrac{\sin[(2n+1)x]}{(2n+1)^2}.$

PROBLEM SET 9

9.1 Compute the Fourier coefficients of the function $f(x) = x^3 + \pi x^2$

9.2 Compute the Fourier coefficients of the function $f(x)$ and its Fourier approximates in the interval $-3 < x < 3$ for $N = 1$. Plot

$$f(x) = \begin{cases} 0 & \text{if } -3 < x < -1 \\ \cosh(x) & \text{if } -1 < x < 1 \\ 0 & \text{if } 1 < x < 3 \end{cases}$$

9.3 Prove the Euler formula for the Fourier series represented in the above definition by $f(t) = a_0 + \sum_{n=1}^{\infty}\left[a_n\cos(n\omega t) + b_n\sin(n\omega t)\right]$, in other words prove the Euler formulas:

$$a_0 = \frac{1}{T}\int_0^T f(t)\,dt$$

$$a_n = \frac{2}{T}\int_0^T f(t)\cos(n\omega t)\,dt \text{ for } n \geq 1$$

$$b_n = \frac{2}{T}\int_0^T f(t)\sin(n\omega t)\,dt \text{ for } n \geq 1.$$

9.4 Expand the function $f(x) = e^x$, $-\pi < x < \pi$ in Fourier series and establish the numerical result $\sum_{n=1}^{\infty}\frac{1}{n^2+1} = \frac{\pi}{2}\left[\coth\pi - \frac{1}{\pi}\right].$

9.5 Construct a Fourier series of the function

$$f(x) = \begin{cases} \sin 2x & -\pi \leq x \leq 0 \\ 0 & 0 \leq x \leq \pi \end{cases}$$

and deduce that $\sum_{n=1}^{\infty}\frac{1}{(2n-1)(2n+3)} = \frac{1}{3}.$

9.6 Compute the Fourier coefficients of the function

$$f(x) = e^{2x} + e^{-x}.$$

9.7   (1) If the series $\dfrac{1}{2}a_0 + \displaystyle\sum_{n=1}^{\infty}(a_n\cos nx + b_n\sin nx)$ is the Fourier series of a continuous and

square integrable function $f(x)$ which has the period, 2π, showing that

(2)   Define the function $F(x) = \int_0^x\left[f(x') - \dfrac{1}{2}a_0\right]dx'$ in terms of the one specific above

and establish the periodicity relation $F(x) = F(x + 2m\pi)$, m, integral.

(3)   Detail the connection between the respective Fourier coefficients of f, F.

(4)   Find the analog of the relation in (2) which involves the integral of F^2.

(5)   Compute the Fourier coefficients of the function $f(x) = 2\cos(\pi x) + 3\sin(2\pi x)$ and write $f(x)$ as Fourier series.

9.8   The deflection z of a uniform plate carrying a load $0 < y < a$ per unit area satisfies,

$$\frac{\partial^4 z}{\partial x^4} + 2\frac{\partial^4 z}{\partial x^2 \partial y^2} + \frac{\partial^4 z}{\partial y^4} = \frac{w}{D}$$

where D is a constant depending on the material and thickness of the plate. For a rectangular plate $0 < x < a$ and $0 < y < a$, simply supported at its edges, w has to satisfy

$$z = \frac{\partial^2 z}{\partial y^2} = 0, \text{when } y = 0 \text{ and } y = b,$$

$$z = \frac{\partial^2 z}{\partial x^2} = 0, \text{when } x = 0 \text{ and } x = a.$$

If w is constant, show that

$$z = \frac{16w}{\pi^6 D}\sum_{m=0}^{\infty}\sum_{n=0}^{\infty}\frac{\sin\left[(2m+1)\pi x/a\right]\sin\left[(2n+1)\pi y/b\right]}{(2m+1)(2n+1)[2m+1)^2/a^2 + (2n+1)^2/b^2]^2}.$$

9.9   Compute the Fourier coefficients of the function

$$f(x) = \begin{cases} x & 0 \le x \le \pi \\ -x & -\pi \le x \le 0 \end{cases}$$

Write down $f(x)$ as a series and prove that the series obtained is uniformly convergent.

9.10.   Compute the Fourier coefficients of the function

$$f(x) = \begin{cases} -1 & -\pi < x < 0 \\ +1 & 0 < x < \pi \end{cases}$$

Write down $f(x)$ as a series and prove that the series obtained is not uniformly convergent.

10 Fourier Transform

> There cannot be a language more universal and more simple, more free from errors and obscurities … more worthy to express the invariable relations of all natural things [than mathematics]. [It interprets] all phenomena by the same language, as if to attest the unity and simplicity of the plan of the universe, and to make still more evident that unchangeable order which presides over all natural causes.
>
> *–Joseph Fourier*

INTRODUCTION

This chapter defines Fourier integrals, Fourier transforms, the properties of the Fourier transforms, the use of Fourier transforms to solve a partial differential equation, the Parseval–Plancherel formula, Dirac combs, Poisson summation formula and energy spectrum.

10.1 FOURIER INTEGRALS

While the Fourier series allows a periodic function to be written as a series of harmonic oscillations at definite frequencies, the Fourier transform allows an aperiodic function to be expressed as an integral sum over a continuous range of frequencies. The following integrals are called Fourier integrals:

$$
\begin{aligned}
f(t) &= \int_0^\infty \left[A(\omega)\cos(\omega t) + B(\omega)\sin(\omega t) \right] d\omega \\
&= \int_0^\infty C(\omega)\cos\left[\omega t + \phi(\omega) \right] d\omega \\
&= \int_0^\infty D(\omega)\sin\left[\omega t + \psi(\omega) \right] d\omega \\
&= \frac{1}{\sqrt{2\pi}} \int_{-\infty}^{+\infty} E(\omega) e^{+i\omega t} d\omega \\
&= \frac{1}{2\pi} \int_{-\infty}^{+\infty} F(\omega) e^{+i\omega t} d\omega \\
&\quad \int_{-\infty}^{+\infty} G(v) e^{+i2\pi v t} d\omega
\end{aligned}
\tag{10.1}
$$

The determination of the functions $A(\omega)$, $B(\omega)$, $C(\omega)$, $D(\omega)$, $E(\omega)$, $F(\omega)$ and $G(v)$ is the central problem of Fourier analysis. Each of the functions $E(\omega)$, $F(\omega)$ and $G(v)$ is known as the Fourier transform of the function $f(t)$. Indeed, there is no universally accepted convention to Fourier transforms. In all the above equations, ω represents an angular frequency in radians per second, while v represents the frequency in cycles per second. The following relationships hold:

$$
\begin{aligned}
A(\omega) &= \frac{1}{\pi} \int_{-\infty}^{+\infty} f(t)\cos(\omega t) dt = \frac{1}{2\pi}\left[F(\omega) + F(-\omega) \right] \\
&= C(\omega)\cos\left[\phi(\omega) \right] = D(\omega)\sin\left[\psi(\omega) \right]
\end{aligned}
\tag{10.2}
$$

DOI: 10.1201/9781003478812-10

$$B(\omega) = \frac{1}{\pi} \int_{-\infty}^{+\infty} f(t)\sin(\omega t)\,dt = \frac{1}{2\pi}\big[F(\omega) + F(-\omega)\big]$$
$$= C(\omega)\sin\big[\phi(\omega)\big] = D(\omega)\cos\big[\psi(\omega)\big]$$

(10.3)

$$E(\omega) = \frac{1}{\sqrt{2\pi}} \int_{-\infty}^{+\infty} f(t)e^{-i\omega t}\,dt = \frac{1}{\sqrt{2\pi}} G(\omega/2\pi)$$

(10.4)

$$F(\omega) = \int_{-\infty}^{+\infty} f(t)e^{-i\omega t}\,dt$$
$$= \pi\big[A(\omega) - iB(\omega)\big] \quad \text{for } \omega > 0$$
$$= \pi\big[A(\omega) - iB(|\omega|)\big] \quad \text{for } \omega > 0$$

(10.5)

$$G(v) = \int_{-\infty}^{+\infty} f(t)e^{-i2\pi vt}\,dt = F(2\pi v)$$

(10.6)

It can also be proved that

$$C^2(\omega) = D^2(\omega) = A^2(\omega) + B^2(\omega) = \frac{1}{\pi^2} F(\omega)F(\omega)$$

$$\tan\big[\phi(\omega)\big] = -\frac{B(\omega)}{A(\omega)} \text{ and } \tan\big[\psi(\omega)\big] = \frac{A(\omega)}{B(\omega)}$$

(10.7)

10.2   DEFINITION OF FOURIER TRANSFORM

We need to understand that there is no universally accepted convention in terms of the definition of the Fourier transform and inverse Fourier transform. If we use the variables x and y instead of t and ω, we can define the Fourier transform and the Fourier inverse transform as follows:

$$F(y) = \int_{-\infty}^{+\infty} f(x)e^{-ixy}\,dx = \text{ Fourier transform of } f(x)$$
$$FT\{f(x)\}$$

(10.8)

$$f(x) = \frac{1}{2\pi} \int_{-\infty}^{+\infty} F(y)e^{+ixy}\,dy = \text{Inverse Fourier transform of } F(y)$$
$$\text{IFT}\{F(y)\}$$

(10.9)

10.2.1   PROPERTIES OF THE FOURIER TRANSFORM

It is useful to know a few properties about a function and its Fourier transform or the effect that various operations on a function would have on the Fourier transform of the function. We would use the symbol FT for the Fourier transform.

10.2.1.1   Addition

The Fourier transform of the sum of two functions, f and g, is the sum of the Fourier transforms of each function.

$$FT\big[f(x) + g(x)\big] = FT\big[f(x)\big] + FT\big[g(x)\big]$$

(10.10)

10.2.1.2   Multiplication by a Constant

Fourier transform of the multiplication of a function f by a constant a is equal to the multiplication of the constant by the Fourier transform of the function.

$$FT\big[af(x)\big]=aFT\big[f(x)\big] \tag{10.11}$$

10.2.1.3   Scaling

If $FT\big[f(x)\big]=F(y)$, then

$$FT\big[f(ax)\big]=\frac{1}{|a|}F\left(\frac{y}{a}\right) \tag{10.12}$$

For a a constant real.

Also,

$$FT\left[\frac{1}{|a|}f\left(\frac{x}{a}\right)\right]=F(ay). \tag{10.13}$$

10.2.1.4   Shifting

If $FT\big[f(x)\big]=F(y)$,

then,

$$FT\big[f(x\pm x_0)\big]=e^{\pm ix_0 y}F(y) \tag{10.14}$$

and

$$FT\big[e^{\pm iy_0 x}f(x)\big]=F(y\mp y_0) \tag{10.15}$$

10.2.1.5   Products of Convolution

The convolution of two functions f and g of a given variable x is written as $f(x)\otimes g(x)$. If $FT\big[f(x)\big]=F(y)$ and $FT\big[g(x)\big]=G(y)$, then the Fourier transform of the convolution of the two functions is given by

$$FT\big[f(x)\otimes g(x)\big]=F(y)\cdot G(y) \tag{10.16}$$

Inversely,

$$FT\big[f(x)\cdot g(x)\big]=\frac{1}{2\pi}F(y)\otimes G(y) \tag{10.17}$$

where the symbol $\otimes$ is used to define the convolution of two functions:

$$f(x)\otimes g(x)=\int_{-\infty}^{+\infty}f(u)g(x-u)\,du. \tag{10.18}$$

10.2.1.6 Area

The area under the curve representing a function is equal to its Fourier transform at the origin:

$$\int_{\infty}^{+\infty} f(x)dx = F(0). \tag{10.19}$$

Conversely, the value of a function at the origin is equal to $\dfrac{1}{2\pi}$, times the area of its Fourier transform:

$$f(0) = \frac{1}{2\pi}\int_{\infty}^{+\infty} F(y)dy. \tag{10.20}$$

10.2.1.7 Derivation

If $FT\left[f(x)\right] = F(y)$ then $FT\left[\dfrac{d^m}{dt^m}f(x)\right] = (iy)^m F(y)$

Inversely,

$$FT\left[(-ix)^m f(x)\right] = \frac{d^m}{dy^m}F(y) \tag{10.21}$$

Examples of the Fourier transformation using *Wolfram Mathematica 7.0*:
  Plot[Exp[...0.04x^2],{x,...10,10}, AxesLabel ... {x, Exp[...0.04x^2]}]
  Plot[%,{y,...1,1},AxesLabel ... {y, FourierTransform[Exp[...0.04x^2]]}]

10.2.2 Use of Fourier Transform to Solve a Partial Differential Equation

The use of Fourier transform can help change a PDE in an ODE. To illustrate this, let us try to solve the wave equation using the Fourier transform.

Consider the wave equation in Cartesian coordinates:

$$\frac{\partial^2 u}{\partial x^2} = \frac{1}{v^2}\frac{\partial^2 u}{\partial t^2} \quad u(x,0) = u(x) \tag{10.22}$$

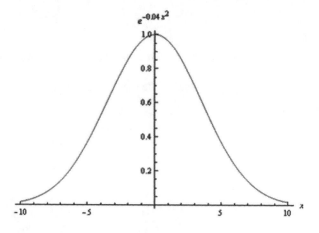

FIGURE 10.1 Graph of [Exp[...0.04x^2].

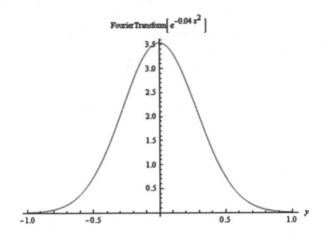

FIGURE 10.2   FourierTransform[Exp[...0.04x^2]].

Recall

$$FT\{f(x)\} = F(y) = \int_{-\infty}^{+\infty} f(x)e^{-ixy}dx \qquad (10.23)$$

Now, let us multiply both sides of the PDE by e^{-ixy}

$$e^{-ixy}\frac{\partial^2 u}{\partial x^2} = \frac{1}{v^2}\frac{\partial^2 u}{\partial t^2}e^{-ixy} \qquad (10.24)$$

and integrate over dx from $-\infty$ to $+\infty$.

$$\int_{\infty}^{+\infty} e^{-ixy}\frac{\partial^2 u}{\partial x^2}\,dx = \frac{1}{v^2}\int_{\infty}^{+\infty}\frac{\partial^2 u}{\partial t^2}e^{-ixy}dx = \frac{1}{v^2}\frac{\partial^2}{\partial t^2}\int_{\infty}^{+\infty} e^{-ixy}udx \qquad (10.25)$$

and using the nth derivatives formula,

$$FT\left[\frac{d^m}{dt^m}f(x)\right] = (iy)^m F(y), \text{ we get}$$

$(iy)^2 U(y,t) = \dfrac{1}{v^2}\dfrac{\partial^2}{\partial t^2}U(y,t)$ or $\ddot{U} + y^2 v^2 U = 0$ which is an ODE whose solution can be written as

$U(y,t) = U(y,0)e^{\pm yvt}$. Now, using the inverse Fourier transform relation

$$FT\left[\frac{d^m}{dt^m}f(x)\right] = (iy)^m F(y) \qquad (10.26)$$

we obtain

$$u(x,t) = \frac{1}{2\pi}\int_{-\infty}^{+\infty} U(y,0)e^{\pm ixy}dy \qquad (10.27)$$

or

$$u(x,t) = \frac{1}{2\pi}\int_{-\infty}^{+\infty} U(y,0)e^{\pm yvt+ixy}dy \tag{10.28}$$

$$u(x,t) = \frac{1}{2\pi}\int_{-\infty}^{+\infty} U(y)e^{iy(x\pm vt)}dy \tag{10.29}$$

which is a function of $x \pm vt$; that is, $u(x,t) = f(x \pm vt)$, the general form of the wave function. Note that $U(y)$ is the Fourier transform of $u(x)$ which appears in the initial condition.

10.2.2.1 Theorem of Moment

Consider a continuous function f of a real value x. The pth moment or moment of p-order is defined by

$$m_p = \int_{\infty}^{+\infty} x^p f(x)dx \tag{10.30}$$

One can prove that

$$m_p = \frac{f^{(p)}(0)}{(-2i\pi)^p}. \tag{10.31}$$

In particular, the zeroth moment is $m_0 = \int_{\infty}^{+\infty} f(x)dx$; the weighted mean by the function f of x is

$$\langle x \rangle = \frac{m_1}{m_0} = -\frac{u'(\rho)}{2i\pi f(0)}$$ and the weighted mean by the function f of x^2 is $\langle x^2 \rangle = \frac{m_2}{m_1} = -\frac{f''(0)}{4\pi^2 f(0)}.$

10.3 PARSEVAL–PLANCHEREL FORMULA

The following relation is established by Parseval and generalized by Plancherel (1910) to Fourier transforms: if $f(x)$ and $g(x)$ are integrable functions, and $\hat{f}(v)$ and $\hat{g}(v)$ their respective Fourier transforms, then

$$\int f(x)\hat{g}(x)dx = \int \hat{f}(v)\,\hat{g}(v)\,dv \tag{10.32}$$

Proof

$$\int f(x)\overline{g}(x)dx = \left[FT[f\cdot\overline{g}]\right]_{v=0} = \hat{f}(v)*\overline{\hat{g}}(-v)$$
$$= \left[\int\hat{f}(t)\overline{\hat{g}}(t-v)dt\right]_{v=0} \tag{10.33}$$
$$= \int\hat{f}(v)\overline{\hat{g}}(v)dv$$

For the case $f = g$, the Parseval–Plancherel formula becomes

$$\int |f(x)|^2\,dx = \int |\hat{f}(v)|^2\,dv \tag{10.34}$$

10.4   DIRAC COMB

Dirac comb is an important notion of distribution defined by

$$\Delta(x) = \sum_{n=-\infty}^{n=+\infty} \delta(x - n) \qquad (10.35)$$

with n an integer
It is a tempered distribution.
Dirac comb vanishes for non-integer
DiracComb[1/2] = 0
DiracComb[3/2] = 0
DiracComb[2.25] = 0

10.4.1   FOURIER TRANSFORM OF DIRAC COMB

The Fourier transform of a Dirac comb is a Dirac comb:

$$FT\left[\sum_{n=-\infty}^{n=+\infty} \delta(x - n)\right] = \sum_{n=-\infty}^{n=+\infty} \delta(v - n) \qquad (10.36)$$

10.5   POISSON SUMMATION FORMULA

For a given function f, we can write the Poisson formula by

$$\left[\sum_{n=-\infty}^{n=+\infty} f(x - nT)\right] = \frac{1}{T} \sum_{n=-\infty}^{n=+\infty} f\left(\frac{n}{T}\right) e^{2imnx/T} \qquad (10.37)$$

For $x = 0$, this formula becomes

$$\left[\sum_{n=-\infty}^{n=+\infty} f(nT)\right] = \frac{1}{T} \sum_{n=-\infty}^{n=+\infty} f\left(\frac{n}{T}\right) \qquad (10.38)$$

and for the period $T = 1$, it becomes

$$\left[\sum_{n=-\infty}^{n=+\infty} f(n)\right] = \sum_{n=-\infty}^{n=+\infty} f(n) \qquad (10.39)$$

This means that each of the series converges.

10.5.1   SCALING OF DIRAC COMB

The scaling property of the Dirac comb follows from the properties of the Dirac delta function:

$$\Delta_T(x/T) = |a|\Delta_{aT}(x) \qquad (10.40)$$

TABLE 10.1
Fourier Transform of Some Functions

Function, $f(t)$	Fourier Transform, $F(\omega)$
Definition of Inverse Fourier Transform $f(t)=\dfrac{1}{2\pi}\int_{-\infty}^{\infty}F(\omega)e^{j\omega\tau}d\omega$	Definition of Fourier Transform $F(\omega)=\int_{-\infty}^{\infty}f(t)e^{-j\omega t}dt$
$f(t-t_0)$	$F(\omega)e^{-j\omega t_0}$
$f(t)e^{j\omega_0}$	$F(\omega-\omega_0)$
$f(\alpha t)$	$\dfrac{1}{|\alpha|}F\left(\dfrac{\omega}{\alpha}\right)$
$F(t)$	$2\pi f(-\omega)$
$\dfrac{d^n f(t)}{dt^n}$	$(j\omega)^n F(\omega)$
$(-jt)^B f(t)$	$\dfrac{d^n F(\omega)}{d\omega^n}$
$\int_{-\infty}^{\infty}f(\tau)d\tau$	$\dfrac{F(\omega)}{j\omega}+\pi F(0)\delta(\omega)$
$\delta(t)$	1
$e^{j\omega d}$	$2\pi\delta(\omega-\omega_0)$
$\mathrm{sgn}(t)$	$\dfrac{2}{j\omega}$
$j\dfrac{1}{\pi}$	$\mathrm{sgn}(\omega)$
$u(t)$	$\pi\delta(\omega)+\dfrac{1}{j\omega}$
$\sum_{n=-\infty}^{\infty}F_n e^{jms_n t}$	$2\pi\sum_{n=-\infty}^{n}F_n\delta(\omega-n\omega_0)$
$\mathrm{rect}\left(\dfrac{t}{\tau}\right)$	$\tau Sa\left(\dfrac{\omega\tau}{2}\right)$
$\dfrac{B}{2\pi}Sa\left(\dfrac{Bt}{2}\right)$	$\mathrm{rect}\left(\dfrac{\omega}{B}\right)$
$tri(t)$	$tri(t)$
$A\cos\left(\dfrac{\pi t}{2\tau}\right)\mathrm{rect}\left(\dfrac{t}{2\tau}\right)$	$\dfrac{A\pi}{\tau}\dfrac{\cos(\omega\tau)}{(\pi/2\tau)^2-\omega^2}$
$\cos(\omega_0 t)$	$\pi\left[\delta(\omega-\omega_0)+\delta(\omega+\omega_0)\right]$
$\sin(\omega_0 t)$	$\dfrac{\pi}{j}\left[\delta(\omega-\omega_0)-\delta(\omega+\omega_0)\right]$

(Continued)

TABLE 10.1 (CONTINUED)
Fourier Transform of Some Functions

$u(t)\cos(\omega_0 t)$	$\dfrac{\pi}{2}\left[\delta(\omega-\omega_0)+\delta(\omega+\omega_0)\right]+\dfrac{j\omega}{\omega_0^2-\omega^2}$
$u(t)\sin(\omega_0 t)$	$\dfrac{\pi}{2j}\left[\delta(\omega-\omega_0)-\delta(\omega+\omega_0)\right]+\dfrac{\omega^2}{\omega_0^2-\omega^2}$
$u(t)e^{-\alpha t}\cos(\omega_0 t)$	$\dfrac{(\alpha+j\omega)}{\omega_0^2+(\alpha+j\omega)^2}$
$u(t)e^{-\alpha t}\sin(\omega_0 t)$	$\dfrac{\omega_0}{\omega_0^2+(\alpha+j\omega)^2}$
$e^{-\alpha|t|}$	$\dfrac{2\alpha}{\alpha^2+\omega^2}$
$e^{-t^2/(2\sigma^2)}$	$\dfrac{\sigma\sqrt{2\pi}\,e^{-t^2\omega^2/2}}{\alpha+j\omega}$
$u(t)e^{-\alpha}$	$\dfrac{1}{\alpha+j\omega}$
$u(t)te^{-\alpha}$	$\dfrac{1}{(\alpha+j\omega)^2}$

10.6 ENERGY SPECTRUM

Fourier transform of a function f of time t gives a function F of the frequency ω by $F(\omega)=\displaystyle\int_{-\infty}^{+\infty}f(t)e^{-i\omega t}dt$.

We can write $F(\omega)$ as a complex function:

$$F(\omega)=R(\omega)+iX(\omega)=A(\omega)e^{i\varphi(\omega)} \tag{10.41}$$

where $R(\omega)$ and $X(\omega)$ are single functions of ω. The amplitude $A(\omega)$ of $F(\omega)$ is called the Fourier spectrum of $f(t)$. The square of the amplitude is called the energy spectrum of $f(t)$ and $\varphi(\omega)$ its phase angle. In general, if we consider a real function $f_1(t)$ as a voltage across an energy source and $f_2(t)$ the current delivered by the source to a given load, the energy delivered by the source is given by the integral:

$$E=\int_{-\infty}^{+\infty}f_1(t)f_2(t)dt. \tag{10.42}$$

Assume $F_1(\omega)$ and $F_2(\omega)$ the respective Fourier transforms of $f_1(t)$ and $f_1(t)$. The Fourier transform of the product $\vec{k}=(k_x,k_y,k_z)$ is given by $F_2(\omega)\otimes F_2(\omega)$, which can be expressed as

$$\int_{-\infty}^{+\infty}f_1(t)f_2(t)dt=\frac{1}{2\pi}\int_{-\infty}^{+\infty}\dot{F}_1(\omega)F_2(\omega)d\omega \tag{10.43}$$

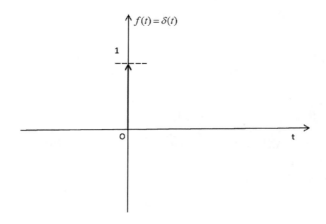

FIGURE 10.3   Dirac delta function at the origin of axis O.

Let us define $E_{12} = \dot{F}_1(\omega)F_2(\omega)$. We can see that the energy is the area under the curve of $\dfrac{\dot{F}_1(\omega)F_2(\omega)}{2\pi}$. Because it involves two functions indexed 1 and 2, this energy is called the cross-energy spectrum of $f_1(t)$ and $f_2(t)$.

Example: Fourier transform of the Dirac delta function, δ.

We recall that for an arbitrary continuous function, $f(t)$, the Dirac delta function, δ, is defined by the equation:

$$\int_{-\infty}^{+\infty} \delta(t-t_o)f(t)dt = f(t_o) \tag{10.44}$$

The Fourier transform of the Dirac-delta function is given by

$$\int_{-\infty}^{+\infty} \delta(t)e^{-i\omega t}dt = 1 \tag{10.45}$$

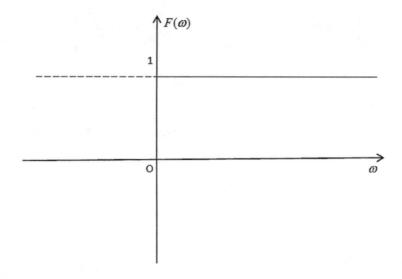

FIGURE 10.4   The Fourier transform of the Dirac delta function at the origin of axis O.

and the inversion formula is

$$\delta(t) = \frac{1}{2\pi}\int_{-\infty}^{+\infty} 1.e^{i\omega t}d\omega = \frac{1}{2\pi}\int_{-\infty}^{+\infty}\cos(\omega t)d\omega \tag{10.46}$$

10.7   FOURIER INTEGRALS IN N-DIMENSIONS

Consider a function of n variables $x_1, x_2, ..., x_n$, that is, $f = f(x_1, x_2, ..., x_n)$. If we can write

$$f(x_1, x_2, ..., x_n) =$$
$$\frac{1}{(2\pi)^n}\int...\int F(y_1, y_2, ..., y_n)e^{+i(x_1y_1+x_2y_2+...+x_ny_n)}dy_1dy_2...dy_n \tag{10.47}$$

then

$$F(y_1, y_2, ..., y_n) = \frac{1}{(2\pi)^n}\int...\int f(x_1, x_2, ..., x_n)$$
$$e^{-i(x_1y_1+x_2y_2+...+x_ny_n)}dx_1dx_2...dx_n \tag{10.48}$$

F is the Fourier transform of f and we can write the Fourier pair:

$$\delta(t) = \frac{1}{2\pi}\int_{-\infty}^{+\infty} 1.e^{i\omega t}d\omega = \frac{1}{2\pi}\int_{-\infty}^{+\infty}\cos(\omega t)d\omega, \text{ with } x_i, y_i, i = 1,2,3,...,n, \text{ conjugate variables.}$$

Note that you can have other conventions with the change in sign on the exponent argument, $+i$, instead of $...i$, or a change of the factor $(2\pi)^{-n/2}$, instead of $(2\pi)^{-n}$. However, in physical science applications in three dimensions, where the position vector used is $\vec{r} = (x, y, z)$, and the wave number used is $\vec{k} = (k_x, k_y, k_z)$, the transformation will be

$$F(\vec{k}) = FT\{f(\vec{r})\} = \iiint f(\vec{r})e^{-i\vec{k}\vec{r}}d\vec{r} \tag{10.49}$$

and inversely

$$f\vec{r} = FT^{-1}\{F(\vec{k})\} = \frac{1}{(2\pi)^3}\iiint F(\vec{k})e^{+i\vec{k}\vec{r}}d\vec{k} \tag{10.50}$$

with the correspondence $f(\vec{r}) \leftrightarrow F(\vec{k})$ of the Fourier pair.

In equation (10.50), $e^{+i\vec{k}}$ is the complex representation of a static harmonic plane wave in three dimensions with the wavefront perpendicular to the wave number, $\vec{k}$. Note that the wavelength of the wave is related to its wave number by the equation:

$$\mu = \frac{2\pi}{|\vec{k}|} \tag{10.51}$$

The function $f(\vec{r})$ is spatial, that is, it only depends on the position, $f(\vec{r},t)$. A four-dimensional transform can be written as follows:

$$F(k,\omega) = FT\{f(r,t)\} = \iint f(r,t)e^{-i(\vec{k}\vec{r}+\omega t)}drdt \tag{10.52}$$

and inversely

$$f\vec{r},t = FT^{-1}\{F(\vec{k},\omega)\} = \frac{1}{(2\pi)^4}\iiint F(\vec{k},\omega)e^{+i(\vec{k}+\omega t)}d\vec{k}d\omega \tag{10.53}$$

ω and t represent the angular frequency and the time, respectively.

$f(\vec{r},t)$ can be regarded as an integral sum of traveling harmonic plane waves. The exponent

term $e^{\pm i(\vec{k}\vec{r}+\omega t)}$ represents a wave traveling in the negative $\vec{k}$ direction and $e^{\pm i(\vec{k}\vec{r}-\omega t)}$ represents a wave traveling in the positive $\vec{k}$ direction.

PROBLEM SET 10

10.1 Calculate the Fourier transform of $H(x)\exp(-\alpha x)$ with α a positive real number and $H(x)$ the Heaviside function. Deduce:

  a. by symmetry the Fourier transform of $\exp(-\alpha|x|)$.

  b. that the Fourier transform of $H(x)\dfrac{x^{n-1}}{(n-1)!}\exp(-\alpha x)$ for n integer is $(\alpha+2i\pi v)^{-n}$.

10.2 a. Use the properties of the Fourier transformation and the integral $\int e^{-\pi x^2}dx = 1$ to obtain a differential equation satisfied by the Fourier transform of $e^{-\pi x^2}$.

  b. Infer the Fourier transform if $\dfrac{1}{\sigma\sqrt{2\pi}}e^{-\pi x^2/2\sigma^2}$.

  c. Use this transformation to calculate the moment m_0, m_1, m_2 of the previous function.

10.3 Use the theorem of Parseval to calculate the following integral:

  a. $\int\left[\dfrac{\sin\pi x}{\pi x}\right]^n dx = 1$, for $n = 2,3,4$

  b. $\int\dfrac{dx}{(1+x^2)^2}dx = 1\cdot$

10.4 Express the Fourier transform of the following distribution:

  a. $\sum_{n=0}^{N-1}\delta(x-nT)$; n integer

  b. $\sum_{n=-\infty}^{n=+\infty}a_n\delta(x-nT)$ with $a_n = f(nT)$, $f(x)$is a integrable function

  c. $\sum_{n=-\infty}^{n=+\infty}b_n\delta(x-nT)$, where the coefficient b_n are related to the coefficients a_n by the relation: $b_n = \dfrac{1}{N}\sum_{k\ni n-N+1}^{n}a_n$

10.5 Verify the continuity of the Fourier transform

10.6 Prove that if T a given distribution is tempered, its derivatives also are tempered.

10.7 Find the Fourier transform of the function $\cos(\omega_o t)$. Graph both the function and its Fourier transform.

Hint. $FT\left[\cos\left(\omega_0 t\right)\right] = \pi\left[\delta\left(\omega - \omega_0\right) + \delta\left(\omega + \omega_0\right)\right].$

10.8   Prove that the Fourier transform of the nth derivative of the Dirac delta function $\delta\left(t\right)$ is $\left(i\omega\right)^n$.

10.9   Using Poisson's sum formula, show that

$$\sum_{n=-\infty}^{n=+\infty} \frac{\sin a\left(t + nT\right)}{t + nT} = \omega_c \frac{\sin\left(2N + 1\right)\omega_c t}{\sin\left(\omega_c t\right)}$$

where $\omega_c = \dfrac{\pi}{T}$ and N is such that $N < \dfrac{aT}{2\pi} < N + 1$

10.10   Show that if $u(t)$ is the solution of the differential equation

$$\frac{d^2 u\left(t\right)}{dt^2} - t^2 u\left(t\right) = \lambda u\left(t\right)$$

then its Fourier transform $U\left(\omega\right)$ is the solution of the same differential equation.

11 Laplace Transform

What we know is not much. What we do not know is immense.

—Pierre-Simon Laplace

INTRODUCTION

This chapter defines Laplace transform and its properties. Even though Laplace and Fourier transforms are two different concepts, they are somehow related.

11.1 DEFINITION

If $f(t)$ is a real or a complex function of a real variable t, the bilateral Laplace transformation of $f(t)$, referred to as $F(p)$ of a complex variable p, is defined by

$$F(p) = \int f(t) e^{-pt} dt \qquad (11.1)$$

Since p is a complex number, assume α and ω the real and imaginary parts of p. Therefore, we can write $p = \alpha + i\omega$ and

$F(\alpha + i\omega) = \int e^{-\alpha t} f(t) e^{-i\omega t} dt$, which is the Fourier transform of the function $f(t) e^{-\alpha t}$.

The unilateral Laplace transformation is defined as

$F(p) = \int_0^\infty f(t) e^{-pt} dt$ and is useful for functions that are null when $t < 0$.

Example 1 (using the following command in Mathematica)

LaplaceTransform[Exp[−t],t,s], we get the Laplace transform of the function e^{-t} as $1/(1 + s)$.

Similarly, for the Laplace transform of te^{-t}, the command LaplaceTransform[tExp[−t],t,s], gives $1/(1 + s)^2$.

11.2 INVERSE LAPLACE TRANSFORM

The inverse of Laplace transformation $F(p)$ is given by

$$f(t) = LT^{-1}\left[F(p)\right] = \frac{1}{2\pi i} \cdot \int_{\alpha_0 - i\omega}^{\alpha_0 + i\omega} F(p) e^{pt} f(p) dp \qquad (11.2)$$

for $f(t) e^{-\alpha t}$, with α_0, α_1 and α_2 all real numbers.

Example 2: Using Mathematica Software
LaplaceTransform[t,t,s]
$1/s^2$
LaplaceTransform[t²,t,s]
$2/s^3$
LaplaceTransform[sin[a*t],t,s]
$a/(a^2 + s^2)$

DOI: 10.1201/9781003478812-11

LaplaceTransform[cos[4*t],t,s]
$s/(16 + s^2)$
Plot[%,{s,0,100},AxesLabel{s,LaplaceTransform cos[4*t]}]

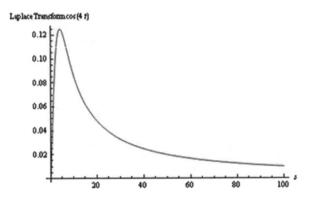

FIGURE 11.1   Graph of the Laplace transform of cos(4t).

LaplaceTransform[cos[b*t],t,s]
$s/(b^2 + s^2)$
InverseLaplaceTransform[1/(1 + s^2),s,t]
sin[t]
Plot[%,{t,0,2 Pi},AxesLabel{t,InverseLaplaceTransform [(1+s^2)$^{-1}$]}]

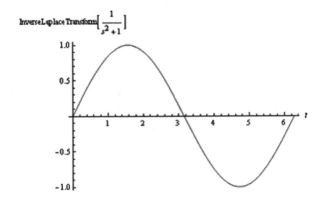

FIGURE 11.2   Graph of the inverse Laplace transform of $1/s^2 + 1$.

11.3   PROPERTIES OF LAPLACE TRANSFORM

11.3.1   ADDITION

$$\text{If } LT\left[f(t)\right] = F(p) \text{ for } \alpha_1 < \alpha < \alpha_2 \text{ and } LT\left[g(t)\right] = G(p)$$
$$\text{for } \beta_1 < \alpha < \beta_2 \text{ then, } LT\left[f(t)+g(t)\right] = F(p)+G(p) \text{ for} \qquad (11.3)$$
$$\text{Sup}(\alpha_1,\beta_1) < \alpha < \inf(\alpha_2,\beta_2).$$

11.3.2 SCALING

If $T\big[f(t)\big]=F(p)$ for $\alpha_1<\alpha<\alpha_2$, with a a complex number, then, $T\big[f(at)\big]=\dfrac{1}{|a|}F\left(\dfrac{p}{a}\right)$ for $|a|\alpha_1<\alpha<|a|\alpha_2$.

11.3.3 TRANSLATION

If $LT\big[f(t)\big]=F(p)$ for $\alpha_1<\alpha<\alpha_2$

$$T\big[f(t-\tau)\big]=e^{-\tau p}F(p)\,\text{for}\,\alpha_1<\alpha<\alpha_2 \tag{11.4}$$

Inversely,

$$LT\big[y'\big]+kLT\big[y\big]=bLT\big[e^{at}\big] \tag{11.5}$$

11.3.4 CONVOLUTION

If $LT\big[f(t)\big]=F(p)$ for $\alpha_1<\alpha<\alpha_2$ and $LT\big[g(t)\big]=G(p)$ for $\beta_1<\alpha<\beta_2$, then $LT\big[f(t)*g(t)\big]$ $=F(p)\cdot G(p)$ for $\text{Sup}(\alpha_1,\beta_1)<\alpha<\inf(\alpha_2,\beta_2)$.

Inversely,

$$LT\big[f(t)\cdot g(t)\big]=\dfrac{1}{2\pi i}\int_{x_0-i\infty}^{t_0+i\infty}F(s)G(p-s)\,ds\ \text{for}\ \alpha_1<\alpha_0<\alpha_2\ \text{and}\ \alpha_1+\beta_1<\alpha<\alpha_2+\beta_2.$$

11.3.5 DERIVATION

Let us differentiate under the integral sign $F(p)$; we obtain:

$$\dfrac{dF(p)}{dp}=\int -tf(t)e^{-pt}dt \tag{11.6}$$

After m successive differentiations one obtains:

$$\dfrac{d^{m}F(p)}{dp^{m}}=\int(-t)^{m}f(t)e^{-pt}dt \tag{11.7}$$

For all the values of p for which $F(p)$ is defined, it is indefinitely differentiable. $F(p)$ is then a holomorphic function at every point within its domain of differentiability. Remember that an analytic function in a complex domain D is indefinitely differentiable at every point of its domain D. An analytic function can be multiform. If it is uniform, it is said to be holomorphic.

11.4 DERIVATION PROPERTIES

If $T\big[f(t)\big]=F(p)$ for $\alpha_1<\alpha<\alpha_2$, then

$$T\left[\dfrac{d^{m}}{dt^{m}}f(t)\right]=p^{m}F(p)\,\text{for}\,\alpha_1<\alpha<\alpha_2 \tag{11.8}$$

Theorem 1

Let $F(t)$ and its first time derivative, $F'(t)$, be continuous for $t \geq 0$, and assume the existence of $LT\left[F(t)\right]$ and $LT\left[F'(t)\right]$ for $s \geq s_0$, then

$$LT\left[F'(t)\right] = sLT\left[F(t)\right] - F(0) \quad s > s_0 \tag{11.9}$$

11.4.1 INTEGRATION

If $LT\left[f(t)\right] = F(p)$ for $\alpha_1 < \alpha < \alpha_2$,

$$LT\left[\int_\infty f(u)du\right] = \frac{F(p)}{p} \text{ for } \sup(\alpha_1, 0) < \alpha < \alpha_2 \tag{11.10}$$

11.5 APPLICATION TO FIRST-ORDER ORDINARY DIFFERENTIAL EQUATIONS (ODE)

Let us use the Laplace transform to solve the initial value problem:

$$y' + ky = be^{at}; y(0) = A; \quad a \neq -k \tag{11.11}$$

To solve this ODE, we will apply the Laplace transform to both sides of the equality:

$$LT\left[y' + ky\right] = LT\left[e^{at}\right] \tag{11.12}$$

Using the linearity property, we have

$$LT\left[y'\right] + kLT\left[y\right] = bLT\left[e^{at}\right] \tag{11.13}$$

But,

$$LT\left[y'(t)\right] = sLT\left[y(t)\right] - y(0); \quad s > s_0 \tag{11.14}$$

and

$$LT\left[e^{at}\right] = \frac{1}{s-a}; \quad s > a \tag{11.15}$$

Therefore,

$$sLT\left[y\right] - y(0) + kLT\left[y\right] = b\frac{1}{s-a}$$

$$(s+k)LT\left[y\right] - y(0) = \frac{b}{s-a}$$

$$LT[y] = \frac{b}{(s+k)(s-a)} + \frac{y(0)}{(s+k)}$$

Let us now use the initial condition $y(0) = A$, then

$$LT[y] = y(s) = \frac{b}{(s+k)(s-a)} + \frac{A}{(s+k)} \qquad (11.16)$$

To perform the inverse operation on the Laplace transform of $y(s)$ to have $y(t)$, we need to change the last expression of the second equality into partial fractions. This will lead to

$$Y(s) = \frac{A}{(s+k)} + \frac{b}{k+a}\left[\frac{1}{s-a} - \frac{1}{s+k}\right]$$

or

$$Y(s) = \frac{1}{(s+k)}\left(A - \frac{b}{k+a}\right) + \frac{b}{k+a}\left(\frac{1}{s-a}\right)$$

Using the inverse Laplace transform, $LT^{-1}\left[\dfrac{1}{s-a}\right] = e^{at}$ and the linearity property, we obtain the solution to the initial value problem:

$$y(t) = e^{-kt}\left(A - \frac{b}{k+a}\right) + \frac{b}{k+a}e^{at} \qquad (11.17)$$

11.6 PROPERTIES OF THE INVERSE LAPLACE TRANSFORM

11.6.1 UNIQUENESS

If $f(t)$ is continuous and has the Laplace transform, $F(s) = L\{f(t)\}$, then $LT^{-1}[F(s)] = f(t)$.

11.6.2 LINEARITY

The linearity property for the inverse Laplace transform is given by

$$LT^{-1}[AG(s) + BH(s)] = Ag(t) + Bh(t) \qquad (11.18)$$

The uniqueness of the property of the Laplace transform means that for a given continuous function, $f(t)$, if we know the Laplace transform $F(s)$ of $f(t)$, we can always express the function $f(t)$ whose Laplace transform is $F(s)$.

Table 11.1 shows some functions and their Laplace transform.

TABLE 11.1
Functions and Their Laplace Transforms

$f(t)$ $t \geq 0$	$F(s)$
1 or $u(t)$ unit step function	$\dfrac{1}{s}$
e^{-at} exponential decay	$\dfrac{1}{s+a}$
$\dfrac{1}{(n-1)!}t^{n-1}e^{-at}$ $n \geq 0$ n integer	$\dfrac{1}{s+a}$
$\sin(bt)$	$\dfrac{1}{s^2+b^2}$ $Re(s)>0$
$\cos(bt)$	$\dfrac{1}{(n-1)!}t^{n-1}e^{-at}$ $n \geq 0$ $Re(s)>0$
$\sinh(bt)$	$\dfrac{b}{s^2-b^2}$ $Re(s)>|b|$
$\cosh(bt)$	$\dfrac{s}{s^2-b^2}$ $Re(s)>|b|$
$e^{at}\sin(bt)$	$\dfrac{b}{(s-a)^2+b^2}$ $Re(s)>a$
$e^{at}\cos(bt)$	$\dfrac{s-a}{(s-a)^2+b^2}$ $Re(s)>a$
$t\sin(bt)v$	$\dfrac{2bs}{(s^2+b^2)^2}$ $Re(s)>0$
$t\cos(bt)$	$\dfrac{s^2-b^2}{(s^2+b^2)^2}$ $Re(s)>0$
$e^{at}f(t)$	$F(s-a)$
$\delta(t-t_0)$	e^{-st_0}
$\delta(t-t_0)f(t)$	$e^{-st_0}f(t_0)$
$f^{(n)}(t)$ $n=1,2,3,\dots$	$s^nF(s)-s^{n-1}f(0)-s^{n-2}f^{(1)}(0)-\dots-f^{(n-1)}(0)$
$(-t)^n f(t)$	$\dfrac{d^n}{ds^n}F(s)$
$\displaystyle\int_0^t f(r)dr$	$\dfrac{1}{s}F(s)$

PROBLEM SET 11

11.1 Find the Laplace transform of each of the function. Determine and sketch the complex plane region in which it is valid

   a. e^{-2t}; b. e^{10t}; c. e^{3-4t}.

11.2 Use Laplace transform to solve the initial value problem

   a. $y' + 3y = 10e^{-2t}$ $y(0) = 1$; b. $y' + 3e^{2t}$ $y(0) = 1$

   c. $y' - y = 10e^{t/5}$ $y(0) = 0$;   d. $y' - 3y = 6e^{3t}$ $y(0) = 0$

   e. $y' - y = 1 + 4e^{3t}$ $y(0) = 5$; f. $y' - y = 3e^{-t} + 4e^{3t}$ $y(0) = 0$

11.3 Show that $LT[\sin bt] = \dfrac{b}{s^2 + b^2}$ $R_e(s) > 0..$

11.4 Show that $LT[\cos bt] = \dfrac{b}{s^2 + b^2}$ $R_e(s) > 0.$

11.5 Show that $LT[\sinh bt] = \dfrac{b}{s^2 - b^2}$ $R_e(s) > |b|.$

11.6 Show that $LT[\cosh bt] = \dfrac{s}{s^2 - b^2}$ $R_e(s) > |b|.$

11.7 Show that $LT[t] = \dfrac{1}{s^2}$ $R_e(s) > 0.$

11.8 Show that $LT[e^{at}\cos 3t] = \dfrac{s-2}{(s-2)^2 + 9}$ $R_e(s) > |2|.$

11.9 Show that $LT[e^{at}t] = \dfrac{1}{(s-\alpha)^2}$ $R_e(s) > |\alpha|.$

11.10 Find $LT^{-1}\left[\dfrac{3s-2}{(s^2 + +6s + 25)}\right].$

11.11 Find $LT^{-1}\left[\dfrac{2s+3}{(s^2 - 2s + 3)}\right].$

11.12 Find $LT^{-1}\left[\dfrac{6}{(s^2 + 2s + 4)}\right].$

11.13 Find $LT^{-1}\left[\dfrac{4s-9}{s^2 + 9}\right].$

11.14   Use the Laplace transform to solve the initial value problem

     a. $y'' + 9y = 0 \quad y(0) = 1, \quad y'(0) = 2$

     b. $y'' - y' - 6y = 0 \; y(0) = 0, \quad y'(0) = 3$

     c. $y'' + y' - 2y = 2e^{4t} \quad y(0) = 0, \; y'(0) = 3$

     d. $y''^{+y} = \sin t \quad y(0) = 2, \quad y'(0) = -1$.

11.15   Show that $LT\{e^{at}t^n\} = \dfrac{n!}{(s-\alpha)^{n+1}}; s > a \; ; n > 0$ and integral.

12 Special Functions

There are things which seem incredible to most men who have not studied mathematics.

–Aristotle

INTRODUCTION

Most of the special functions are studied in this chapter. We start by introducing Bessel differential equations and then we define and study Legendre polynomials and the associated Legendre polynomials, the Laguerre functions and their differential equations, the Chebyshev polynomials, the Hermite polynomials, the Gamma functions and the Beta functions, the factorial functions and the hypergeometric functions.

12.1 BESSEL DIFFERENTIAL EQUATION

Bessel functions were first discovered by Daniel Bernoulli. However, it is Friedrich Wilhelm Bessel (July 22, 1784–March 17, 1846) who worked on those functions while he was analyzing the motion of the planets due to gravitational interactions. Bessel was a German mathematician and astronomer. In fact, the asteroid 1552 Bessel was named in his honor. In 1810, at the age of 26, Bessel was appointed the director of the Königsberg Observatory by King Frederick William III of Prussia. There, he published tables of atmospheric refractions based on Bradley's observations, which won him the Lalande Prize from the French Academy of Sciences in 1811. Bessel was able to pin down the position of over 50,000 stars during his time at Königsberg.

Bessel's ordinary differential equation (ODE) can be written as

$$x^2 \frac{d^2 Z_v}{dx^2} + x \frac{dZ_v}{dx} + \left(x^2 - v\right)^2 Z_v = 0. \tag{12.1}$$

The solution to Bessel's ODE, $J_n(x)$, where n is an integer replacing v can also be defined by the generating function, $g(x,t) = e^{(x/2)[t-1/t]}$. The expansion of the generating function $g(x,t)$ in the Laurent series gives

$$e^{(x/2)[t-1/t]} = \sum_{n=-\infty}^{n=+\infty} J_n(x) t^n. \tag{12.2}$$

$J_n(x)$ is called the Bessel function of the first kind. The expansion of the generating function in the MacLaurin series gives the expression for the Bessel function of the first kind as

$$J_n(x) = \sum_{s=0}^{\infty} \frac{(-1)^s}{s!(n+s)!} \left(\frac{x}{2}\right)^{n+2s} \tag{12.3}$$

The Bessel functions oscillate but are not periodic except in the limit where $n \to \infty$. Also, the amplitude of the function decreases exponentially.

DOI: 10.1201/9781003478812-12

Example 1 of Bessel Equation

Let us solve the following Bessel equation,

$$x^2 y'' + xy' + \left(x^2 - n^2\right) y = 0 \tag{12.4}$$

using the following technique called the indicial method also known as Frobenius' method. This method uses a generalized Laurent series and searches for the most general solution on the form:

$$y = \sum_{\lambda=0}^{\infty} a_\lambda x^{\lambda+k} \tag{12.5}$$

with $a_0 \neq 0$ and k a constant real number.

To solve the Bessel differential equation we need to find the first and second derivative of y, that is, $y' = \sum_{\lambda=0}^{\infty} a_\lambda \left(\lambda + k\right) x^{\lambda+k-1}$ and

$$y'' = \sum_{\lambda=0}^{\infty} a_\lambda \left(\lambda + k\right)\left(\lambda + k - 1\right) x^{\lambda+k-2}.$$

Now, we plug y, y' and y'' back onto the differential equation. Hence,

$$\sum_{\lambda=0}^{\infty} a_\lambda \left(\lambda + k\right)\left(\lambda + k - 1\right) x^{\lambda+k} + a_\lambda \left(\lambda + k\right) x^{\lambda+k}$$

$$+ a_\lambda x^{\lambda+k+2} - n^2 a_\lambda x^{\lambda+k} = 0.$$

By setting $\lambda = 0$, we obtain the coefficient of x^k, the lowest power of x, which is $a_0 k\left(k - 1\right) + a_0 k - n^2 a_0 = 0$ or $a_0 \left[k\left(k - 1\right) + k - n^2\right] = 0$.

$a_0 \neq 0$; therefore, $k^2 - n^2 = 0 \Leftrightarrow k = \pm n$.

Let's now examine the coefficient of x^{k+1}. We obtain:

$$a_1 \left(k + 1\right)\left(k\right) + a_1 \left(k + 1\right) - n^2 a_1 = 0$$

$$a_1 \left[\left(k + 1\right)\left(k\right) + \left(k + 1\right) - n^2\right] = 0$$

If $a_1 \neq 0$ then $\left(k + 1\right) - n^2 = 0$

$$\left(k + 1 + n\right)\left(k + 1 - n\right) = 0$$

$$k = -1 - n \text{ or } k = -1 + n$$

Since the values of k are different from $\pm n$, we should rather have $a_1 = 0$.

Let us now look at the coefficient of $x^{\lambda+k+2}$, and using $k = n$ we obtain

$$a_{\lambda+2} = -\frac{a_\lambda}{\left(\lambda + 2\right)\left(2n + \lambda + 2\right)} \tag{12.6}$$

λ varies from 0 to ∞. Let us calculate the successive values of a_λ:

for $\lambda = 0$, $a_2 = -\dfrac{a_0}{2(2n+2)} = \dfrac{-a_0}{2^2(n+1)}$

for $\lambda = 1$, $a_3 = 0$, and for all $\lambda = 2k+1$, $a_\lambda = 0$

for $\lambda = 2$

$$a_4 = -\frac{a_2}{4(2n+4)} = \frac{a_0}{4(2n+4)2^2(n+1)}$$

$$= (-1)^2 \frac{a_0}{2^{2\times 2}2!(n+1)(n+2)} = \frac{a_0 n!}{(n+2)!}$$

In general, for $\lambda = 2p-2$, $a_{2p} = (-1)^p \dfrac{a_0 n!}{2^{2p} p!(n+p)!}$

We can now write the form of the solution to the Bessel differential equation:

$$y(x) = \sum_{p=0}^{\infty} 2^n (-1)^p \frac{a_0 n!}{2^{2p} p!(n+p)!} \left(\frac{x}{2}\right)^{n+2p} \tag{12.7}$$

The previous summation is also referred to as the Bessel function $J_n(x)$.

For $k = -n$ the solution is going to be $J_{-n}(x) = (-1)^n J_n(x)$ with n an integer.

For n negative, Bessel's function given by equation (12.7) becomes

$$J_{-n}(x) = \sum_{s=0}^{\infty} \frac{(-1)^s}{s!(s-n)!} \left(\frac{x}{2}\right)^{2s-h}. \tag{12.8}$$

$$\text{Plot}[\{\text{BesselJ}[0,x], \text{BesselJ}[1,x], \text{BesselJ}[2,x],$$

$$\text{BesselJ}[3,x]\}, \{x,0,20\}, \text{AxesLabel} \rightarrow \{x, \text{BesselFunctions}\}, \text{PlotLegend} \rightarrow \{J_0, J_1, J_2, J_3\}].$$

12.1.1 Properties of Bessel Function of the First Kind

$$P_1. \text{ For all } n, J_{-n}(x) = (-1)^n J_n(x) \tag{12.9a}$$

Recurrence relations:

$$P_2. \ J_{n-1}(x) + J_{n+1}(x) = \frac{2n}{x} J_n(x) \tag{12.9b}$$

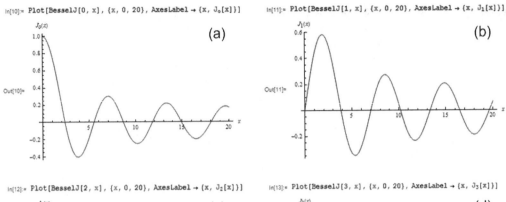

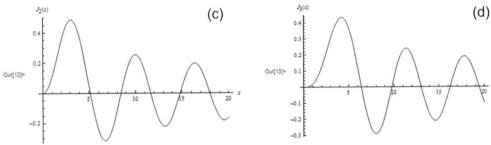

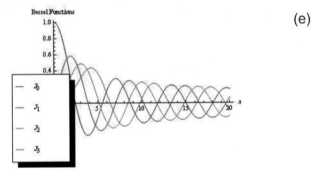

FIGURE 12.1 Example of the graph of various Bessel functions.

$$P_3. \quad J_{n-1}(x) - J_{n+1}(x) = 2J'_n(x) \tag{12.9c}$$

$$P_4. \quad J_{n+1}(x) = \frac{n}{x} J_n(x) - J'_n(x) \tag{12.9d}$$

$$P_5. \quad \frac{d}{dx}\left[x^{-n} J_n(x)\right] = -x^{-n} J_{n+1}(x) \tag{12.9e}$$

12.1.2 INTEGRAL REPRESENTATION OF BESSEL FUNCTION

By replacing t by $x > 0$, the generating function becomes

$$e^{ix\sin\theta} = J_0(x) + 2\left[J_2(x)\cos2\theta + J_4(x)\cos4\theta + \cdots\right]$$

$$+2i\left[J_1(x)\sin\theta + J_3(x)\sin3\theta + \cdots\right] \tag{12.10}$$

Now, by identifying the real and imaginary parts of equation (12.10) we have the following expression with the summation sign:

$$\cos(x\sin\theta) = J_0(x) + 2\sum_{n=1}^{\infty} J_{2n}(x)\cos(2n\theta) \tag{12.11}$$

$$\sin(x\sin\theta) = 2\sum_{n=1}^{\infty} J_{2n-1}(x)\sin(2n-1)\theta \tag{12.12}$$

12.1.3 BESSEL FUNCTIONS OF THE FIRST KIND FOR INTEGER ORDERS N = 0,1,2,...

The general formula is given by

$$J_n(x) = \frac{1}{\pi}\int^{\pi} \cos\cos(n\theta - x\sin\theta)d\theta = \frac{1}{\pi}\int^{\pi} \cos\cos(x\sin\theta - n\theta)d\theta \tag{2.13}$$

We can find the integral forms of the Bessel functions of the 0th and first order, that is:

$$J_0(x) = \frac{1}{\pi}\int^{\pi} \cos(x\sin\theta)d\theta = \frac{1}{\pi}\int^{\pi} \cos(x\sin\theta)d\theta \tag{12.14}$$

$$J_1(x) = \frac{1}{\pi}\int^{\pi} \cos(\theta - x\sin\theta)d\theta = \frac{1}{\pi}\int^{\pi} \cos(x\sin\theta - \theta)d\theta$$

$$= \frac{1}{\pi}\int_0^{\pi} \cos\theta\sin(x\cos\theta)d\theta \tag{12.15}$$

12.1.4 BESSEL FUNCTIONS OF THE SECOND KIND

The constant v involved in the Bessel equation is the order of the Bessel functions found in the solution to Bessel's differential equation and can take any real number value. For cylindrical problems, the order of the Bessel function is an integer value, $v = n$, while for spherical problems, the order of the Bessel function is of half integer value, $v = n + 1/2$. Now since Bessel's differential equation is a second-order differential equation, there must be two linearly independent solutions. Typically, the general solution is given by $y = AJ_v(x) + BY_v(x)$, where the special functions $J_v(x)$ are Bessel functions of the first kind which are finite at $x = 0$ for all real values v and the special functions $Y_v(x)$ are Bessel functions of the second kind (also known as Weber or Neumann functions) which are singular at $x = 0$.

12.1.5 BESSEL FUNCTIONS OF THE SECOND KIND FOR INTEGER ORDERS $N = 0, 1, 2, \ldots$

We can find the integral forms of the Bessel functions of the zeroth and first order, that is:

$$Y_n(x) = -\frac{2(x/2)^{-n}1}{\sqrt{\pi}\,{}^{\text{``}}\left(\frac{1}{2}-n\right)} \int \frac{\cos(xt)\,dt}{(t^2-1)^{n+1/2}} \quad x > 0 \tag{12.16}$$

$$Y_n(x) = \frac{1}{\pi}\int_0^\pi \sin(x\sin\theta - n\theta)\,d\theta - $$
$$\frac{1}{\pi}\int_0^\pi [e^{nt} + e^{-nt}\cos(n\pi)]\exp(-x\sinh t)\,dt \tag{12.17}$$

$$Y_0(x) = \frac{4}{\pi^2}\int^{\pi/2} \cos(x\cos\theta)\left[\gamma + \ln\left(2x\sin^2\theta\right)\right]d\theta \quad x > 0 \tag{12.18}$$

$$Y_0(x) = -\frac{2}{\pi}\int^\infty \cos(x\cosh t)\,dt \quad x > 0 \tag{12.19}$$

12.1.6 RECURRENCE RELATIONS ON THE BESSEL FUNCTIONS IN THE GENERAL CASE

Here are some relations about the Bessel functions of higher order for all real values, v:

$$J_{v+1}(x) = \frac{2v}{x}J_v(x) - J_{v-1}(x) \tag{12.20}$$

$$J'_{v+1}(x) = \frac{1}{2}\left[J_{v-1}(x) - J_{v+1}(x)\right]$$

$$J'_v(x) = J_{v-1}(x) - \frac{v}{x}J_v(x)$$

$$J'_v(x) = \frac{v}{x}J_v(x) + J_{v+1}(x) \tag{12.21}$$

$$\frac{d}{dx}\left[x^v J_v(x)\right] = x^v J_{v-1}(x)$$

$$\frac{d}{dx}\left[x^{-v}J_v(x)\right] = -x^{-v}J_{v+1}(x)$$

For n and m nonnegative positive integers, if we use the orthogonality properties such as

$$\int^\pi \cos n\theta\cos m\theta\,d\theta = \frac{\pi}{2}\delta_{nm} \tag{12.22}$$

$$\int_0^\pi \sin n\theta\sin m\theta\,d\theta = \frac{\pi}{2}\delta_{nm}, \tag{12.23}$$

we get

$$\frac{1}{\pi}\int^{\pi}\cos\left(x\sin\theta\right)\cos n\theta \, d\theta = \begin{cases} J_n\left(x\right) & \text{for } n \quad \text{even} \\ 0 & \text{for } n \quad \text{odd} \end{cases} \tag{12.24}$$

$$\frac{1}{\pi}\int_0^{\pi}\sin\left(x\sin\theta\right)\sin n\theta \, d\theta = \begin{cases} 0 & \text{for } n \quad \text{even} \\ J_n\left(x\right) & \text{for } n \quad \text{odd} \end{cases} \tag{12.25}$$

Finally, a general integral representation of the Bessel's function is written as

$$J_n\left(x\right) = \frac{1}{\pi}\int^{\pi}\cos\left(n\theta - x\sin\theta\right)d\theta$$

$$\text{for } n = 0,1,2,\ldots \tag{12.26}$$

Remember that equation (11.26) is not the only integral representation for Bessel's function. There are many other integral representations.

12.1.7 The Fourier–Bessel Integral

The Fourier integrals of the function $f(x)$ of one variable and the function $f(x,y)$ of two variables have the successive forms:

$$f\left(x\right) = \frac{1}{2\pi}\int_{-\infty}^{+\infty}d\mu \int_{\infty}^{+\infty} f\left(\xi\right)e^{i\mu\left(x-\xi\right)}d\xi \tag{12.27}$$

$$f\left(x,y\right) = \frac{1}{\left(2\pi\right)^2}\int_{-\infty}^{+\infty}\int_{\infty}^{+\infty}d\mu d\mu' \int_{\infty}^{+\infty}\int_{\infty}^{+\infty} f\left(\xi,\eta\right)e^{i\mu\left(x-\xi\right)+i\mu'\left(y-\eta\right)}d\xi d\eta \tag{12.28}$$

Using the polar coordinates:

$$\begin{aligned} & x = r\cos\varphi, \, y = r\sin\varphi; \\ & \xi = \rho\cos\psi, \eta = \rho\sin\psi \\ & \mu = \lambda\cos\psi, \mu' = \lambda\sin\psi, \end{aligned} \tag{12.29}$$

then we obtain $d\xi d\eta = \rho d\rho d\psi, d\mu d\mu' = \lambda d\lambda d\xi$.

Assuming that $f\left(x,y\right) = f\left(r\right)e^{in\varphi}$ with n an integer, we obtain the Bessel–Fourier integral:

$$f\left(r\right) = \int_0^{+\infty}\int^{+\infty} f\left(\rho\right)J_n\left(\lambda\rho\right)J_n\left(\lambda r\right)\lambda d\lambda \rho d\rho. \tag{12.30}$$

It is essential that the function $f(r)$ has the following properties:

- be defined and continuous in the interval $\left(0,+\infty\right)$;
- has a finite number of maxima and minima in any finite interval;
- the integral $f\left(r\right) = \int\rho\left|f\left(\rho\right)\right|d\rho$ converges. The above conditions are necessary for the expansion of the Bessel–Fourier integral to be possible.

12.1.8   NEUMANN FUNCTIONS

Neumann function can be regarded as a linear combination of Bessel's function of first and second kind: it can be written as

$$N_v(x) = \frac{\cos(v\pi)J_v(x) - J_{-v}(x)}{\sin(v\pi)} \tag{12.31}$$

or

$$N_v(x) = \frac{J_v(x)}{\tan(v\pi)} - \frac{J_{-v}(x)}{\sin(v\pi)} \tag{12.32}$$

12.2   LEGENDRE POLYNOMIALS

12.2.1   INTRODUCTION TO LEGENDRE POLYNOMIALS

The Legendre polynomials are especially important in physics and engineering. We define them here first for real variables x comprises between '1 and +1. We define Legendre polynomials as follows:

$$P_0(x) = 1;$$

$$P_n(x) = \frac{1}{2^n n!} \frac{d^n(x^2-1)^n}{dx^n}; n = 1,2,3,\dots \tag{12.33}$$

We need to prove that these polynomials are of degree n and that they are orthogonal for all $x \in [-1,1]$.

Using the binomial expansion we have

$$(x^2-1)^n = \sum_{r=0}^{n} \frac{n!(-1)^{n-r}}{(n-r)!r!} x^{2r} \tag{12.34}$$

Legendre polynomial can then be written as

$$
\begin{aligned}
P_n(x) &= \frac{1}{2^n n!} \frac{d^n(x^2-1)^n}{dx^n} \\
&= \frac{1}{2^n} \sum_{r=0}^{n} \frac{(-1)^{n-r}2r(2r-1)(2n-2)\dots(2r-n+1)x^{2r-n}}{(n-r)!r!} \\
&\quad \times \frac{(2r-n)(2r-n-1)\dots3.2.1}{(2r-n)(2r-n-1)\dots3.2.1} \\
&= \frac{1}{2^n} \sum_{r=0}^{n} \frac{(-1)^{n-r}(2r)!x^{2r-n}}{(n-r)!r!(2r-n)!} = \sum_{r=0}^{n} \frac{(-1)^{n-r}1.3.5\dots(2r-1)x^{2r-n}}{(n-r)!(2r-n)!2^{n-r}}
\end{aligned} \tag{12.35}
$$

Since $(-m)! \equiv \infty$, the first term gives

For n odd, $\dfrac{(-1)^{(n-1)/2}1.3.5...n}{2.4.6...(n-1)} x$, and $\dfrac{(-1)^{n/2}1.3.5...(n-1)}{2.4.6...n} x$ for n even.

We also have

$$P_n(1) = 1 \ \forall n \ge 0 \tag{12.36}$$

The Legendre polynomials are polynomials of degree n and are odd or even depending on whether n is odd or even so that

$$P_n(x) = (-1)^n P_n(-x) \ \ \forall n \ge 0 \tag{12.37}$$

The first-order Legendre polynomials are

$$P_0(x) = 1; P_1(x) = x; P_2(x) = \frac{3}{2}x^2 - \frac{1}{2}; P_3(x) = \frac{5}{2}x^3 - \frac{3}{2}x; P_4(x) = \frac{35}{8}x^4 - \frac{15}{4}x^2 + \frac{3}{8}$$

We now need to prove that Legendre polynomials are orthogonal that is:

$$\int_1^l P_m(x) P_n(x) = 0, \text{ for } m \ne n,$$

and that in general,

$$\int_1 P_m(x) P_n(x) dx = \frac{2}{2n+1} \delta_{mn}.$$

and the required orthogonal set is given by

$$\varphi(x) = \sqrt{\frac{2n+1}{2}} P_n(x); \ < \varphi_m | \varphi_n = \delta_{mn} \tag{12.38}$$

We need to prove that these polynomials are of degree n and that they are orthogonal for all $x \in [-1,1]$.

If $m < n$ then the integration by part yields

$$\int_1^1 P_n(x) x^m dx = \frac{1}{2^n n!} \int_1^b \frac{d^n (x^2-1)^n}{dx^n} x^m dx \tag{12.39}$$

which gives

$$\int_1^b P_m(x) P_n(x) = 0. \tag{12.40}$$

12.2.1.1   Coulomb's Expansion

It is possible to have the Legendre polynomials in the following series:

$$\frac{1}{\sqrt{1 - 2\mu x + \mu^2}} = \sum_{n=0}^{\infty} \mu^n P(x); \ |\mu| < 1; \ -1 < x < +1 \tag{12.41}$$

This is known as the Coulomb's expansion; it allows us to find the distance between a point at r_0 on the polar axis and any other point (r,θ) not on the sphere of radius, r_0.

For $r < r_0$,

$$\frac{1}{R} \equiv \frac{1}{r_0\sqrt{1+(r/r_0)^2 - 2(r/r_0)\cos(\theta)}} = \frac{1}{r_0}\sum_{n=0}^{\infty}(r/r_0)^n P_n\left[\cos(\theta)\right] \qquad (12.42)$$

For $r > r_0$,

$$\frac{1}{R} \equiv \frac{1}{r\sqrt{1+(r_0/r)^2 - 2(r_0/r)\cos(\theta)}} = \frac{1}{r}\sum_{n=0}^{\infty}(r_0/r)^n P_n\left[\cos(\theta)\right] \qquad (12.43)$$

But these are not valid expansion for $r = r_0$.

12.2.1.2   Recursion Formula

The recursion formula is given by

$$(n+1)P_{n+1}(x)-(2n+1)xP_n(x)+nP_{n-1}(x)=0,\ n\geq 1. \qquad (12.44)$$

12.2.1.3   Differential Equation

The differential equation related to Legendre polynomial is given by

$$(x^2-1)\frac{d^2}{dx^2}P_n(x)+2x\frac{d}{dx}P_n(x)-n(n+1)P_n(x)=0. \qquad (12.45)$$

with the software Mathematica, the command to obtain Legendre polynomial of degree n is Legendre $P[n,x]$.

For $n = 10$ we have
Legendre$P[10,x]$
$1/256('63 + 3465x^{2'} 30030x^4 + 90090x^{6'} 109395x^8 + 46189x^{10})$

For $n = 1$
Legendre$P[1,x]$
x

For $n = 0$
Legendre $P[0,x]$
1

For $n=1$
Legendre$P[1,1]$
1

Plot[LegendreP[1,x],{x,-1,1}]

Plot[LegendreP[2,x],{x,-1,1}]

Plot[LegendreP[3,x],{x,-1,1}]

Plot[LegendreP[4,x],{x,-1,1}]

Plot[LegendreP[5,x],{x,-1,1}]

Plot[LegendreP[6,x],{x,-1,1}]

Plot[LegendreP[7,x],{x,-1,1}]

Plot[LegendreP[8,x],{x,-1,1}]

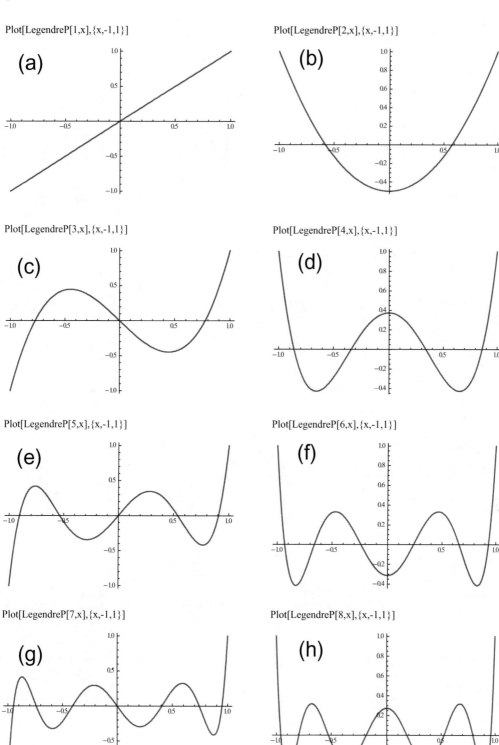

FIGURE 12.2   Example of the graphs of various Legendre polynomials.

Plot[LegendreP[9,x],{x,-1,1}]   Plot[LegendreP[10,x],{x,-1,1}]

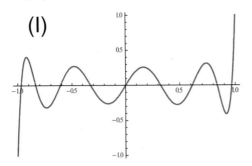   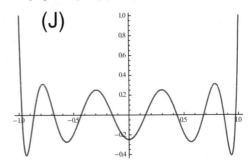

FIGURE 12.2   (Continued).

12.2.1.4   Legendre Polynomial of the Second Kind

The nth-order Legendre polynomial of the second kind is obtained by the use of the command: LegendreQ[n,x].

The following are the expressions of the first 11 orders of Legendre polynomial of the second kind.

LegendreQ[0,x]
$(1/2)\text{Log}[1\,'x] + 1/2\text{Log}[1 + x]$

LegendreQ[1,x]
$1 + x\,((1/2)\,\text{Log}[1\,'x] + 1/2\,\text{Log}[1 + x])$

LegendreQ[2,x]
$((3x)/2) + 1/2('1 + 3x^2)((1/2)\text{Log}[1\,'x] + 1/2\text{Log}[1 + x])$

LegendreQ[3,x]
$2/3\,'(5x^2)/2\,'1/2x(3\,'5x^2)((1/2)\text{Log}[1\,'x] + 1/2\text{Log}[1 + x])$

LegendreQ[4,x]
$(55x)/24\,'(35x^3)/8 + 1/8\,(3\,'30x^2 + 35x^4)\,((1/2)\text{Log}[1\,'x] + 1/2\,\text{Log}[1 + x])$

LegendreQ[5,x]
$(8/15) + (49x^2)/8\,'(63x^4)/8 + 1/8x\,(15\,'70x^2 + 63x^4)\,((1/2)\,\text{Log}[1\,'x] + 1/2\,\text{Log}[1 + x])$

LegendreQ[6,x]
$((231x)/80) + (119x^3)/8\,'(231x^5)/16 + 1/16\,('5 + 105x^2\,'315x^4 + 231x^6)\,((1/2)\,\text{Log}[1\,'x] + 1/2\,\text{Log}[1 + x])$

LegendreQ[7,x]
$16/35\,'(849x^2)/80 + (275x^4)/8\,'(429x^6)/16\,'1/16x\,(35\,'315x^2 + 693x^4\,'429x^6)\,((1/2)\,\text{Log}[1\,'x] + 1/2\,\text{Log}[1 + x])$

LegendreQ[8,x]
$(15159x)/4480\,'(4213x^3)/128 + (9867x^5)/128\,'(6435x^7)/128 + 1/128\,(35\,'1260x^2 + 6930x^4\,'12012x^6 + 6435\,x^8)\,('(1/2)\,\text{Log}[1\,'x] + 1/2\,\text{Log}[1 + x])$

LegendreQ[9,x]
$-(128/315) + (14179x^2)/896$ ′ $(11869x^4)/128 + (65065x^6)/384$ ′ $(12155x^8)/128 + 1/128x$ $(315$ ′ $4620x^2 + 18018x^4$ ′ $25740x^6 + 12155x^8)$ $(′(1/2)$ Log$[1$ ′ $x] + 1/2$ Log$[1 + x])$

LegendreQ[10,x]
$-((61567x)/16128) + (26741x^3)/448$ ′ $(157157x^5)/640 + (70499x^7)/192$ ′ $(46189x^9)/256 + 1/256$ $(′63 + 3465x^2$ ′ $30030x^4 + 90090x^6$ ′ $109395x^8 + 46189x^{10})$ $(′(1/2)$ Log$[1$ ′ $x] + 1/2$ Log$[1 + x])$

12.2.1.5   Graph of the First 11th Order of the Legendre Polynomials of the Second Kind

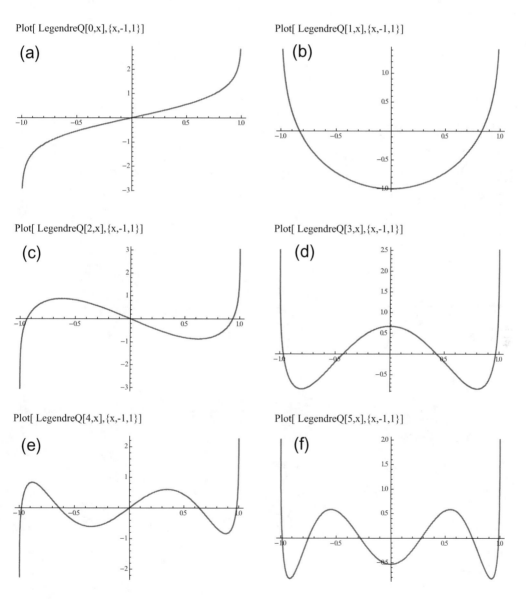

FIGURE 12.3   Graph of the first 11th order of the Legendre polynomials of the second kind.

Plot[LegendreQ[6,x],{x,-1,1}]

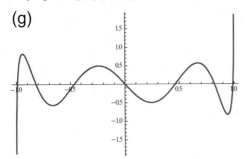

Plot[LegendreQ[7,x],{x,-1,1}]

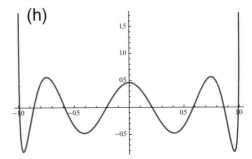

Plot[LegendreQ[8,x],{x,-1,1}]

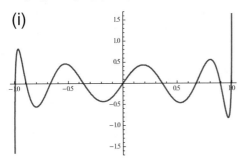

Plot[LegendreQ[9,x],{x,-1,1}]

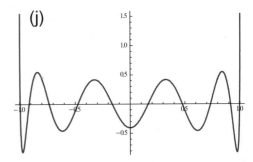

Plot[LegendreQ[10,x],{x,-1,1}]

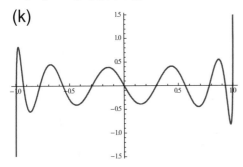

FIGURE 12.3 (Continued).

12.2.2 Associated Legendre Polynomials

The associated Legendre polynomials are defined as follows:

$$P_n^m(x) \equiv \frac{1}{2^n n!}(1-x^2)^{m/2}\frac{d^m P_n(x)}{dx^m}; \quad 0 \le m \le n. \tag{12.46}$$

$$P_n^0(x) \equiv P_n(x).$$

They form an infinite orthogonal set of functions over $[-1,1]$, that is:

$$\int_1^1 P_n^m(x)P_{n'}^m(x)dx = \frac{2}{2n+1}\frac{(n+m)!}{(n-m)!}\delta_{nn'} \tag{12.47}$$

for $m > n$, $P_n^m(x) = 0$.

The set $\{P_n^m(x)\}$ forms a complete set of functions at least for bounded piecewise continuous functions on the interval,
$$[-1,1].$$

12.3 LAGUERRE FUNCTIONS

Laguerre ODE derives from the radial ODE of the Schrodinger equation for the hydrogen atom. The equation can be written as

$$xy''(x) + (1-x)y'(x) + ny(x) = 0$$

where n is a nonnegative integer; $n = 0,1,2,3, \dots$.

The solutions to the Laguerre equation are referred to as Laguerre polynomials, $L_n(x)$.

The contour integral $y_n(x) = \dfrac{1}{2\pi i}\int \dfrac{e^{-xz/1-z}}{(1-z)^{n+1}}\, dz \equiv L_n(x)$ is solution to Laguerre ODE. The contour includes $z = 0$ but does not include the point $z = 1$. As a result, a generating function for the Laguerre polynomials is

$$g(x,z) = \frac{e^{-xz/1-z}}{1-z} = \sum_{n=0}^{\infty} L_n(x) z^n; \ |z| < 1 \tag{12.48}$$

12.3.1 RODRIGUES FORMULA FOR LAGUERRE POLYNOMIALS

Laguerre polynomials can be expressed by Rodrigues' formula, that is:

$$L_n(x) = \frac{e^x}{n!}\frac{d^n}{dx^n}\left(x^n e^{-x}\right); \ n = 0,1,2,3,\dots \tag{12.49}$$

$L_n(x)$ can also be written as a series on the form:

$$L_n(x) = \frac{(-1)^n}{n!}\left[x^n - \frac{n^2}{1!}x^{n-1} + \frac{n^2(n-1)^2}{2!}x^{n-2} + \dots + (-1)^n n!\right] \tag{12.50}$$

The Laguerre polynomials can also be written as the following series:

$$L_n(x) = \sum_{m=0}^{n} \frac{(-1)^m n!}{(n-m)!m!m!} x^m \tag{12.51}$$

12.3.2 RECURRENCE RELATIONS

$$(n+1)L_{n+1}(x) = (2n+1-x)L_n(x) - nL_{n-1}(x) \tag{12.52}$$

$$xL_{n+1}'(x) = nL_n(x) - nL_{n-1}(x) \tag{12.53}$$

$$g(0,z) = \sum_{m=0}^{\infty} L_m(x) z^m = \frac{1}{1-z} = \Sigma_{m=0}^{\infty} z^m \tag{12.54}$$

This implies that $L_n(0) = 1$.

12.3.3 ORTHOGONALITY CONDITIONS

Laguerre polynomials $L_n(x)$, with $n = 0, 1, 2, 3...$ form a complete orthogonal set on the interval $[0,\infty]$ with respect to the weighting function, e^{-x}. It can be shown that

$$\int e^{-x} L_m(x) L_n(x) dx = \delta_{mn} = \begin{cases} 0, & m \neq n \\ 1, & m = n \end{cases} \tag{12.55}$$

The nth-order Laguerre polynomial is represented in Mathematica by the command, LaguerreL[n,x]

LaguerreL[0,x]
1

LaguerreL[1,x]
1 − x

LaguerreL[2,x]
1/2 (2 − 4x + x^2)

LaguerreL[3,x]
1/6 (6 − 18x + 9x^2 − x^3)

LaguerreL[4,x]
1/24 (24 − 96x + 72x^2 − 16x^3 + x^4)

LaguerreL[5,x]
1/120 (120 − 600x + 600x^2 − 200x^3 + 25x^4 − x^5)

LaguerreL[6,x]
1/720 (720 − 4320x + 5400x^2 − 2400x^3 + 450x^4 − 36x^5 + x^6)

LaguerreL[7,x]
(5040 − 35280x + 52920x^2 − 29400x^3 + 7350x^4 − 882x^5 + 49x^6 − x^7)/5040

LaguerreL[8,x]
(40320 − 322560x + 564480x^2 − 376320x^3 + 117600x^4 − 18816x^5 + 1568x^6 − 64x^7 + x^8)/40320

LaguerreL[9,x]
(1/362880)(362880 − 3265920x + 6531840x^2 − 5080320x^3 + 1905120x^4 − 381024x^5 + 42336x^6 − 2592x^7 + 81x^8 − x^9)

LaguerreL[10,x]
(1/3628800)(3628800 − 36288000x + 81648000x^2 − 72576000x^3 + 31752000x^4 − 7620480x^5 + 1058400x^6 − 86400x^7 + 4050x^8 − 100x^9 + x^{10})

12.3.4   GRAPH OF LAGUERRE POLYNOMIAL OF ORDER N; N = 1,2,3,...,10.

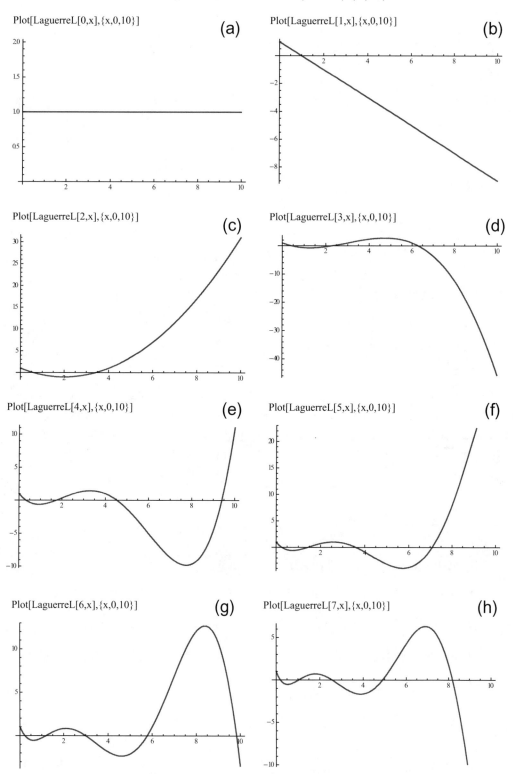

FIGURE 12.4   Graphs of the Laguerre polynomial of order n, $n = 1,2,3,..., 10$.

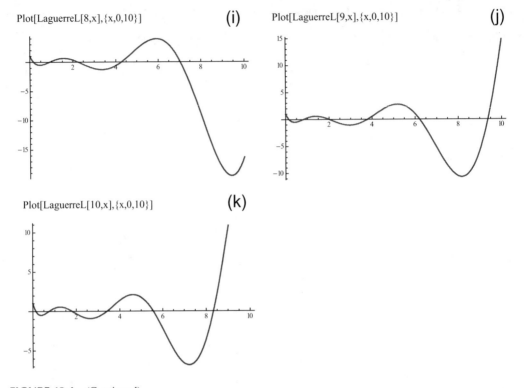

FIGURE 12.4   (Continued).

Laguerre ODE is not self-adjoint and the solutions do not form by themselves an orthonormal set. The orthonormality condition was given with the weighting function, e^{-x}. However, defined with a new weighting function, $e^{-x/2}$, that is, obtaining other new orthogonal Laguerre functions, $\ell_n(x) = e^{-x/2}L_n(x)$, we obtain the following ODE for $\ell_n(x)$:

$$x\ell_n''(x) + \ell_n'(x) + \left(n + \frac{1}{2} - \frac{x}{4}\right)\ell_n(x) = 0 \tag{12.56}$$

where ℓ_n is self-adjoint.

12.3.5   ASSOCIATED LAGUERRE DIFFERENTIAL EQUATIONS

The associated Laguerre differential equation is defined by the differential equations:

$$xL_n^{k''}(x) + (k + 1 - x)L_n^{k'}(x) + nL_n^k(x) = 0 \tag{12.57}$$

The solution is given by

$$L_n^k(x) = (-1)\frac{d^k}{dx^k}L_{n+1}(x) \tag{12.58}$$

In general,

$$L_n^k(x) = \sum_{m=0}^{n}(-1)^m\frac{(n+k)!}{(n-m)(k+m)!m!}x^m, k > -1 \tag{12.59}$$

12.3.6 Rodrigues' Representation of the Associated Laguerre ODE

Rodrigues' representation of the associated Laguerre ODE is defined by

$$L_n^k(x) = \frac{e^x x^{-k}}{n!} \frac{d^n}{dx^n}\left(e^{-x} x^{n+k}\right)$$
(12.60)

12.3.7 Generating Function

A generating function for the Laguerre polynomials is

$$g(x,z) = \frac{e^{-xz/1-z}}{(1-z)^{k+1}} = \sum_{n=0}^{\infty} L_n^k(x) z^n; \ |z| < 1$$
(12.61)

The associated Laguerre polynomials are obtained from Mathematica by the command LaguerreL[n,k,x]. Here are some examples of the associated Laguerre polynomials and graphs:

LaguerreL[0,k,x]
1

LaguerreL[1,k,x]
$1 + k - x$

LaguerreL[n,0,x]

LaguerreL[n,x]

LaguerreL[1,1,0]
2

LaguerreL[1,1,x]
$2 - x$

12.3.8 Orthogonality Conditions

The orthogonality conditions with the weighting function, $e^{-x} x^k$ is given by

$$\int^{\infty} e^{-x} e^{-x} x^k L_n^k(x) L_m^k(x)\, dx = \frac{(n+k)!}{n!} \delta_{mn}$$
(12.62)

12.4 CHEBYSHEV POLYNOMIALS

Chebyshev polynomials of the first kind are solutions to the differential equations:

$$(1-x^2) y'' - xy' + n^2 y = 0$$
(12.63)

where n is a positive integer
  Chebyshev polynomials of the second kind are solutions to the differential equations:

$$(1-x^2) y'' - 3xy' + n(n+2) y = 0.$$
(12.64)

where n is a positive integer.
  These equations are special cases of the Sturm–Liouville differential equations.

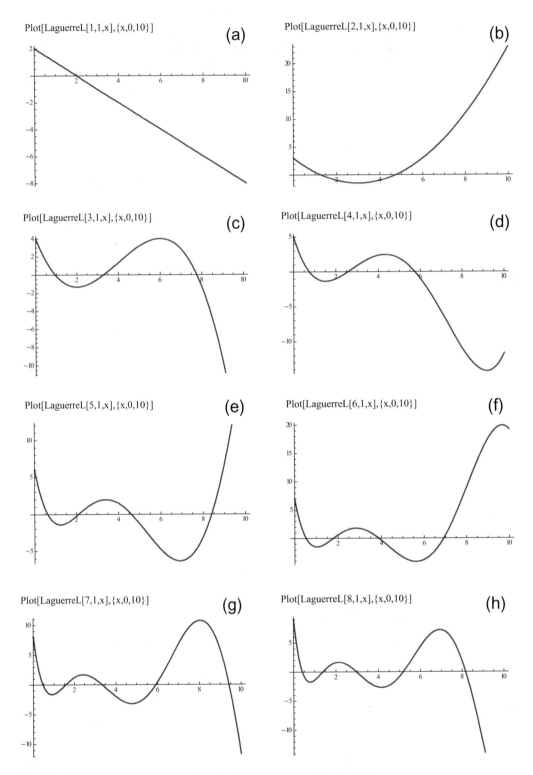

FIGURE 12.5 Examples of the associated Laguerre polynomials and graphs.

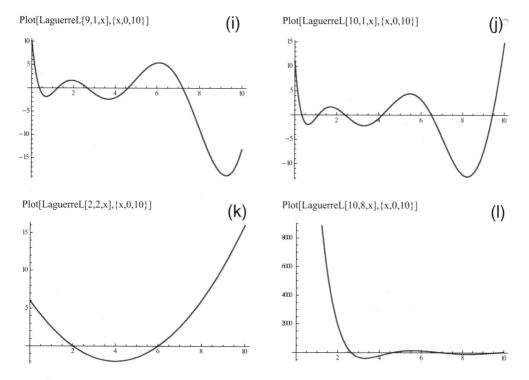

FIGURE 12.5 (Continued).

The generating function of the Chebyshev polynomials of the first kind is

$$g(x,t) = \frac{1-tx}{1-2tx+t^2} = \sum_{n=0}^{\infty} T_n(x)t^n \tag{12.65}$$

For $n = 0$, $T_0(x) = 1$

For $n = 1$, $T_1(x) = x$

For $n = 0$, $T_0(x) = 1$

And the recurrence relation is

$$n = 0, \quad T_{n+1}(x) = 2xT_n(x) - T_{n-1}(x) \tag{12.66}$$

Example of Chebyshev polynomial of the first kind: Chebyshev T[5,x]

$$5x' \, 20x^3 + 16x^5$$

12.4.1 GRAPH OF CHEBYSHEV POLYNOMIALS OF THE FIRST KIND USING THE MATHEMATICA COMMAND

Plot[{ChebyshevT[3, x], ChebyshevT[4, x], ChebyshevT[5, x]}, {x, ′1, 1}, AxesLabel -> {x, T}, PlotLegend -> {"T[3,x]", "T[4,x]", "T[5,x]"}, LegendPosition -> {1.1, ′0.4}]

The generating function of the Chebyshev polynomials of the second kind is

$$g(x,t) = \frac{1}{1-2tx+t^2} = \sum_{n=0}^{\infty} U_n(x)t^n \tag{12.67}$$

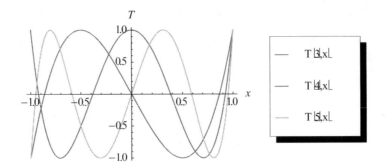

FIGURE 12.6 Graph of Chebyshev polynomials of the first kind, T, with Mathamatica.

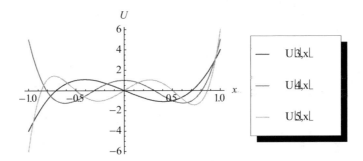

FIGURE 12.7 Graph of Chebyshev polynomials of the second kind, U, with Mathematica.

For $n = 0$, $T_0(x) = 1$

For $n = 1$, $T_1(x) = x$

For $n = 0$, $T_0(x) = 1$

And the recurrence relation is

$$n = 0, T_{n+1}(x) = 2xT_n(x) - T_{n-1}(x) \tag{12.68}$$

ChebyshevU[10,x]

$$-1 + 60x^2 {}'560x^4 + 1792x^6 {}'2304x^8 + 1024x^{10}$$

12.4.2 GRAPH OF CHEBYSHEV POLYNOMIALS OF THE SECOND KIND USING THE MATHEMATICA COMMAND

Plot[{ChebyshevU[3,x],ChebyshevU[4,x],ChebyshevU[5,x]},{x,1,1},AxesLabel□{x,U},PlotLegend
□{"U[3,x]","U[4,x]","U[5,x]"},LegendPosition□{1.1,′0.4}]

12.5   HERMITE POLYNOMIAL

Charles Hermite (1822–1901) is a French mathematician who worked on the theory of functions such as applications of the elliptic functions. He proved that e is a transcendental number representation of integers in Hermitian forms.

12.5.1   APPLICATIONS OF HERMITE POLYNOMIAL

Hermite polynomial defines the eigenstates of quantum simple harmonic oscillators, constructs filter banks for perfect reconstruction. Hermite functions are a set of eigenfunctions of the continuous Fourier transform. They are also involved in Koopman's Hilbert space formulations of classical mechanics and stochastic processes.

Hermite polynomials are in the form of

$$H_k(x) = (-1)^k \frac{d^k}{dx^k}\left(e^{-x^2}\right)e^{x^2}, \, k = 0,1,2,\ldots \tag{12.69}$$

The related generating functions are given by

$$g(x,w) = \sum_{k=0}^{\infty} \frac{H_k(x)}{k!} w^k = e^{2xw-w^2} \tag{12.70}$$

Now let us derive the generating function of Hermite polynomial

$$
\begin{aligned}
H_k(x) &= (-1)^k e^{x^2} \frac{d^k}{dx^k}\left(e^{-x^2}\right) \\
&= (-1)^k e^{x^2} \frac{k!}{2\pi i} \oint \frac{e^{-z^2}}{(z-x)^{k+1}} dz \\
\sum_{k=0}^{\infty} \frac{H_k(x)}{k!} w^k &= \sum_{k=0}^{\infty} \frac{w^k}{k!}(-1)^k e^{x^2} \frac{k!}{2\pi i} \oint \frac{e^{-z^2}}{(z-x)^{k+1}} dz \\
\sum_{k=0}^{\infty} \frac{(-w)^k}{2\pi i} e^{x^2} &\oint \frac{e^{-z^2}}{(z-x)^{k+1}} dz
\end{aligned}
\tag{12.71}
$$

Let us evaluate $\oint \dfrac{e^{-z^2}}{(z-x)^{k+1}} dz$ at the pole of order $k + 1$ at $z = x$.

$$
\begin{aligned}
a_{-1} &= \lim_{z\to x} \frac{1}{(k)!} \frac{d^k}{dz^k}\left[(z-x)^{k+1} \frac{e^{-z^2}}{(z-x)^{k+1}}\right] \\
&= \frac{1}{(k)!} \lim_{z\to x} \frac{d^k}{dz^k}\left[e^{-z^2}\right] \\
&= \frac{1}{(k)!} \lim_{z\to x}(-2z)^k e^{-z^2} = \frac{(-2x)^k e^{-x^2}}{(k)!} \\
\oint \frac{e^{-z^2}}{(z-x)^{k+1}} dz &= 2\pi i \frac{(-2x)^k e^{-x^2}}{(k)!} = 2\pi i \frac{(-2x)^k e^{-x^2}}{(k)!} \\
\sum_{k=0}^{\infty} \frac{H_k(x)}{k!} w^k &= \sum_{k=0}^{\infty} \frac{(-w)^k}{(k)!} = \sum_{k=0}^{\infty} \frac{(2xw)^k}{k!}
\end{aligned}
\tag{12.72}
$$

12.5.2   RECURRENCE RELATIONS

$$H_{k+1}(x) = (2x)H_k(x) - H'_k(x)$$
$$H'_k(x) = 2kH_k(x)$$
$$H_1(x) = (2x)H_0(x) - H'_0(x) = 2x$$
$$H_2(x) = 4x^2 - 2$$
$$H_3(x) = 8x^3 - 4x - 8x = 8x^3 - 12x \qquad (12.73)$$
$$H_4(x) = 16x^4 - 48x^2 + 12$$
$$H_5(x) = 32x^5 - 160x^3 + 120x$$
$$H_6(x) = 64x^6 - 480x^4 + 720x^2 - 120$$

These relations are found by taking the derivative of $H_k(x)$ with respect to x and taking the derivative of the generating function with respect to x.

The first recurrence relation is a simple way of finding the terms of the Hermite polynomial.

Deriving the recurrence relations:

$$H_k(x) = (-1)^k e^{x^2} \frac{d^k}{dx^k}\left(e^{-x^2}\right)$$

$$\frac{d}{dx}H_k(x) = (-1)^k \left[\frac{d^k}{dx^k}\left(e^{-x^2}\right)e^{x^2}(2x) + e^{x^2}\frac{d^{k+1}}{dx^{k+1}}\left(e^{-x^2}\right) \right]$$

$$H'_k(x) = (2x)H_k(x) - H_{k+1}(x)$$

or

$$H_k(x) = (-1)^k e^{x^2} \frac{d^k}{dx^k}\left(e^{-x^2}\right) \qquad (12.74)$$

$$g(x,w) = \sum_{k=0}^{\infty} \frac{H_k(x)}{k!} w^k = e^{2xw - w^2}$$

$$\frac{dg}{dx} = \sum_{k=0}^{\infty} \frac{H'(x)}{k!} w^k$$

$$\frac{dg}{dx} = e^{2xw - w^2}(2w) = 2w\sum_{k=0}^{\infty}\frac{H_k(x)}{k!}w^k = 2w\sum_{k=0}^{\infty}\frac{H_k(x)}{k!}w^{k+1} \qquad (12.75)$$

Setting $k = k - 1$,

$$2\sum_{k=0}^{\infty}\frac{H_k(x)}{k!}w^{k+1} \Rightarrow 2\sum_{k=0}^{\infty}\frac{H_{k-1}(x)}{(k-1)!}w^k$$

$$2\sum_{k=0}^{\infty}\frac{H_{k-1}(x)}{(k-1)!}w^k = \sum_{k=0}^{\infty}\frac{H_k^{'}(x)}{k!}w^k$$

$$\frac{2H_{k-1}(x)}{(k-1)!} = \frac{H_k^{'}(x)}{k!}$$

$$H_k^{'}(x) = \frac{2H_{k-1}(x)k!}{(k-1)!} = 2H_{k-1}(x) \tag{12.76}$$

Let us now prove the differential equation

$$H''_k(x) - 2xH'_k(x) + 2kH_k(x) = 0 \tag{12.77}$$

Starting from $H_{k+1}(x) = (2x)H_k(x) - H'_k(x)$ and differentiating both sides, gives successively:

$$H'_{k+1}(x) = 2xH'_k(x) + 2H_k(x) - H''_k(x)$$
$$H'_k(x) = 2kH_{k-1}(x)$$
$$H'_{k+1}(x) = 2(k+1)H_k(x)$$
$$2(k+1)H_k(x) = 2xH'(x) + 2H_k(x) - H''_k(x)$$
$$H''_k(x) - 2xH'_k(x) + 2H_k(x) = 0 \tag{12.78}$$

12.5.3 ORTHOGONALITY CONDITION

$H_k(x)$ is orthogonal when integrated over the interval $(-\infty, \infty)$, after being weighted by the function, $w(x) = e^{-x^2}$; therefore,

$$\int_{-\infty}^{\infty} H_m(x)H_k(x)w(x)dx = 0$$

$$\text{when } m \neq k \tag{12.79}$$

Proof of the Orthogonality

$$\int_{-\infty}^{\infty} w(x)H_k(x)H_m(x)dx = \int_{-\infty}^{\infty} e^{-x^2}(-1)^k\frac{d^k}{dx^k}\left(e^{-x^2}\right)e^{x^2}H_m(x)dx$$

$$= \int_{-\infty}^{\infty} w(x)H_k(x)H_m(x)dx \tag{12.80}$$

Integrating by parts gives

$$\int_{-\infty}^{\infty} w(x)H_k(x)H_m(x)dx$$

$$= (-1)^k\frac{d^{k-1}}{dx^{k-1}}\left(e^{-x^2}\right)H_m(x)\bigg|_{-\infty}^{\infty} - (-1)^k\int_{-\infty}^{\infty}\frac{d^{k-1}}{dx^{k-1}}\left(e^{-x^2}\right)\frac{d}{dx}H_m(x)dx$$

$$(-1)^k \frac{d^{k-1}}{dx^{k-1}}\left(e^{-x^2}\right)H_m(x)\bigg|_{-\infty}^{\infty} = 0$$

$$\int_{-\infty}^{\infty} w(x)H_k(x)H_m(x)dx = -(-1)^k \int_{-\infty}^{\infty} \frac{d^{k-1}}{dx^{k-1}}\left(e^{-x^2}\right)\frac{d}{dx}H_m(x)dx$$

$$= (-1)^k \frac{d^{k-2}}{dx^{k-2}}\left(e^{-x^2}\right)H_m(x)\bigg|_{-\infty}^{\infty} + \int_{-\infty}^{\infty} \frac{d^{k-2}}{dx^{k-2}}\left(e^{-x^2}\right)\frac{d^2}{dx^2}H_m(x)dx$$

$$(-1)^k \frac{d^{k-2}}{dx^{k-2}}\left(e^{-x^2}\right)H_m(x)\bigg|_{-\infty}^{\infty} = 0 \tag{12.81}$$

This leads to the following relations:

$$\int_{-\infty}^{\infty} w(x)H_k(x)H_m(x)dx = \int_{-\infty}^{\infty} \frac{d^{k-2}}{dx^{k-2}}\left(e^{-x^2}\right)\frac{d^2}{dx^2}H_m(x)dx \tag{12.82}$$

$$\int_{-\infty}^{\infty} w(x)H_k(x)H_m(x)dx = \int_{-\infty}^{\infty} \frac{d^{k-m-1}}{dx^{k-m-1}}\left(e^{-x^2}\right)\frac{d^{m+1}}{dx^{m+1}}H_m(x)dx$$

$$\frac{d^{m+1}}{dx^{m+1}}H_m(x)dx = 0, \text{ proving}$$

$$\int_{-\infty}^{\infty} w(x)H_k(x)H_m(x)dx = 0\# \tag{12.83}$$

The Mathematica command for the fifth Hermite polynomial is given by

 HermiteH[5,x]

 $120x'160x^3 + 32x^5$.

12.5.4 Graph of the Third, Fourth and the Fifth Hermite Polynomials

Plot[{HermiteH[3,x],HermiteH[4,x],HermiteH[5,x]},{x,1,1},AxesLabel{x,H},PlotLegend{"H[3,x]", "H[4,x]","H[5,x]"},LegendPosition{1.1,'0.4}].

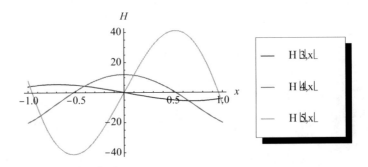

FIGURE 12.8 Graph of the third, fourth and fifth, Hermite polynomials.

12.6   THE GAMMA FUNCTION AND THE BETA FUNCTION

12.6.1   THE GAMMA FUNCTIONS

12.6.1.1   Definition

The Gamma function was introduced by the Swiss mathematician Leon-Hard Euler and studied by Adrian-Marie Legendre, Gauss, Joseph Liouville, Hermite and others. It is defined in the sense of Euler in 1930 by the function:

$$\Gamma(x) = \int [-\log(t)]^{x-1} dt, \ \ for \ x > 0 \tag{12.84}$$

or

$$\Gamma(x) = \int_0^\infty t^{x-1} e^{-t} dt \tag{12.85}$$

or

$$\Gamma(x) = 2\int^\infty t^{2x-1} e^{-t^2} dt. \tag{12.86}$$

For $x = 1$, $\Gamma(1) = \int_0^\infty e^{-t} dt = 1$.

For $x > 0$ and proceeding by integration by parts, we can prove that

$$\Gamma(x+1) = \int_0^\infty t^x e^{-t} dt = \left[-t^{x-1} e^{-t} dt = x\Gamma(x). \right. \tag{12.87}$$

For x integer, say $x = n =$ integer, the Γ function leads to the definition of the factorial function, that is $\Gamma(n+1) = n$!
  For $n = 0$, $\Gamma(1) = 0! = 1$
  One property of the Γ function is that

$$\Gamma(x) = \frac{\Gamma(x+n)}{x(x+1)...(x+n-1)} \qquad x+n > 0. \tag{12.88}$$

12.6.1.2   Generalization of the Gamma function: Integral Representation

The integral representation of the Gamma function is given by the expression:

$$\Gamma(z) = \int_0^\infty t^{z-1} e^{-t} dt = (z-1)! \tag{12.89}$$

The following definitions are given by Euler in 1729 and by Gauss in 1811: For $x > 0$, the function $\Gamma_p(x)$ can be defined by

$$"_p(x) = \frac{p! p^x}{x(x+1)...(x+p)} = \frac{p^x}{x(1+x/1)...(1+x/p)} \tag{12.90}$$

Also, $\Gamma(x) = \lim_{p \to \infty} \Gamma_p(x)$.

Theorem 1 (of Weierstrass)

For all real number, x, except the negative integer, $0,'1,'2,....$, $\dfrac{1}{\Gamma(x)}$ can be expressed as

$$\frac{1}{\Gamma(x)} = xe^{\gamma x} \prod_{p=1}^\infty \left[\left(1 + \frac{x}{p} \right) e^{-x/p} \right] \tag{12.91}$$

where γ is the Euler–Mascheroni constant defined by

$$\gamma = \lim_{p \to \infty} \left(1 + \frac{1}{2} + \ldots + \frac{1}{p} - \log(p) \right) = 0.57721 \tag{12.92}$$

A better approximation of γ is $\gamma = \frac{1}{2}\left(\sqrt[3]{10} - 1 \right) = 0.5772173$.

Gamma function can be represented in the plane as follows:

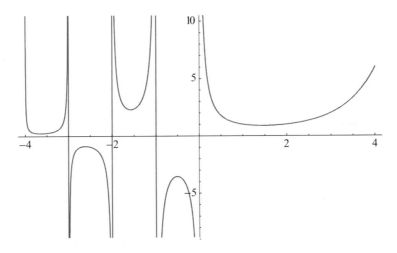

FIGURE 12.9   Graph of the Gamma function.

12.6.1.3   Derivatives of the Gamma Functions

The first derivative of the Gamma function can be derived by

$$\Gamma'(x) = \int_0^\infty t^{x-1} t^{x-1} e^{-t} \log(t)\,dt \tag{12.93}$$

The nth derivative of the Gamma function is expressed by

$$\Gamma^{(n)}(x) = \int_0^\infty t^{x-1} e^{-t} \log^n(t)\,dt \tag{12.94}$$

12.6.1.4   Properties of the Gamma Function

One of the key properties of the Gamma function relates $\dfrac{1}{\Gamma(x)}$ and $\dfrac{1}{\Gamma(-x)}$, and is given by the following relation:

$$\frac{1}{\Gamma(x)}\frac{1}{\Gamma(-x)} = -x^2 e^{\gamma x} e^{-\gamma x} \prod_{p=1}^{\infty}\left[\left(1 + \frac{x}{p}\right) e^{-x/p}\left(1 - \frac{x}{p}\right) e^{x/p}\right] \tag{12.95}$$

12.6.1.5   Absolute Value of the Gamma Function in the Complex Plane

The following command is used to represent in Mathematica the graph of the Absolute value of the Gamma function in the complex plane:

Plot3D[Abs[Gamma[$x + I\,y$]],{x,'5,2},{y,'0.5,0.5}].

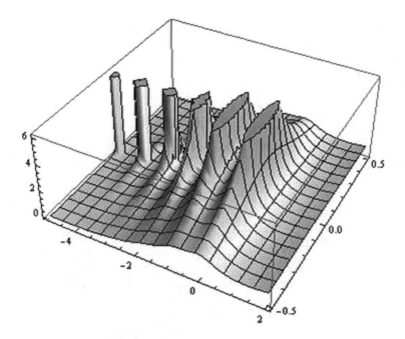

FIGURE 12.10 Graph of the absolute value of the Gamma function in the complex plane.

12.6.1.6 The Incomplete Gamma Functions

The incomplete Gamma function, $\gamma(z,x)$, is defined in the relation:
$\Gamma(z) = \gamma(z,x) + \Gamma(z,x)$ and is written as follows:

$$\gamma(z,x) = \int^x e^{-t} t^{z-1} dt \qquad x > 0$$

$\Gamma(z,x)$ is defined: $\Gamma(z,x) = \int_t^\infty e^{-t} t^{z-1} dt \quad x > 0$

12.6.2 THE BETA FUNCTIONS

The Beta function is defined by the integral of $t^{y-1}(1-t)^{z-1}$ in the range [0,1]:

$$B(y,z) = \int^b t^{y-1}(1-t)^{z-1} dt = \frac{\Gamma(y)\Gamma(z)}{\Gamma(z+y)} \tag{12.96}$$

12.6.2.1 The Incomplete Beta Functions

For the incomplete Beta function x is not fixed and varies between 0 and 1.

$$B_x(y,z) = \int_0^x t^{y-1}(1-t)^{z-1} dt \qquad 0 < x < 1 \tag{12.97}$$

The regularized normalized form of the incomplete Beta function is given by

$$I_x(y,z) = \frac{B_x(y,z)}{B(y,z)} \tag{12.98}$$

Beta[5,4] = 1/280
Beta[1,1] = 1

12.6.2.2 Graph of 3-D Beta Function for Real Positive Values

The Graph of 3-D Beta function for real positive values can be done by using the Mathematica command:

Plot3D[Beta[*a*,*b*],{*a*,0,2},{*b*,0,2}]

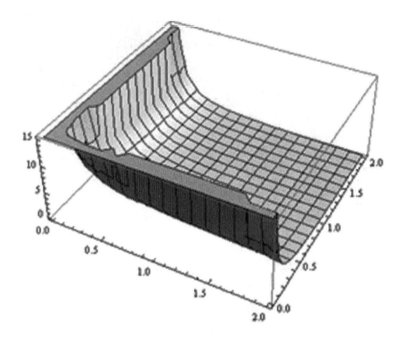

FIGURE 12.11   Graph of 3-D Beta function for real positive values.

Plot3D[Abs[Beta[*x* + *Iy*,′5.1]],{*x*,′10,4},{*y*,′5,5}]

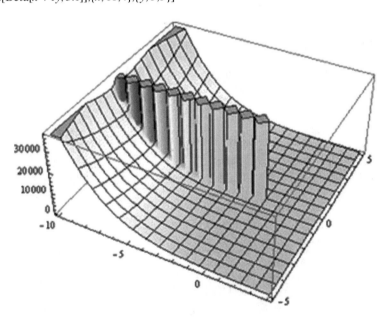

FIGURE 12.12   Graph of the absolute value of Beta in the complex plane.

12.6.3 THE FACTORIAL FUNCTION

The factorial function is defined for integers by

$$n! = \begin{cases} 1.2 \cdot 3 \ldots (n-1) \cdot n & n = 1,2,3\ldots \\ 1 & n = 0 \end{cases} \tag{12.99}$$

Pochhammer's polynomial is given by

$$(z)_n = \begin{cases} z.(z+1)(z+2)\ldots(z+n-1) = \dfrac{\Gamma(z+n)}{\Gamma(z)} & n > 0 \\ \qquad\qquad = \dfrac{(z+n-1)!}{(z-1)!} & \\ 1 = 0! & n = 0 \end{cases} \tag{12.100}$$

where z is a complex number.

The Gamma function is also defined for z complex by

$$\Gamma(z) = (z-1)! \tag{12.101}$$

$$\text{Plot3D}\left[\text{Abs}\left[\text{Factorial}\left[x + Iy\right]\right], \{x, -8, 1\}, \{y, -1.5, 1\}\right]$$

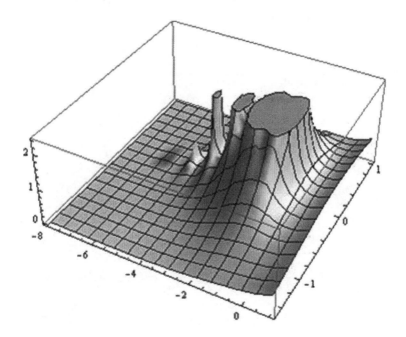

FIGURE 12.13 Graph of the absolute value of the factorial function of $x + Iy$.

12.6.3.1 Double Factorial

Until now, single factorial has been defined; but double factorial is also important and is given in terms of the values of the integer n.

$$n!! = \begin{cases} n(n-2)(n-4)\ldots.5.3.1 & n > 0 \ n \text{ odd} \\ n(n-2)(n-4)\ldots.6.4.2 & n > 0 \ n \text{ even} \\ 1 \ n = -1,0 \end{cases} \qquad (12.102)$$

Using the software Mathematica, the following commands give a list of the double factorial of n when n varies from 1 to 10 and from 10 to 20.

Table[$n!!$,{n,10}]
{1,2,3,8,15,48,105,384,945,3840}

Table[$n!!$,{n,10,20}]
{3840,10395,46080,135135,645120,2027025,10321920,34459425,185794560,654729075,3715891200}

12.6.3.2   Identity
The identity relation is given by

$$n! = n!(n-1)!! \qquad (12.103)$$

12.7   HYPERGEOMETRIC FUNCTION

12.7.1   GAUSS'S HYPERGEOMETRIC FUNCTION

Gauss's hypergeometric function is of great importance in mathematical physics and participates in the construction of Daubechies and Chebyshev-associated polynomials. Gauss's hypergeometric function is the solution to the differential equation:

$$z(z-1)F'' + \left[(a+b+1)z-c\right]F' + abF = 0 \qquad (12.104)$$

We will assume holomorphic solutions about $z = 0$. These solutions are on the form:

$$F\left(a,b|c|z\right) = \sum_{n=0}^{\infty} \alpha_n z^n. \qquad (12.105)$$

Let us now differentiate F twice; we get

$$F'\left(a,b|c|z\right) = \sum_{n=0}^{\infty} n\alpha_n z^{n-1} = \sum_{n=0}^{\infty} (n+1)\alpha_{n+1} z^n \qquad (12.106)$$

$$F''\left(a,b|c|z\right) = \sum_{n=0}^{\infty} n(n-1)\alpha_n z^{n-2} = \sum_{n=0}^{\infty} (n+2)(n+1)\alpha_{n+2} z^n \qquad (12.107)$$

We now substitute these derivatives back into the differential equation (12.104). We will then have

$$\sum_{n=0}^{\infty} n(n-1)\alpha_n - n(n+1)\alpha_{n+1} + (1+a+b)n\alpha_n - c(n+1)\alpha_{n+1} + ab\alpha_n \Big] z^n \equiv 0 \qquad (12.108)$$

Equating the coefficient to zero leads us to obtain,

$$(n+1)(n+c)\alpha_{n+1} = (n+a)(n+b)\alpha_n \tag{12.109}$$

Consider the first term $\alpha_0 = 1$, then

$$\alpha_1 = \frac{ab}{c}\alpha_0 = \frac{ab}{c} \tag{12.110}$$

$$\alpha_{n+1} = \frac{(n+a)(n+b)}{(n+c)(n+1)}\alpha_n \tag{12.111}$$

which can also be put into the Gamma form:

$$\alpha_n = k\frac{``(n+a)``(n+b)}{``(n+c)n!} \tag{12.112}$$

where k is a constant and can be found by using the normalization condition, $\alpha_0 = 1: k = \frac{\Gamma(c)}{\Gamma(a)\Gamma(b)}$. Therefore,

$$\alpha_n = \frac{\Gamma(c)}{\Gamma(a)\Gamma(b)}\frac{\Gamma(n+a)\Gamma(n+b)}{\Gamma(n+c)n!} \tag{12.113}$$

The solution (12.105) can be written in the form:

$$F(a,b|c|z) = 1 + \frac{a(a+1)b(b+1)}{2!c(c+1)}z^2 + \cdots$$

$$= \frac{\Gamma(c)}{\Gamma(a)\Gamma(b)}\sum_{n=0}^{\infty}\frac{\Gamma(n+a)\Gamma(n+b)}{\Gamma(n+c)n!}z^n \tag{12.114}$$

The series represented by equation (12.114) does not converge for $c = 0$, or $c = ´1, ´2, \ldots$, that is, for c equals a negative integer. Also, the series will terminate for a finite number of terms if a or b is a negative integer.

The hypergeometric function $F(a,b|c|z) = \sum_{n=0}^{\infty}\alpha_n z^n$ is also represented by the series $2F1[a,b,c,z]$. Using Mathematica software, we can write the following command, Series XE "Series" $\left[\text{Hypergeometric2F1}[a,b,c,x],\{x,0,3\}\right]$, to retrieve the series to the third order.

Series[Hypergeometric2F1[a,b,c,x],{x,0,3}]

1 + (a b x)/c + (a (1 + a) b (1 + b) x²)/(2 c (1 + c)) + (a (1 + a) (2 + a) b (1 + b) (2 + b) x³)/(6 c (1 + c) (2 + c)) + O[x]⁴

$$\text{Plot}\Big[\text{Hypergeometric2F1}\big[1/3,1/3,2/3,x\big],\{x,-1,1\}\Big]$$

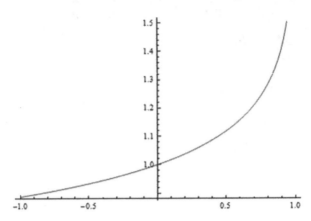

FIGURE 12.14   Graph of the hypergeometric function $F(x)$ for $a = 1/3$, $b = 1/3$, $c = 2/3$ with x ranging between -1 and $+1$.

$$\text{Plot}\Big[\text{Hypergeometric2F1}\big[1/3,1/3,2/3,x\big],\{x,-1,1\}\Big]$$

$$\text{Plot}\Big[\text{Hypergeometric2F1}\big[1,1,1,x\big],\{x,-100,100\}\Big]$$

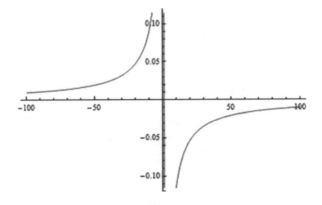

FIGURE 12.15   Graph of the hypergeometric function $F(x)$ for $a = 1$, $b = 1$, $c = 1$ with x ranging between -100 and $+100$.

$$\text{Plot}\Big[\text{Hypergeometric2F1}\big[1,1,1,x\big],\{x,-100,100\}\Big].$$

$$\text{Plot}\Big[\text{Hypergeometric2F1}\big[2,1,1,x\big],\{x,-100,100\}\Big]$$

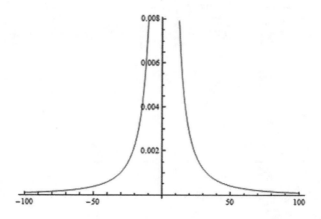

FIGURE 12.16 $\text{Plot}\Big[\text{Hypergeometric2F1}\big[2,1,1,x\big],\{x,-100,100\}\Big].$

12.7.2 THE KUMMER CONFLUENT HYPERGEOMETRIC FUNCTION

The Kummer confluent hypergeometric function is given by $1\text{F}1\big[a,b,z\big]$
    The change of variable $z \to z/b$ in the hypergeometric equation (12.104) gives

$$bz\left(\frac{z}{b}-1\right)F'' + \left[(a+b+1)\frac{z}{b}-c\right]bF' + abF = 0 \tag{12.115}$$

By simplifying by b and taking the limit $b \to \infty$, we obtain the differential equation associated with the confluent hypergeometric function:

$$zF''\big(a|c|z\big) + (c-z)F'\big(a|c|z\big) - aF\big(a|c|z\big) = 0 \tag{12.116}$$

12.7.2.1 Integral Representation

$$
\begin{aligned}
F\big(a|c|z\big) &= \lim_{b\to\infty} F\big(a,b+c|c|z/b\big) \\
&= \lim_{b\to\infty}\frac{\Gamma(c)}{\Gamma(a)\Gamma(c-a)}\int (t-z/b)^{-b-c}t^{b}(t-1)^{c-a-1}dt \\
&= \lim_{b\to\infty}\frac{\Gamma(c)}{\Gamma(a)\Gamma(c-a)}\int 1-zt/b)^{-b-c}t^{a-1}(1-t)^{c-a-1}dt \\
&= \lim_{b\to\infty}\frac{\Gamma(c)}{\Gamma(a)\Gamma(c-a)}\int (1-zt/b)^{-(b/zt)zt-c}t^{a-1}(1-t)^{c-a-1}dt \\
&= \lim_{b\to\infty}\frac{\Gamma(c)}{\Gamma(a)\Gamma(c-a)}\int e^{zt}t^{a-1}\big(1-t\big)^{c-a-1}dt \\
&= \text{Re}(c) > \text{Re}(a) > 0
\end{aligned}
\tag{12.117}
$$

With Mathematica software, the command,
Series[Hypergeometric1F1[a, b, x],{x,0,3}]
gives:

$$1 + (ax)/b + (a(1 + a)x^2)/(2b (1 + b)) + (a(1 + a)(2 + a)x^3)/(6b (1 + b) (2 + b)) + O[x]^4$$

Graph of confluent hypergeometric functions:

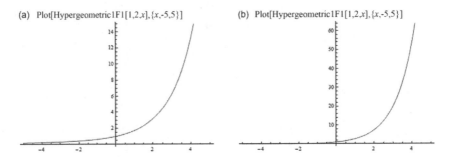

(a) Plot[Hypergeometric1F1[1,2,x],{x,-5,5}]

(b) Plot[Hypergeometric1F1[1,2,x],{x,-5,5}]

FIGURE 12.17   Graph of some confluent hypergeometric functions.

PROBLEM SET 12

12.1   Prove the following two equations

$$\frac{d}{dx}\left[x^{v}J_{v}(x)\right] = x^{v}J_{v-1}(x)$$

$$\frac{d}{dx}\left[x^{-v}J_{v}(x)\right] = -x^{-v}J_{v+1}(x)$$

12.2   Prove the following integral representation form of Bessel's function:

$$J_{v}(x) = \frac{2}{\sqrt{\pi}\,(v-1/2)!}\left(\frac{x}{2}\right)^{v}\int_{0}^{\pi/2}\cos(x\sin\theta)\cos^{2v}\theta\,d\theta \text{ , where } v > -\frac{1}{2}.$$

12.3   Prove Property P_1
    1.   Prove Property P_2 (Use the generating function)
    2.   Prove Property P_3
    3.   Prove Property P_4
    4.   Prove Property P_5

12.4   Using the software Mathematica command, plot the Bessel function of first kind $J_n(x)$, for $n = 5$ and for $n = 6$. Make a comparison.

12.5   Derive the following expression of Bessel's function:

$$J_n(x) = (-1)^n x^n \left(\frac{1}{x}\frac{d}{dx}\right)^n J_0(x)$$

12.6   Show that an Hermite polynomial of degree n, $H_n(x)$, is solution to the differential equation, $\dfrac{d^2}{dx^2}H_n(x) - 4\pi x \dfrac{d}{dx}H_n(x) + 4\pi n H_n(x).$

Infer the Hermite functions $h_n(x)$ are the eigenfunctions of the operators,

$$\frac{d^2}{dx^2} - 4\pi x^2 \frac{d}{dx}, \text{ because } \frac{d^2}{dx^2} h_n(x) - 4\pi x \frac{d}{dx} h_n(x) - \delta_n h_n(x).$$

a. What are the corresponding eigenvalues?
b. What is the Fourier transformation of the above operator? Deduce that Hermite functions are orthogonal and invariant (at a constant factor) by Fourier transformation.

12.7 Consider the hypergeometric equation

$$z(z-1)u''(z) + \left[(a+b+1)z - c\right]u'(z) + abu(z) = 0 \equiv L_z\left[u\right]$$

Where a, b and c are real parameters such that $c > b > 0$. Construct an integral representation for a solution as

$$u(z) = \int_C (z-t)^\mu w(t)dt$$

where the integral path C, the index μ, and the function $w(t)$, must be determined as follows:

a. Chose the index μ such that

$$L_z\left[u\right] \equiv \int_t L_z\left[(z-t)^\mu w(t)\right]dt = \int_t \ell_t\left[(z-t)^\mu w(t)\right]dt$$

where $\ell_t = \alpha(t)\dfrac{d^2}{dt^2} + \beta(t)\dfrac{d}{dt}$

that is, you must determine $\mu, \alpha(t), \beta(t)$ to make the last equality hold.

b. Now integrate by part to obtain

$$\int L_z\left[u\right] = [\text{OUT INTEGRATED TERMS}]_{\text{ends of } C} + \int(z-t)^\mu M_t\left[w(t)\right]dt = 0$$

and set $M_t\left[w(t)\right] = 0$
to solve for nonzero $w(t)$

12.8 Prove that

$$F\left(a|c|z\right) = \lim_{b\to\infty} F\left(a, b+c|c|z/b\right)$$

$$= \lim_{b\to\infty} \frac{\Gamma(c)}{\Gamma(a)\Gamma(c-a)} \int e^{zt} t^{a-1}(1-t)^{c-a-1}dt$$

$$\text{Re}(c) > \text{Re}(a) > 0$$

12.9 Problem: Prove equation 12.7.

13 Green's Function

Donnez-moi un point fixe et un levier et je soulèverai la Terre.

–Archimède

INTRODUCTION

Green's functions are often used in physics, engineering and related sciences. In this chapter, we define Green's function and delta Dirac function and we give some applications.

13.1 DEFINITION

The concept of Green's function was developed for the first time by the British mathematician George Green in the 1830s. Green's function is a type of function used to solve inhomogeneous differential equations subject to specific initial conditions or boundary conditions. The Green functions are of foremost importance in physics, especially in classical and quantum physics. They participate in many ways in relativistic and nonrelativistic quantum physics. We would introduce Green's function by evoking the momentum operator in quantum mechanics represented by $p \equiv -i\nabla$. In one dimension, we can write, $p_x \equiv -i\dfrac{d}{dx}$.

Consider now that we are trying to solve the following ordinary differential equation:

$$-i\frac{dy}{dx} = f(x) \qquad (13.1)$$

Therefore,

$$dy = if(x)dx$$

$$[y]_{y_0}^{y} = i\int_t^x f(x')dx'$$

For all $x' \in [a,x]$ and the boundary condition, $f(a) = y_o$.

We can find

$$y(x) = y_0 + i\int_i^k f(x')dx' \qquad (13.2)$$

Let us now assume that $x \in [a,b]$ and introduce the Heaviside function $\theta(x)$ defined as the following step function:

$$\theta(x) = \begin{cases} 1 & x > 0 \\ 0 & x < 0 \end{cases} \qquad (13.3)$$

DOI: 10.1201/9781003478812-13

The graph of the Heaviside function is the following:

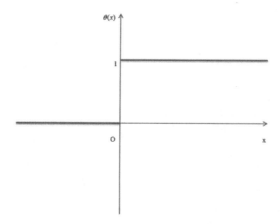

FIGURE 13.1 Heaviside function representing a step function with only two values: 0 and 1.

The Heaviside function can help rewrite the solution $y(x)$ of the ordinary differential equation (13.1) as follows:

$$y(x) = y_0 + i\int_a^b \theta(x - x') f(x') dx' \tag{13.4}$$

For $x = x'$, $x - x' = 0$, $\theta(0) = 1$ and we can think of the integral as on the form:

$$y(x) = y_0 + Kf(x) \tag{13.5}$$

where K is defined by

$$Kf(x) = i\int_i^k \theta(x - x') f(x') dx' \tag{13.6}$$

$i\theta(x - x')$ is the kernel of the integral operator, K.

When the kernel is from the solution of an equation involving a differential operator, it is often referred to as Green's function of that differential. In this case, the differential operator is the operator, $i\dfrac{d}{dx}$. We can then write the expression of the Green function related to $i\dfrac{d}{dx}$ by $G_1(x, x') \equiv i\theta(x - x')$, with the boundary condition, $y(a) = y_0$.

13.1.1 EXAMPLE OF THE SECOND-ORDER DIFFERENTIAL EQUATION

Consider the second-order differential equation,

$$\frac{d^2 y}{dx^2} = f(x) \tag{13.7}$$

with the boundary conditions $y(a) = y_0, y'(a) = \tilde{y}_0$

The first integration of equation (13.7) gives

$$\frac{dy}{dx} = \int_a^x f(x) dx + y'(a) \tag{13.8}$$

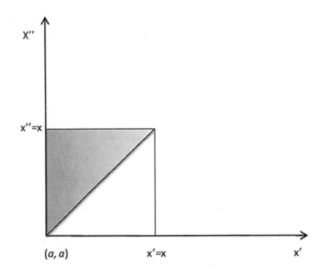

FIGURE 13.2 Integration figure of a second-order differential equation.

$$y = \int_i^b \left[\int_i^{x''} f(x') dx' \right] dx'' + \int_i^b y'(a) dx'' + y(a) \tag{13.9}$$

$$y = \int_i^r dx'' \int_i^{x''} f(x') dx' \right] dx' + \tilde{y}_0 (x-a) + y_0 \tag{13.10}$$

Consider now the figure:

Let us integrate over the shaded region, from $x' = a$ to $x' = x$ and from $x'' = a$ to $x'' = x$. As a result, one obtains by interchanging the order of limits of integration,

$$y(x) = y_0 + \tilde{y}_0 (x-a) + \int_i^b dx' f(x') \int_u^b dx'' \tag{13.11}$$

or

$$y(x) = y_0 + \tilde{y}_0 (x-a) + \int_i^r (x-x') f(x') dx' \tag{13.12}$$

By restricting x to the interval, $[a,b]$, we can write

$$y(x) = y_0 + \tilde{y}_0 (x-a) + \int_l^b (x-x') \theta(x-x') f(x') dx' \tag{13.13}$$

Now, by analogy with what was done with Green's function G_1, we can introduce a new Green's function G_2, defined by $G_2(x,x') = (x-x')\theta(x-x')$ which corresponds to the operator $\dfrac{d^2}{dx^2}$ for the boundary conditions, $y(a) = y_0$ and $y'(a) = \tilde{y}_0$. It is important to recall that $G_2(x,x')$ is continuous, while $G_1(x,x')$ is discontinuous at $x = x'$.

13.2 GREEN'S FUNCTION AND DELTA DIRAC FUNCTION

Consider the equation, $Ly = f$, with L some linear ordinary differential operator and f a given function of x. Suppose now that L has a complete set of eigenfunctions, $\{\phi_n(x)\}$, such that

$$L\phi_n(x) = \lambda_n \phi_n(x) \tag{13.14}$$

with λ_n the eigenvalues of L whose eigenfunctions are, respectively, $\phi_n(x)$. By analogy, we can write $y(x)$ and $f(x)$ as linear combinations of $\phi_n(x)$, that is:

$$y(x) = \sum_{n=1}^{\infty} \alpha_n \phi_n(x) \text{ and } f(x) = \sum_{n=1}^{\infty} \beta_n \phi_n(x) \tag{13.15}$$

If we substitute $y(x)$ and $f(x)$, separately, in the equation $Ly = f$, we get

$$L \sum_{n=1}^{\infty} \alpha_n \phi_n(x) = \sum_{n=1}^{\infty} \beta_n \phi_n(x) \tag{13.16}$$

$$\sum_{n=1}^{\infty} \alpha_n L \phi_n(x) = \sum_{n=1}^{\infty} \beta_n \phi_n(x) \tag{13.17}$$

$$\sum_{n=1}^{\infty} \alpha_n \lambda_n \phi_n(x) = \sum_{n=1}^{\infty} \beta_n \phi_n(x) \tag{13.18}$$

$$\sum_{n=1}^{\infty} (\alpha_n \lambda_n - \beta_n) \phi_n(x) = 0 \tag{13.19}$$

The functions $\phi_n(x)$ form a complete set of eigenfunctions; they are linearly independent and, as a result, are elements of the Hilbert space. The previous equation implies the identity:

$\alpha_n \lambda_n - \beta_n = 0$; so $\alpha_n = \dfrac{\beta_n}{\lambda_n}$ and $y(x)$ is given by

$$y(x) = \sum_{n=1}^{\infty} \frac{\beta_n}{\lambda_n} \phi_n(x) \text{ with } \lambda_n \neq 0 \text{ and } \beta_n = (\phi_n, f); \text{ we then have}$$

$$y(x) = \sum_{n=1}^{\infty} \frac{1}{\lambda_n} (\phi_n, f) \phi_n(x) = \sum_{n=1}^{\infty} \frac{1}{\lambda_n} \phi_n(x) \int \phi_n{}^*(x'), f(x') dx' = \int G(x,x') f(x') dx' \tag{13.20}$$

with $G(x,x') = \sum_{n=1}^{\infty} \dfrac{1}{\lambda_n} \phi_n(x) \phi_n{}^*(x')$, Green's function corresponding to the linear ordinary differential operator L.

If λ is real then $G(x,x')$ is symmetric, that is $G(x,x') = G(x',x)^*$.

Example

Consider the operator L as the second-order derivative; that is, $L = \dfrac{d^2}{dx^2}$, for $x \in [0,1]$. The normalized eigenfunctions of L on $[0,1]$ which vanish at 0 and 1 are $\phi_n(x) = \sqrt{2}\sin(n\pi x)$, $n = 1, 2, \ldots$ with the eigenvalues $\lambda = -n^2\pi$.

In fact,

$$\frac{d\phi_n(x)}{dx} = \sqrt{2}n\pi\cos(n\pi x), \tag{13.21}$$

and

$$\frac{d^2\phi_n(x)}{dx^2} = -\sqrt{2}n^2\pi^2\sin(n\pi x) = -n^2\pi^2\phi_n(x) = \lambda_n \phi_n(x). \tag{13.22}$$

So, Green function in this case is

$$G(x,x') = \frac{2}{\pi^2} \sum_{n=1}^{\infty} \frac{\sin(n\pi x)\sin(n\pi x')}{n^2} \phi_n(x)\phi_n^*(x') \tag{13.23}$$

which is a uniformly convergent series to a continuous function. We can now calculate $G(x,x')$ at a few points, for example, $x = x' = \dfrac{1}{2}$, then

$$G(1/2,1/2) = -\frac{2}{\pi^2} \sum_{n=1}^{\infty} \frac{1}{n^2} = -\frac{2}{\pi^2} \frac{\pi^2}{8} = -\frac{1}{4} \tag{13.24}$$

Let us now reconsider Green's function,

$$G(x,x') = \sum_{n=1}^{\infty} \frac{1}{\lambda_n} \phi_n(x)\phi_n^*(x') \tag{13.25}$$

and operate L on it. This leads to

$$LG(x,x') = L \sum_{n=1}^{\infty} \frac{1}{\lambda_n} \phi_n(x)\phi_n^*(x') \tag{13.26}$$

$$LG(x,x') = \sum_{n=1}^{\infty} \frac{L\phi_n(x)\phi_n^*(x')}{\lambda_n} = \sum_{n=1}^{\infty} \phi_n(x)\phi_n^*(x') = I(x,x') \tag{13.27}$$

Also, for all $f(x)$, and with $\{\phi_n\}$ forming a complete set of orthonormal functions, we have

$$\int I(x,x')f(x')dx' = \int \sum \phi_n(x)\phi_n^*(x')f(x')dx' \tag{13.28}$$

$$\int I(x,x')f(x')dx' = \sum \phi_n(x)\int \phi_n^*(x')f(x')dx' \tag{13.29}$$

$$\int I(x,x')f(x')dx' = \sum_{n=1}^{\infty} \phi_n(x)(\phi_n, f) = f(x). \tag{13.30}$$

$\int I(x,x')f(x')dx'$ is the property obtained with the δ – Dirac function, and as a result, the function $I(x,x')$ can be identified as the δ – Dirac function: $(x,x') = \delta(x-x')$; we can also write $LG(x,x') = \delta(x-x')$, which could also be written as $LK = \delta I$, with K an integral operator defined by

$$Kf(x) = \int G(x,x')f(x')dx' \tag{13.31}$$

If the operator L is invertible, then $Ly = f$ implies that

$$y = L^{-1}f = Kf = \int G(x,x')f(x')dx' \tag{13.32}$$

Now, using Green's function, $G_1(x,x')$, we have

$$-i\frac{d}{dx}G_1(x,x') = \delta(x-x') \tag{13.33}$$

But, $G_1(x,x') = +i\theta(x-x')$, which implies that

$$\frac{d}{dx}\theta(x-x') = \delta(x-x'). \tag{13.34}$$

with the change of variable $t = x - x'$; $dt = dx$, we have $\dfrac{d}{dx}\theta(t) = \delta(t)$.

13.3   PROPERTIES

$$\forall t \in \left[-\varepsilon, \varepsilon\right] \int_{-\varepsilon}^{\varepsilon} \frac{d}{dt}\theta(t)\,dt = [\theta(t)]_{-\varepsilon}^{\varepsilon} = \theta(\varepsilon) - \theta(-\varepsilon) = 1 - 0 = 1 \qquad (13.35)$$

which is equivalent to δ – Dirac property, $\int_{-\varepsilon}^{\varepsilon}\delta(t)\,dt = 1$.

Let us now apply $\dfrac{d}{dx}\theta(t)$ to a function, $f(t)$:

$\forall t \in \left[-\varepsilon, \varepsilon\right]$, let us calculate

$$I = \int_{-\varepsilon}^{\varepsilon} \frac{d}{dt}\theta(t)f(t)\,dt = \int_{-\varepsilon}^{\varepsilon} d\theta(t)f(t) = [\theta(t)f(t)]_{-\varepsilon}^{\varepsilon} - \int_{-\varepsilon}^{\varepsilon}\theta(t)f'(t)\,dt.$$

$$I = \theta(\varepsilon)f(\varepsilon) - \theta(-\varepsilon)f(-\varepsilon) - \int_{\varepsilon}^{\varepsilon}\theta(t)f'(t)\,dt$$

$$= f(\varepsilon) - \int_{\varepsilon}^{\varepsilon}\theta(t)f'(t)\,dt \qquad (13.36)$$

$$= f(\varepsilon) - [f(t)]_{0}^{\varepsilon}$$

$$= f(\varepsilon) - [f(\varepsilon) - f(0)] = f(0)$$

which is the other property of the δ – Dirac function, that is

$$\int_{-\varepsilon}^{\varepsilon}\delta(t)f(t)\,dt = f(0) \qquad (13.37)$$

Remember how we define the function $\theta(t)$:

$$(x) = \begin{cases} 1 & x > 0 \\ 0 & x \le 0 \end{cases} \qquad (13.38)$$

We can now define $\theta(at)$. Suppose $at = u$, then $t = \dfrac{u}{a} > 0$ for $a > 0$; so

$$\theta(at) = \theta(u) = 1 \equiv \theta(t). \qquad (13.39)$$

For $a < 0$, $t < 0$ and $\theta(t) = 0$. We can therefore rewrite $\theta(at)$ as follows:

$$\theta(at) = \begin{cases} \theta(t) & a > 0 \\ 1 - \theta(t) & a < 0 \end{cases} \qquad (13.40)$$

By differentiating, we obtain

$$\begin{cases} a\dfrac{d\theta(at)}{adt} = a\dfrac{d\theta(at)}{d(at)} = \dfrac{d\theta(t)}{d(t)} & a > 0 \\[3mm] \dfrac{d\theta(at)}{d(t)} = \dfrac{d(1-\theta(t))}{d(t)} = -\dfrac{d\theta(t)}{d(t)} & a < 0 \end{cases} \qquad (13.41)$$

13.4   GREEN'S FUNCTION IN ONE DIMENSION: EXAMPLE OF THE FORCED DAMPING HARMONIC OSCILLATION

Let us start by solving the linear ODE:

$$\ddot{x}(t) + 2\gamma \dot{x}(t) + \omega_o^2 x(t) = F(t) \tag{13.42}$$

where $F(t)$ represents the external driving force. The coefficient of the first term is of unit mass, $m = 1$. In this equation, γ represents the coefficient of the frictional force and ω_0^2 the spring constant. Assume that $F(t)$ has a Fourier transform,

$$\hat{F}(k) = \frac{1}{\sqrt{2\pi}} \int_{\infty}^{+\infty} e^{ikt} F(t) dt \tag{13.43}$$

For the Fourier transform to exist, we will consider that $F(t), x(t), \dot{x}(t), \ddot{x}(t)$ all tending to zero for sufficiently large t. Let us then apply $\frac{1}{\sqrt{2\pi}} e^{ikt} dt$ to both sides of equation (13.42) and take the integral. We have

$$\frac{1}{\sqrt{2\pi}} \left[\ddot{x}(t) + 2\gamma \dot{x}(t) + \omega_o^2 x(t) \right] e^{ikt} dt = \frac{1}{\sqrt{2\pi}} F(t) e^{ikt} dt \tag{13.44}$$

$$\frac{1}{\sqrt{2\pi}} \int_{\infty}^{+\infty} \left[\ddot{x}(t) + 2\gamma \dot{x}(t) + \omega_o^2 x(t) \right] e^{ikt} dt = \frac{1}{\sqrt{2\pi}} \int_{\infty}^{+\infty} F(t) e^{ikt} dt \tag{13.45}$$

After some integrations by parts and using the boundary conditions, we obtain that

$$\left[-k^2 - 2i\gamma k + \omega_0^2 \right] \hat{x}(k) = \hat{F}(k) \tag{13.46}$$

So,

$$\hat{x}(k) = \frac{\hat{F}(k)}{\left[-k^2 - 2i\gamma k + \omega_0^2 \right]} \tag{13.47}$$

The inverse Fourier transform of $\hat{x}(k)$ will give

$$x(t) = \frac{1}{\sqrt{2\pi}} \int_{-\infty}^{+\infty} \hat{x}(k) e^{-ikt} dk \tag{13.48}$$

or

$$x(t) = \frac{1}{\sqrt{2\pi}} \int_{-\infty}^{+\infty} \frac{\hat{F}(k)}{\left[-k^2 - 2i\gamma k + \omega_0^2 \right]} e^{-ikt} dk \tag{13.49}$$

But we already knew the expression of $\hat{F}(k)$; then,

$$x(t) = \frac{1}{2\pi} \int_{\infty}^{+\infty} \left\{ \frac{e^{-ikt}}{\left[-k^2 - 2i\gamma k + \omega_0^2 \right]} \int_{\infty}^{+\infty} e^{ikt'} F(t') dt' \right\} dk \qquad (13.50)$$

$$x(t) = \frac{1}{2\pi} \int_{\infty}^{\infty} F(t') dt' \left\{ \int_{\infty}^{+\infty} \frac{e^{-ik(t-t')}}{\left[-k^2 - 2i\gamma k + \omega_0^2 \right]} dk \right\} \qquad (13.51)$$

which could be written as

$$x(t) = \int_{\infty}^{\infty} G(t,t') F(t') dt' \qquad (13.52)$$

with the Green function, $G(t,t') = \frac{1}{2\pi} \int_{-\infty}^{+\infty} \frac{e^{-ik(t-t')}}{\left[-k^2 - 2i\gamma k + \omega_0^2 \right]} dk$.

The general solution will consider the solution of the homogeneous equation $\ddot{x}(t) + 2\gamma \dot{x}(t)$ $+\omega_0^2 = 0$ whose characteristics equation is $r^2 + 2\gamma r + \omega_0^2 = 0$ with the discriminant $\Delta' = \gamma^2 - \omega_0^2$.

If $\gamma^2 - \omega_o^2 < 0$, then $\Delta' < 0$ and there are two zeros:

$r = -\gamma \pm i\sqrt{\omega_0^2 - \gamma^2}$ and the general solution of the differential equation (13.42) will be written as

$$x(t) = Ae^{\left(-\gamma - i\sqrt{\omega_0^2 - \gamma^2} \right)t} + Be^{\left(-\gamma + i\sqrt{\omega_0^2 - \gamma^2} \right)t} + \int_{-\infty}^{+\infty} G(t,t') F(t') dt' \qquad (13.53)$$

A and B can be determined by the initial conditions.

13.5 GREEN'S FUNCTION IN THREE DIMENSIONS

Green's function can be calculated in three dimensions in a number of situations. The case we are going to use here is related to the Hamiltonian used in quantum mechanics. We will replace the potential energy by a given real number, λ. Therefore, the Hamiltonian will be given by

$$H_0 = \vec{\nabla}^2 + \lambda \qquad (13.54)$$

We consider the coefficient of $\vec{\nabla}^2$ to be 1 to simplify the expression of H_0.
    Consider the equation to be solved:

$$H\phi(r) = F(r) \qquad (13.55)$$

and assume that $\phi(r)$ and $F(r)$ are Fourier transformable. From their Fourier transform, we understand that the coefficient $\frac{1}{\sqrt{2\pi}}$ in one dimension becomes $\left(\frac{1}{\sqrt{2\pi}} \right)^3$ in three dimensions. We can then write successively:

$$\hat{F}(k) = \frac{1}{(2\pi)^{3/2}} \int_{\infty}^{+\infty} e^{-ikr} F(r) d^3r$$

$$\phi(k) = \frac{1}{(2\pi)^{3/2}} \int_{\infty}^{+\infty} e^{-ikr} \phi(r) d^3r$$

Let us now apply Fourier transform to both sides. We have

$$\int_{-\infty}^{+\infty}\vec{\nabla}^2 e^{-ikr}\phi(r)d^3r+\lambda\int_{-\infty}^{+\infty}e^{-ikr}\phi(r)d^3r\right)=\int_{-\infty}^{+\infty}e^{-ikr}F(r)d^3r \tag{13.56}$$

We will now use Green's theorem:

$$\int_{\text{volume}}\left(F\vec{\nabla}^2 G-G\vec{\nabla}^2 F\right)d^3r=\int_{\text{surface}}\left(F\vec{\nabla}G-G\vec{\nabla}F\right)\vec{n}\cdot d\vec{S} \tag{13.57}$$

where $\vec{n}$ is the unit vector normal to the surface; $\vec{n}$ is an outward vector and S is the surface enclosing the volume V.

$$\frac{1}{(2\pi)^{3/2}}\int_{\text{volume}}\left(e^{-i\vec{k}\vec{r}}\vec{\nabla}^2\phi(r)d^3r-\frac{1}{(2\pi)^{3/2}}\int_{\text{volume}}\phi(r)\vec{\nabla}^2 e^{-i\vec{k}\vec{r}}d^3r\right.$$
$$=\frac{1}{(2\pi)^{3/2}}\int_{\text{surface}}\left(e^{-i\vec{k}\vec{r}}\vec{\nabla}\phi(r)\vec{n}dS-\frac{1}{(2\pi)^{3/2}}\int_{\text{surface}}\phi(r)\vec{\nabla}e^{-i\vec{k}\vec{r}}\vec{n}dS\right. \tag{13.58}$$

To evaluate the surface integral we will use the solid angle $d\Omega=\frac{dS}{r^2}$; we will consider the surface to be a surface of radius R. Then, we will have to tend R to ∞. Thus, the surface term becomes

$$\frac{1}{(2\pi)^{3/2}}\lim_{R\to\infty}R^2\left[\int\left(e^{-ikr}\frac{d\phi}{dr}-\phi\frac{d}{dr}e^{-ikr}\right)d\Omega\right]_{r=R} \tag{13.59}$$

If $\phi(r)$ tends to 0 as $|\vec{r}|\to\infty$, then the previous term will tend to 0 for $R\to\infty$. Green's theorem reduces to

$$\frac{1}{(2\pi)^{3/2}}\int_{\text{volume}}\left(e^{-i\vec{k}\vec{r}}\vec{\nabla}^2\phi(r)d^3r=\frac{1}{(2\pi)^{3/2}}\int_{\text{volume}}\phi(r)\vec{\nabla}^2 e^{-i\vec{k}\vec{r}}d^3r \tag{13.60}$$

Also,

$$\frac{1}{(2\pi)^{3/2}}\int_{\text{volume}}e^{-i\vec{k}\vec{r}}\vec{\nabla}^2\phi(r)d^3r=\frac{1}{(2\pi)^{3/2}}k^2\int_{\text{volume}}\phi(r)e^{-i\vec{k}\vec{r}}d^3r$$
$$=k^2\hat{\phi}(k) \tag{13.61}$$

And substituting this in equation (13.60) gives

$$\left(-k^2+\lambda\right)\hat{\phi}(\vec{k})=\hat{F}(\vec{k}) \tag{13.62}$$

Now, if $\lambda<0\exists\kappa\in R/\lambda=-\kappa^2$ and then $\hat{\phi}(\vec{k})$ could be written as

$$\hat{\phi}(\vec{k})=-\frac{\hat{F}(\vec{k})}{k^2+\kappa^2} \tag{13.63}$$

Fourier inverse transformation will help to find $\phi\left(\vec{k}\right)$ in the form of

$$\phi\left(\vec{r}\right)=\xi\left(\vec{r}\right)-\frac{1}{\left(2\pi\right)^{3/2}}\int_{\text{volume}}\frac{\widehat{F}\left(\vec{k}\right)}{k^2+\kappa^2}e^{i\vec{k}\vec{r}}d^3\vec{k} \tag{13.64}$$

where $\xi\left(\vec{r}\right)$ is an arbitrary linear combination of solutions of the homogeneous equation,

$$H_0\phi\left(\vec{r}\right)=\left(\nabla^2-\kappa^2\right)\phi\left(\vec{r}\right)=0 \tag{13.65}$$

Overall, we can write $\phi\left(\vec{r}\right)$ in the form,

$$\phi\left(\vec{r}\right)=\xi\left(\vec{r}\right)+\int_{\text{volume}}G\left(\vec{r},\vec{r}'\right)F\left(\vec{r}'\right)d^3\vec{r}' \tag{13.66}$$

with Green's function $G\left(\vec{r},\vec{r}'\right)$ given by

$$G\left(\vec{r},\vec{r}'\right)=-\frac{1}{\left(2\pi\right)^{3/2}}\int_{\text{volume}}\frac{e^{i\vec{k}\left(\vec{r}-\vec{r}'\right)}}{k^2+\kappa^2}d^3\vec{k} \tag{13.67}$$

PROBLEM SET 13

13.1   Prove that equation (13.5) can be written:

$$G\left(\vec{r},\vec{r}'\right)=-\frac{1}{4\pi}\frac{e^{-\kappa\left|\vec{r}-\vec{r}'\right|}}{\left|\vec{r}-\vec{r}'\right|}$$

13.2   Use complex analysis to evaluate Green's function:

$$G\left(t,t'\right)=\frac{1}{2\pi}\int_{-\infty}^{+\infty}\frac{e^{-ik\left(t-t'\right)}}{\left[-k^2-2i\gamma k+\omega_0^2\right]}dk$$

of the forced damped harmonic oscillator. Assume $\omega_o>\gamma$; prove that

$$\begin{cases}G\left(t,t'\right)=\dfrac{e^{-\gamma\left(t-t'\right)}\sin\left[\sqrt{\omega_o^2-\gamma^2}\left(t-t'\right)\right]}{\sqrt{\omega_o^2-\gamma^2}} & t<t'\\[4mm] G\left(t,t'\right)=0 & t<t'\end{cases}$$

and summarize:

$$\left(t,t'\right)=\theta\left(t-t'\right)\frac{e^{-\gamma\left(t-t'\right)}\sin\left[\sqrt{\omega_o^2-\gamma^2}\left(t-t'\right)\right]}{\sqrt{\omega_o^2-\gamma^2}}.$$

13.3   Show that $\left[\dfrac{d^2}{dt^2}+2\gamma\dfrac{d}{dt}+\omega_0^2\right]G\left(t,t'\right)=\delta\left(t-t'\right)$; study the case where the driving force

is $F\left(t\right)=F_0e^{-\alpha t}$ and find the general solution of the forced damped harmonic oscillator
ODE (1).

13.4   Solve the problem

$$\frac{1}{x}u'' + \frac{1}{x^2}u' = f(x), \text{ with } x \in [0,1] \text{ and } u(0) + u'(1) = 1; u'(0) + u(1) = 0.$$

Discuss the existence and the uniqueness of the solution.

13.5   Construct the solution of the problem

$$u'' - \frac{2}{x^2}u = f(x) \text{ with } x \in [0,1]$$

where $f(x)$ is a square integrable function and $u(0) = 0$; $u(1) = 1$.

13.6   An operator in $\mathbb{R}^3$ is defined by

$$Lu = \frac{\partial^2 u}{\partial x \partial y} + \gamma^2 u \text{ where } \gamma^2 \text{ is a positive constant.}$$

a.   Write Green's formula for L.

b.   Show that $J_0\left(2\gamma\sqrt{xy}\right)$ is a classical solution of $Lu = 0$, for x, $y > 0$.

c.   Show that $E = H(x)H(y)J_0\left(2\gamma\sqrt{xy}\right)$ is a fundamental solution for L with a pole at the origin.

13.7   Let $0 < a < b$; consider the Boundary Value Problem:

$$\left(xg'\right)' + \left(x + \frac{v^2}{x}\right)g = \delta(x-\xi), a < x, \xi < b, g(a,\xi) = g(b,\xi) = 0$$

Using information about Bessel's functions, determine $g(x,\xi)$. Do the same for the problem:

$$\left(xg'\right)' - \left(x + \frac{v^2}{x}\right)g = \delta(x-\xi), a < x, \xi < b, g(a,\xi) = g(b,\xi) = 0$$

Are there any values of v^2, for which any either construction fails?

13.8   The boundary value problem

$$-u'' = f, 0 < x < 1, u(0) = u(1), u'(0) = u'(1)$$

describes heat flow in a thin ring of unit circumference. Show that the modified Green's function for this problem is

$$g_M(x,\xi) = \frac{1}{12} + \frac{(x-\xi)^2}{2} - \frac{1}{2}|x-\xi|$$

13.9   Consider the differential operator L such that

$$LG\left[(x,x')\right] = \left(\frac{\partial^2}{\partial x^2} - k^2\right)G(x,x') = 0, \quad x \neq x', \quad k > 0, 0 < x < l$$

Obtain Green's function corresponding to the boundary conditions:

a. $G(0,x')=0\,G(l,x')=0,\;\dfrac{\partial}{\partial x}G\bigg|_{\substack{x=x'+0\\x=x'-0}}=1$

b. $G(0,x')=0,\left(\dfrac{\partial}{\partial x}+\alpha\right)G(x,x')=0,\;x=l$

and discuss the limit $\alpha=\infty,\alpha=0$.

13.10   A nonhomogeneous ODE is defined by

$$\left[\frac{d^2y}{dx^2}+\left(2x-\frac{1}{x}\right)\frac{dy}{dx}\right]=-f(x),1<x<2$$

Find a Green's function, $G(x,x')$, such that

$$y(x)=\int^2 G(x,x')f(x')dx',\text{and }y(1)=y(2)=0.$$

14 Integral Equations

Everything should be made as simple as possible, but not simpler.

–Albert Einstein

INTRODUCTION

Integral equations are encountered in various physical phenomena such as nonlinear phenomena. In this chapter, we introduced the notion of integral equations and their classifications and solved some basic examples.

14.1 INTRODUCTION TO INTEGRAL EQUATIONS

An integral equation is all equations whose unknown is under the sign sum. Assume, for instance, that we want to solve a first-order differential equation

$$y' = f(x, y) \tag{14.1}$$

with the initial condition

$$y(x_0) = y_0 \tag{14.2}$$

where $f(x,y)$ is a function of x and y the unknown. Taking the integral, of both sides of the differential equation leads to the integral equation,

$$y(x) = \int_0^x f(x, y) \, dx + y(x_0) \tag{14.3}$$

Similarly, if we try to solve the second-order differential equation $y'' = f(x, y)$, with the initial conditions $y(x_0) = y_0$ and $y'(x_0) = y_0'$, we are led to the integral equations:

$$y'(x) = \int_0^x f(x, y) \, dx + y'(x_0) \tag{14.4}$$

and

$$y(x) = \int_0^x dx \int_0^x f(z, y(z)) \, dz + y(x_0) + y_0'(x - x_0). \tag{14.5}$$

A transformation of the double integral into a simple integral, gives

$$y(x) = \int_0^x (x - z) f(z, y(z)) \, dz + y_0 + y_0'(x - x_0). \tag{14.6}$$

Thus, the general solution of the differential equation $y'' = f(x, y)$ is obtained from the integral equation on the form:

$$y(x) = \int_{x_0}^x (x - z) f(z, y(z)) \, dz + a_1 x + a_0 \tag{14.7}$$

Where a_1 and a_2 are arbitrary constant real numbers.

DOI: 10.1201/9781003478812-14

14.2 CLASSIFICATION OF THE INTEGRAL EQUATIONS

Let us classify the linear integral equation in which the unknown function is defined only on the x-axis. The integral equation can be written as

$$y(x) = \int_l^r K(x,z) y(z) dz + f(x) \tag{14.8}$$

where $K(x,z)$ is the kernel, $y(x)$ is the unknown function, $f(x)$ is a given function and a is a real number.

Equation (14.8) is called the Volterra equation of the second kind. The same equation with constant limits, that is:

$$y(x) = \int_a^b K(x,z) y(z) dz + f(x) \tag{14.9}$$

where a is a given constant, is called Fredholm equation of the second kind. If the unknown function is only under the sign sum, the integral equations,

$$\int_a^x K(x,z) y(z) dz = f_1(x) \tag{14.10}$$

and

$$\int_a^k K(x,z) y(z) dz = f_1(x) \tag{14.11}$$

are successively called the Volterra equation of the first kind and the Fredholm equation of the first kind.

As an example of the Volterra equation of the first kind, we have the Abel equation of the first kind:

$$\varphi(h) = \frac{1}{\sqrt{2g}} \int_0^h \frac{u(y) dy}{\sqrt{h-y}} \tag{14.12}$$

14.2.1 AN EXAMPLE OF FREDHOLM EQUATION OF THE FIRST KIND

Assuming $u(x)$ is the static deformation of a string and $p(z)$ is the linear charge deformation, that is $p(z) = \dfrac{dq}{dz}$. The distribution is continuous and $dq = p(z) dz$. Each charge will create the deformation:

$$du = \frac{1}{T_0} K(x,z) p(z) dz \tag{14.13}$$

where

$$K(x,z) = \begin{cases} \dfrac{z(l-x)}{l} & z \le x \\ \dfrac{x(l-z)}{l} & x \le z \end{cases} \tag{14.14}$$

The static deformation with a continuous distribution of charge will be $u(x) = \frac{1}{T_0} \int_0^l K(x,z) p(z) dz$,

which is the Fredholm equation of the first kind. The theory of integral equations presents a lot of affinities with linear algebra problems. It is important to recall that a linear transformation in a space of n dimensions is in the form:

$$y_i = a_{i1}u_1 + a_{i2}u_2 + \ldots + a_{in}u_n, \text{ with } i = 1,2,3,\ldots,n$$

and is represented by the matrix (a_{ik}). Therefore, the transformation can be written in the matrix form: $Y = Au$, where $u = (u_1, u_2, u_3, \ldots, u_n)$ is the initial vector and $Y = (y_1, y_2, y_3, \ldots, y_n)$ the transformed vector; A is the matrix whose elements are the a_{ik}. In the case of integral equations, instead of vectors, we will operate on functions defined in some interval $[a,b]$. The matrix also will be replaced by the kernel $K(x,z)$ and the integral summation. As a result, the linear transformation will be represented by

$$y(x) = \int_a^b K(x,z) u(z) dz$$

where $u(z)$ is the original or input function and the transformed or $y(x)$ the output function.

At this stage, we need to recall that the eigenvalues of the matrix A are the values μ for which the eigenvalue equation, $Ax = \mu x$, has non null solutions. We would also call eigenvalue of the kernel, the values μ for which the homogeneous integral equation, $\int_a^b K(x,z) y(z) dz = \mu y(x)$, has non-null solutions. In the theory of integral equations, it is useful to introduce the characteristic values $\lambda = \mu^{-1}$. We will say that λ is a characteristic value if the equation $y(x) = \lambda \int_a^b K(x,z) y(z) dz$ also has non-null solutions.

14.3   KERNEL DEGENERACY

A kernel is said to be degenerate if it is in the form:

$$K(x,y) = \sum_{i=1}^{n} \varphi_i(x) \psi_i(y) \tag{14.15}$$

Example 1
Solve the integral equation:

$$f(x) = x + \lambda \int_0^1 \left(xy^2 + x^2 y \right) f(y) dy$$

Solution
To solve this integral equation, let us define

$$A = \int_0^1 y^2 f(y) dy \tag{14.16}$$

$$B = \int_0^1 yf(y) dy \tag{14.17}$$

We can now write $f(x)$ as follows:

$$f(x) = x + \lambda A x + \lambda B x^2 \tag{14.18}$$

Using the expression (14.18) of $f(x)$ back into (14.16) and (14.17), one obtains

$$A = \frac{1}{4} + \frac{1}{4}\lambda A + \frac{1}{5}\lambda B \qquad (14.19)$$

$$B = \frac{1}{3} + \frac{1}{3}\lambda A + \frac{1}{4}\lambda B \qquad (14.20)$$

which gives

$$A = \frac{60 + \lambda}{240 - 120\lambda - \lambda^2} \qquad (14.21)$$

$$B = \frac{80}{240 - 120\lambda - \lambda^2} \qquad (14.22)$$

and the solution of the integral equation is

$$f(x) = \frac{(240 - 60\lambda)x + 80\lambda x^2}{240 - 120\lambda - \lambda^2} \qquad (14.23)$$

As we can see, there are two values of λ for which the denominator of $f(x)$ is zero. In other words, there are two values of λ for which the function is infinite. These two values are called the eigenvalues of the integral equation. If λ is one of these eigenvalues, then the homogeneous integral equation has a nontrivial solution. These solutions will be called eigenfunctions of the kernel K. Some theorems that we don't need to prove are important to solve some questions. Some of these theorems are stated below:

Theorem 1a
If λ is an eigenvalue, the inhomogeneous equation
$f(x) = g(x) + \lambda \int_a^b K(y,x) f(y) dy$ has a solution if and only if $\int_a^b \varphi(x) g(x) dx = 0$, for every $g(x)$
that obeys the transposed homogenous equation, $f(x) = \lambda \int_a^b K(y,x) f(y) dy$.

Theorem 1b
If λ is not an eigenvalue, then λ is also not the eigenvalue of the transposed equation $f(x) = g(x) + \lambda \int_a^b K(y,x) f(y) dy$.

Theorem 1c
If λ is an eigenvalue, then λ is also the eigenvalue of the transposed equation $f(x) = g(x) + \lambda \int_a^b K(y,x) f(y) dy$,

that is, $f(x) = \lambda \int_a^b K(y,x) f(y) dy$, has at least one nontrivial solution.

14.4   INTEGRAL OPERATOR

14.4.1   DEFINITION

An integral operator, Θ, is an integral equation of the form $\int_a^b K(x,y) f(y) dy = g(x)$, where K is a real-valued function of two variables, x and y. We know that $K(x,y)$ is the kernel; we can also write: $\Theta f = g$, where f maps the function g in the interval $[a,b]$. Furthermore,

$$\Theta:[a,b]\to[a,b] \qquad (14.24)$$

$$f\mapsto g$$

14.4.2 LINEARITY OF THE INTEGRAL OPERATOR Θ

The integral operator is linear, that is, for every complex number,

$$\alpha,\beta,\Theta(\alpha f+\beta g)=\alpha\Theta(f)+\beta\Theta(g). \qquad (14.25)$$

An example of an integral operator is the Laplace transform. Indeed, $LT(f)=\int_0^\infty f(x)e^{-sx}dx=\overline{f}(s)$, in which the kernel is $K(x,s)=e^{-sx}=K(s,x)$ by symmetry.

Let us now define the Volterra equations of the first kind as

$$\int_a^x K(x,y)\phi(y)dy=f(x),\ x\in[a,b] \qquad (14.26)$$

and Volterra equations of the second kind as

$$\phi(x)-\lambda\int_a^x K(x,y)\phi(y)dy=f(x),\ x\in[a,b] \qquad (14.27)$$

14.5 SOLUTION TO THE VOLTERRA EQUATION OF THE SECOND KIND

Theorem 2

If $f(x)$ and $K(x,y)$ are continuous functions of x and y of $[a, b]$, then the Volterra equation of the second kind has one and only one solution given by the uniformly convergent series:

$$\phi(x)=f(x)+\sum_{n=1}^\infty\lambda^n\int_a^x(x,y)f(y)dy \qquad (14.28)$$

where $K_1(x,y)=K(x,y)$

$$K_{n+1}(x,y)=\int_y^x K(x,s)K_n(s,y)ds, n\ge1 \qquad (14.29)$$

Proof

$\phi(x)-\lambda\int_a^x K(x,y)\phi(y)dy=f(x)$ can be rewritten using the operator's properties: $(I-\lambda K)\phi(x)=f(x)$.

The inverse of $(I-\lambda K)$ is given by the Newman series,

$$(I-\lambda K)^{-1}=I+\sum_{j=1}^\infty(-1)^j(-A)^j. \qquad (14.30)$$

Therefore, the solution would be given by

$$\phi(x)=f(x)+\sum_{k=1}^n\lambda^k K^k f(x). \qquad (14.31)$$

14.5.1   CONVERGENCE OF THE SERIES GIVEN BY (14.31)

For the proof, let us define: k_m such that $\kappa_m{}^2(x) = \left| K(x)^m f(x) \right|^2$; Thus,

$$\kappa_m^2(x) = \left| \int_a^x K(x,y) \left[K^m f(y) dy \right] \right|^2$$

$$\leq \int_a^x |K(x,y)|^2\, dy \int_a^x \left| K^{m-1} f(y) \right|^2 dy \qquad (14.32)$$

$$\leq k^2(x) \int_a^x \kappa_{m-1}^2(y)\, dy$$

where

$$k^2(x) = \int_a^x \left| K(x,y) \right|^2 dy \leq \int_a^b \left| K(x,y) \right|^2 dy \qquad (14.33)$$

We need to find a bound on $\kappa_m{}^2(x)$.

For $m = 1$.

$$\kappa_1^2(x) = \left| K^1 f(x) \right|^2 = \left| \int_a^x K(x,y) f(y) dy \right|^2$$

$$\leq \int_a^x |K(x,y)|^2\, dy \int_a^x \left| f(y) \right|^2 dy \qquad (14.34)$$

$$\leq k^2(x) \int_a^b |f(y)|^2\, dy$$

For $m = 2$

$$\mathcal{K}_2^2(x) = \left| K^2 f(x) \right|^2 = \left| \int_a^x K(x,y) K f(y) dy \right|^2$$

$$= \left| \int_a^x K(x,y) dy \int_a^y K(y,s) f(s) ds \right|^2$$

$$\leq \int_a^x |K(x,y)|^2\, dy \int_a^x dy \left| \int_a^y ds K(y,s) f(s) \right|^2 \qquad (14.36)$$

$$\leq k^2(x) \int_a^x dy \left[\int_a^y ds \left| K(y,s) \right|^2 \int_a^y \left| f(s) \right|^2 ds \right]$$

$$\leq k^2(x) \int_a^x dy\, k^2(y) \int_a^y | f(s) |^2\, ds$$

$$\leq k^2(x) \int_a^x dy\, k^2(y) \int_a^b | f(y) |^2\, dy$$

Let us now use the method of induction to prove that, for all m,

$$\kappa_m{}^2(x) \leq \frac{k^2(x)}{(m-1)^2} \left[\int_a^x k^2(y) dy \right]^{m-1} \int_a^b | f(y) |^2\, dy. \qquad (14.37)$$

We have proved that the previous proposition is true for $m = 1$ and $m = 2$. Assume that the proposition is true at the order $m - 1$ and let us prove that it is true at the order m. We have

$$
\begin{aligned}
\kappa_m^2(x) &\leq k^2(x)\int_a^x \kappa_{m-1}^2(y)\,dy \\
&\leq k^2(x)\int_a^x \frac{k^2(y)}{(m-2)!}\,dy\left[\int_a^y k^2(s)\,ds\right]^{m-2}\int_a^b |f(s)|^2\,ds \\
&= \frac{k^2(x)}{(m-2)!}\int_a^x\left[\int_a^y k^2(s)\,ds\right]^{m-2} d\left[\int_a^y k^2(s)\,ds\right]\int_a^b |f(s)|^2\,ds \\
&= \frac{k^2(x)}{(m-1)!}\left[\int_a^x k^2(s)\,ds\right]^{m-1}\int_a^b |f(y)|^2\,dy
\end{aligned}
\tag{14.38}
$$

which proves that proposition (14.37) is true for all m.

Let us now find a bound for the general term $\lambda^n K(x)^n f(x)$, which appears in (14.32):

$$
\begin{aligned}
\left|\lambda^n K(x)^n f(x)\right| &= \left|\lambda^n\right|\left\|K(x)^n f(x)\right\| \\
&\leq \left|\lambda^n\right|\left\{\frac{k^2(x)}{(n-1)!}\left[\int_a^x k^2(y)\,dy\right]^{n-1}\right\}^{1/2}\|f\| \\
&\leq \left|\lambda^n\right|k(x)\left(\frac{\|K\|^{n-1}}{\sqrt{(n-1)!}}\right)\|f\|
\end{aligned}
\tag{14.39}
$$

where $\|K\|^2 = \int_a^b dx \int_a^b dy\, |K(x,y)|^2$.

Since $\dfrac{\lambda^{n+1}\|K\|^n}{\sqrt{n!}} \Big/ \dfrac{\lambda^n\|K\|^{n-1}}{\sqrt{(n-1)!}} = \dfrac{\lambda\|K\|}{\sqrt{n}}$ tends to zero as n tends to ∞, then the series $\displaystyle\sum_{n=1}^{\infty}\frac{\lambda^n\|K\|^{n-1}}{\sqrt{(n-1)!}}$ is

absolutely convergent, which implies using the Weierstrass M test that the series $\displaystyle\sum_{n=1}^{\infty}\lambda^n K^n f(x)$ is

uniformly convergent.

We also need to prove that equation (14.32) is a solution to equation (14.27). We have

$$
\begin{aligned}
f(x)&+\sum_{n=1}^{\infty}\lambda^n K^n f(x)-\lambda\int K\left[f(y)+\sum_{n=1}^{\infty}\lambda^n K^n f(y)\right]dy \\
&= f(x)+\sum_{n=1}^{\infty}\lambda^n K^n f(x)-\lambda\int_a^x Kf(y)\,dy-\Sigma\int_a^x \lambda^{n+1}K^n f(y)\,dy \\
&= f(x)+\sum_{n=1}^{\infty}\lambda^n K^n f(x)-\lambda Kf(x)-\sum_{n=1}^{\infty}\lambda^{n+1}K^{n+1}f(x) \\
&= f(x)
\end{aligned}
\tag{14.40}
$$

Therefore, $I + \sum_{n=1}^{\infty} \lambda^n K^n$ is the inverse of $I - \lambda K$ and always exists. The solution to equation (14.27) is unique.

14.6   DEFINITION OF THE RESOLVENT

The resolvent represented by $\mathfrak{R}(x, y; \lambda)$ is defined as follows:

$$\mathfrak{R}(x, y; \lambda) = \sum_{n=1}^{\infty} \lambda^{n-1} K_n(x, y). \tag{14.41}$$

$\mathfrak{R}(x, y; \lambda)$ is a uniformly convergent series of continuous functions.

Since $\mathfrak{R}(x, y; \lambda)$ is continuous in x and y, the solution to the integral equation (14.27) can be expressed as

$$\phi(x) = f(x) + \lambda \int_a^x \mathfrak{R}(x, y; \lambda) f(y) dy \tag{14.42}$$

14.7   CONNECTION TO ORDINARY DIFFERENTIAL EQUATIONS

Suppose we want to solve a nonsingular linear inhomogeneous ordinary differential equation of the form:

$$a(x) \frac{d^2 u(x)}{dx^2} + a(x) \frac{du(x)}{dx} + c(x) u(x) = f(x) \tag{14.43}$$

with the initial conditions, $u(x_0) = c_1$, $u'(x_0) = c_2$, $x_0 \in [a, b]$. If we assume $\phi(x) = \dfrac{d^2 u(x)}{dx^2}$,

$K(x, y) = \dfrac{b(x)}{a(x)} + \dfrac{c(x)}{a(x)} (x - y)$ and

$$F(x) = \frac{1}{a(x)} \left\{ f(x) - c_2 b(x) - c(x) \left[c_1 + c_2 (x - x_0) \right] \right\}$$

Then, the integral equation

$$\phi(x) + \int_{x_0}^{x} K(x, y) \phi(y) dy = F(x) \tag{14.44}$$

is satisfied by equation (14.43).

Conversely, the function of the form:

$$u(x) = c_1 + c_2 (x - x_0) + \int_{x_0}^{x} (x - y) \phi(y) dy = F(x) \tag{14.45}$$

is a solution to equation (14.43) with the initial conditions:

$$u(x_0) = c_1 \quad u'(x_0) = c_2.$$

Overall, we can admit the following theorem:

Theorem 3:
If for $x \in [a,b]$, the functions $a(x)$, $b(x)$, $c(x)$ and $f(x)$ are piecewise, continuous and bounded, while $a(x)$ is of one sign there, then for the differential equation:

$$a(x)\frac{d^2u(x)}{dx^2} + a(x)\frac{du(x)}{dx} + c(x)u(x) = f(x)$$

subjects to the initial conditions:

$$u(x_0) = c_1, \quad u'(x_0) = c_2, \quad x_0 \in [a,b],$$

there exists a unique solution such that $u(x)$ and $\dfrac{du(x)}{dx}$ are continuous, for all $x \in [a,b]$.

PROBLEM SET 14

14.1   Solve the integral equations

   a. $f(x) = e^{2x} + \lambda \int_0^1 xt f(t)\,dt$

   b. $g(x) = \lambda \int_0^\pi \sin(x-t) g(t)\,dt$.

14.2   Solve the integral equations

   a. $f(x) = x + \lambda \int_0^1 xy f(y)\,dy$

   b. $f(x) = x^4 + \lambda \int_0^1 x^2 y^2 f(y)\,dy$.

14.3   Solve the integral equations

   a. $h(x) = x + \lambda \int_0^\infty e^{-y} h(y)\,dy$

   b. $v(x) = \lambda \int_{-1}^1 v(y)\cos(y)\,dy$.

14.4   Solve the integral equations

   a. $u(x) = x + \lambda \int_0^1 e^{x-y} u(y)\,dy$.

   b. $v(x) = \lambda \int_0^T v(y) e^{y/T}\,dy$

14.5   Solve the integral equations

$$f(x) = x + \lambda \int_0^x xy f(y)\,dy.$$

14.6   Write the initial value problem:

$$u''(y) + yu(y) = 2u(y) = 0$$

with $u(0) = \alpha$ and $u'(0) = \beta$ as a Volterra integral equation. In this contest, prove the following result: $u(x) + x\int_0^x u(y)\,dy = \alpha + \beta x$.

15 Introduction to Dynamical Systems

INTRODUCTION

In this chapter, we review some ordinary differential equations and their analysis in dynamical systems; we introduce systems of ordinary differential equations and how to solve them; we explain the diagonalization and define operators and their convergences. The chapter also contains the exponential of an operator, phase portraits, separatrices, Jordan forms and stable and unstable spaces. Moreover, we also introduce the concept of chaos and Lorenz's equations which have applications in meteorology and atmospheric physics.

15.1 DEFINITION

Assume that we want to solve the linear system of equations:

$$x = Ax \tag{15.1}$$

where $x \in \mathbb{R}$ and A a $n \times n$ matrix. We can also write the system of equations as follows:

$$
\begin{pmatrix} x_1 \\ x_2 \\ x_3 \\ \cdot \\ \cdot \\ \cdot \\ x_n \end{pmatrix} = \begin{bmatrix} a_{11} & \cdots\cdots\cdots & a_{1n} \\ \cdot & & \\ & & \\ & \cdot & \\ & & \\ \cdot & & \\ a_{n1} & \cdots & a_{nn} \end{bmatrix} \begin{pmatrix} x_1 \\ x_2 \\ \\ \cdot \\ \\ \\ x_n \end{pmatrix} \tag{15.2}
$$

If the initial condition is $x(0) = x_0$, then the solution to the system of equations would be $x(t) = x_0 e^{At}$, where e^{At} is a $n \times n$ matrix defined by the Taylor series.

Example 1
Consider the uncoupled system of linear differential equations:

$$
\begin{cases} \dot{x}(t) = -x(t) \\ \dot{y}(t) = 2y(t) \end{cases} \tag{15.3}
$$

The solution to the first differential equation is exponential: $x(t) = x_0 e^{-t}$ and the solution to the second differential equation is $y(t) = y_0 e^{2t}$. So, the solution could be written in the matrix form:

$$
\begin{bmatrix} x \\ y \end{bmatrix} = \begin{bmatrix} x_0 \\ y_0 \end{bmatrix} \begin{bmatrix} e^{-t} & 0 \\ 0 & e^{2t} \end{bmatrix} \tag{15.4}
$$

DOI: 10.1201/9781003478812-15

We see that the system could be written in terms of matrix such as

$$\begin{bmatrix} \dot{x} \\ \dot{y} \end{bmatrix} = \begin{bmatrix} -1 & 0 \\ 0 & 2 \end{bmatrix} \begin{bmatrix} x \\ y \end{bmatrix} \tag{15.5}$$

The dynamical system defined by the previous linear system of equations (15.1) is the mapping $\phi : \mathbb{R} \times \mathbb{R}^2 \to \mathbb{R}^2$ defined by the solution $x(t,c)$ with $c = \begin{bmatrix} x_0 \\ y_0 \end{bmatrix}$. The phase portrait of the system of differential equations (15.1) is the set of all solutions' curves of the system in the phase space, $\mathbb{R}^n$. The function $f(x) = Ax$ defines the mapping $f : \mathbb{R}^2 \to \mathbb{R}^2$, which defines a vector field on $\mathbb{R}^2$. At each point x in the phase space $\mathbb{R}^2$, the solution curves (15.4) are tangent to the vectors in the vector field defined by $f(x) = Ax$.

Example 2
Consider the following uncoupled system of linear equations in $\mathbb{R}^3$:

$$\begin{cases} \dot{x}(t) = x(t) \\ \dot{y}(t) = y(t) \\ \dot{z}(t) = -z(t) \end{cases} \tag{15.6}$$

From $\dot{x}(t) = x(t)$, we have $\dfrac{\dot{x}(t)}{x} = 1$ or $\begin{cases} \dot{x}(t) = x(t) \\ \dot{y}(t) = y(t) \\ \dot{z}(t) = -z(t) \end{cases}$. Taking the integral on both sides gives $\int \dfrac{dx}{x} = \int dt$;

that is $\ln(x) = t + k$ where k is an arbitrary constant. The result of the integration gives $x = e^{t+k} = ce^t$ where $c = e^k$ is also an arbitrary constant.

By proceeding the same way with for the differential equations $\dot{y}(t) = y(t)$ and $\dot{z}(t) = z(t)$, the general solution of the system of differential equation (15.6) is given by

$$\begin{cases} x(t) = c_1 e^t \\ y(t) = c_2 e^t \\ z(t) = c_2 e^{-t} \end{cases} \tag{15.7}$$

15.2   DIAGONALIZATION

A linear system $x = Ax$ can be reduced to an uncoupled system of linear equations by the diagonalization techniques.

Theorem 1
If the eigenvalues $\lambda_1, \lambda_2 \ldots, \lambda_n$ of a $n \times n$ matrix are real or distinct, then any set of corresponding eigenvectors, $\{v_1, v_2, \ldots, v_n\}$, forms a basis for $\mathbb{R}^n$, the matrix $P = [v_1, v_2, \ldots, v_n]$ is invertible and $P^{-1}AP = \text{diag}[\lambda_1, \lambda_2, \ldots \lambda_n]$ is a diagonal matrix.

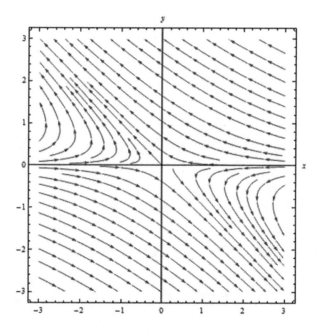

FIGURE 15.1   Phase portrait for the system, $\begin{cases} \dot{x} = -x - 3y \\ \dot{y} = 2y \end{cases}$.

Example 4

Solve the linear system and represent the phase portrait.

$$\begin{cases} \dot{x} = -x - 3y \\ \dot{y} = 2y \end{cases} \tag{15.8}$$

Solution

We can write this system in matrix form:

$$\begin{bmatrix} \dot{x} \\ \dot{y} \end{bmatrix} = \begin{bmatrix} -1 & -3 \\ 0 & 2 \end{bmatrix} \begin{bmatrix} x \\ y \end{bmatrix} \tag{15.9}$$

The matrix associated with this system is

$$A = \begin{bmatrix} -1 & -3 \\ 0 & 2 \end{bmatrix} \tag{15.10}$$

We need now to find the eigenvalues and the eigenvectors of the matrix A. The characteristic equation is given by

$$\det(A - \lambda I) = \begin{vmatrix} -1 - \lambda & -3 \\ 0 & 2 - \lambda \end{vmatrix} = (\lambda + 1)(\lambda - 2) = 0. \tag{15.11}$$

The eigenvalues are $\lambda_1 = -1$ and $\lambda_2 = 2$; the associated eigenvectors are $u_1 = \begin{bmatrix} 1 \\ 0 \end{bmatrix}$ and $u_2 = \begin{bmatrix} -1 \\ 1 \end{bmatrix}$; the matrix of eigenvectors is the transformation matrix, $P = \begin{bmatrix} 1 & -1 \\ 0 & 1 \end{bmatrix}$. Matrix P is invertible and has the inverse $P^{-1} = \begin{bmatrix} 1 & 1 \\ 0 & 1 \end{bmatrix}$.

$P^{-1}AP = \begin{bmatrix} -1 & 0 \\ 0 & 2 \end{bmatrix}$ is a diagonal matrix as expected. Let us now use the linear transformation,

$Y = P^{-1}X \Leftrightarrow X = PY$.

So, $\dot{Y} = P^{-1}\dot{X} = P^{-1}AX = \left(P^{-1}AP \right)Y = \operatorname{diag}\left[\lambda_1, \lambda_2 \right]Y$; then we have the system:

$\begin{bmatrix} \dot{x} \\ \dot{y} \end{bmatrix} = \begin{bmatrix} -1 & 0 \\ 0 & 2 \end{bmatrix}\begin{bmatrix} x \\ y \end{bmatrix}$; the corresponding equation is $\begin{cases} \dot{x} = -x \\ \dot{y} = 2y \end{cases}$

and the solutions are:

$$\begin{cases} x_1 = k_1 e^{-t} \\ y_1 = k_2 e^{2t} \end{cases} \tag{15.12}$$

But we also know that $X = PY$; therefore,

$$\begin{bmatrix} x \\ y \end{bmatrix} = \begin{bmatrix} 1 & -1 \\ 0 & 1 \end{bmatrix}\begin{bmatrix} k_1 e^{-t} \\ k_2 e^{2t} \end{bmatrix}$$

which gives the general solution of the system

$$\begin{cases} x = k_1 e^{-t} - k_2 e^{2t} \\ y = k_2 e^{2t} \end{cases} \tag{15.13}$$

From Mathematica command, we can have the stream plot (Figure 15.1):

$$\text{StreamPlot}\left[\{-x - 3y, 0x + 2y\}, \{x, -3, 3\}, \{y, -3, 3\}, \text{Axes} \rightarrow \text{True}, \text{AxesLabel} \rightarrow \{x, y\} \right]$$

Example 4

Consider the matrix $A = \begin{bmatrix} 1 & 3 \\ 3 & 1 \end{bmatrix}$

a. Find the eigenvalues and the eigenvectors of A. Form the transformation matrix P. Prove that $B = P^{-1}AP$ is a diagonal matrix.
b. Solve $Y = BY$.
c. Find the solution to $X = AX$ or $X = PY$.

Solution

a. The characteristic equation is written as

$$\det(A - \lambda I) = \begin{vmatrix} 1-\lambda & 3 \\ 3 & 1-\lambda \end{vmatrix} = (\lambda + 2)(\lambda - 4) = 0$$

and yields two eigenvalues: $\lambda_1 = -2$ and $\lambda_2 = 4$.

A given vector $V = \begin{pmatrix} x \\ y \end{pmatrix}$ is eigenvector associated with the eigenvalues λ if $A\vec{V} = \lambda\vec{V}$. For $\lambda_1 = -2$

we have

$\begin{bmatrix} 1 & 3 \\ 3 & 1 \end{bmatrix}\begin{bmatrix} x \\ y \end{bmatrix} = -2\begin{bmatrix} x \\ y \end{bmatrix}$, which yield $y = -x$; so $\begin{bmatrix} x \\ y \end{bmatrix} = \begin{bmatrix} x \\ -x \end{bmatrix} = x\begin{bmatrix} 1 \\ -1 \end{bmatrix}$; the eigenvector $V_1 = \begin{bmatrix} 1 \\ -1 \end{bmatrix}$ can

be chosen to be the one associated with the eigenvalue $\lambda_1 = -2$. Similarly, the eigenvector $V_1 = \begin{bmatrix} 1 \\ 1 \end{bmatrix}$

corresponds to the eigenvalue $\lambda_2 = 4$. The transformation matrix P is the matrix of the eigenvectors:

$P = \begin{bmatrix} 1 & 1 \\ -1 & 1 \end{bmatrix}$ and the inverse of P is $P^{-1} = \begin{bmatrix} 1/2 & -1/2 \\ 1/2 & 1/2 \end{bmatrix}$.

$P^{-1}AP = \begin{bmatrix} -2 & 0 \\ 0 & 4 \end{bmatrix} = B$ is a diagonal matrix. We can notice that the entries of B are nothing but

eigenvalues of matrix A.

a. Solve $\dot{Y} = BY$

We have $\begin{bmatrix} \dot{y}_1 \\ \dot{y}_2 \end{bmatrix} = \begin{bmatrix} -2 & 0 \\ 0 & 4 \end{bmatrix}\begin{bmatrix} y_1 \\ y_2 \end{bmatrix}$, which gives the system of linear equations: $\begin{cases} \dot{y}_1 = -2y_1 \\ \dot{y}_2 = 4y_2 \end{cases}$

whose solutions are $\begin{cases} y_1 = c_1 e^{-2t} \\ y_2 = c_2 e^{4t} \end{cases}$;

b. Find the solution to $X = AX$ or $X = PY$

$\begin{bmatrix} x_1 \\ x_2 \end{bmatrix} = \begin{bmatrix} 1 & 1 \\ -1 & 1 \end{bmatrix}\begin{bmatrix} c_1 e^{-2t} \\ c_2 e^{4t} \end{bmatrix}$ implies

$\begin{cases} x_1 = c_1 e^{-2t} + c_2 e^{4t} \\ x_2 = -c_1 e^{-2t} + c_2 e^{4t} \end{cases}$, where c_1 and c_1 are arbitrary real constants (Figure 15.2).

StreamPlot$\left[\{x+3y, 3x+y\}, \{x, -3, 3\}, \{y, -3, 3\}, \text{Axes} \rightarrow \text{True}, \text{AxesLabel} \rightarrow \{x, y\}\right]$

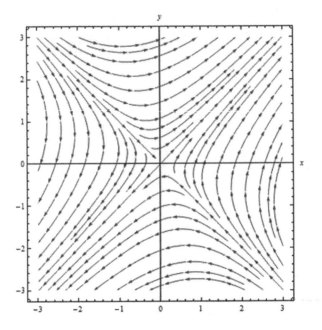

FIGURE 15.2   Phase portrait for the system $\begin{cases} \dot{x} = x + 3y \\ \dot{y} = 3x + y \end{cases}$.

15.3   DEFINITION OF OPERATORS

Consider a linear operator T defined as: $T : \mathbb{R}^n \to \mathbb{R}^n$. Let us call $L(\mathbb{R}^n)$ the linear space of linear operators on $\mathbb{R}^n$. We define the norm of the operator as follows:

$$\|T\| = \max_{|x| \leq 1} |T(x)|$$

where $|x|$ denotes the Euclidean norm of $x \in \mathbb{R}^n$ that is: $|x| = \sqrt{x_1^2 + x_2^2 + x_3^2 + \ldots x_n^2}$.

15.3.1   PROPERTIES OF THE OPERATOR NORM

For $A, B \in L(\mathbb{R}^n)$

  a. $\|T\| \geq 0, \|T\| = 0$ if $T = 0$

  b. $\|kT\| = k|T|, \|kT\| = |k|\|T\|$ for $k \in \mathbb{R}$

  c. $\|A + B\| \leq \|A\| + \|B\|$

15.4   CONVERGENCE

A sequence of linear operators $T_k \in L(\mathbb{R}^n)$ is said to converge to a linear operator $T \in L(\mathbb{R}^n)$ as $k \to \infty$, that is $\lim_{k \to \infty} T_k = T$ if the following condition is satisfied:

$$\forall \varepsilon > 0, \ \exists N \, / \, k \geq N \Rightarrow T - T_k < \varepsilon \tag{15.14}$$

Lemma

For $A, B \in L(\mathbb{R}^n)$ and $x \in \mathbb{R}^n$

    a. $|T(x)| \leq \|T\| |x|$

    b. $|TS| \leq \|T\| \|S\|$

    c. $\|T^k\| \leq \|T\|^k$ for $k = 0, 1, 2, \ldots$

Theorem 2

Given $T \in L(\mathbb{R}^n)$ and $t_0 > 0$, the series $\displaystyle\sum_{k=0}^{\infty} \frac{T^k t^k}{k!}$ is absolutely and uniformly convergent for all

$|t| \leq t_0$.

Proof

Let $\|T\| = a$. From the above lemma, for $|t| \leq t_0$ we have $\left\| \dfrac{T^k t^k}{k!} \right\| \leq \dfrac{\|T\|^k |t|^k}{k!} \leq \dfrac{a^k t_0^{\ k}}{k!}$.

But $\displaystyle\sum_{k=0}^{\infty} \frac{a^k t_0^{\ k}}{k!} = e^{at_0}$; it then follows from the Weierstrass theorem that the series $\displaystyle\sum_{k=0}^{\infty} \frac{T^k t^k}{k!}$ is

absolutely and uniformly convergent for all $|t| \leq t_0$.

15.5   EXPONENTIAL OF AN OPERATOR

The exponential of an operator can be written as $e^T = \displaystyle\sum_{k=0}^{\infty} \frac{T^k}{k!}$. It can be seen that e^T is a linear

operator on $\mathbb{R}^n$. Also, from the above theorem, $e^T \leq e^{\|T\|}$.

In the rest of this chapter, we assume that a linear transformation T on $\mathbb{R}^n$ is represented by the $n \times n$ matrix A with respect to the standard basis for $\mathbb{R}^n$. We can then define the exponential of matrix At as follows:

$$e^{At} = \sum_{k=0}^{\infty} \frac{A^k t^k}{k!}. \tag{15.15}$$

e^{At} is also an $n \times n$ matrix just like A, and e^{At} can be computed in terms of the eigenvalues and eigenvectors of A.

Proposition 1

If P and T are linear transformations on $\mathbb{R}^n$, and $S = PTP^{-1}$, then $e^S = Pe^T P^{-1}$.

Proof

$$e^S = \lim_{n \to \infty} \sum_{k=0}^{n} \frac{\left(PTP^{-1}\right)^k}{k!} = P \sum_{k=0}^{n} \frac{T^k}{k!} P^{-1} = Pe^T P^{-1}$$

Corollary 1

If $P^{-1}AP = \text{diag}\left[\lambda_i\right]$, then $e^{At} = P\text{diag}\left[e^{\lambda_i t}\right]P^{-1}$

Proposition 2

If P and T are linear transformations on $\mathbb{R}^n$ which commute, that is, $ST = TS$, then

$$e^{S+T} = e^S e^T. \tag{15.16}$$

Proof

From the binomial theorem, if we assume that S and T commute, that is $ST = TS$, then we have

$$(S+T)^n = n! \sum_{j+k=n} \frac{S^j T^k}{j!k!} \tag{15.17}$$

$$e^{S+T} = \sum_{n=0}^{\infty} \sum_{j+k=n} \frac{S^j T^k}{j!k!} = \sum_{j=0}^{\infty} \frac{S^j}{j!} \sum_{k=0}^{\infty} \frac{T^k}{k!} = e^S e^T \tag{15.18}$$

Corollary 2

If T is a linear transformation on $\mathbb{R}^n$, then the inverse of the linear transformation, e^T, is given by $\left(e^T\right)^{-1} = e^{-T}$, which commutes; that is, $ST = TS$, then $e^{S+T} = e^S e^T$.

Corollary 3

Consider the matrix $A = \begin{bmatrix} a & -b \\ b & a \end{bmatrix}$; then

$$e^A = e^a \begin{bmatrix} \cos b & -\sin b \\ \sin b & \cos b \end{bmatrix} \tag{15.19}$$

Corollary 4

Consider the matrix $A = \begin{bmatrix} a & b \\ 0 & a \end{bmatrix}$, then

$$e^A = e^a \begin{bmatrix} 1 & b \\ 0 & 1 \end{bmatrix} \tag{15.20}$$

Generalization

For a given matrix $B = P^{-1}AP$ having the forms:

$$B = \begin{bmatrix} \lambda & 0 \\ 0 & \mu \end{bmatrix}, B = \begin{bmatrix} \lambda & 1 \\ 0 & \lambda \end{bmatrix}_{\text{or}} B = \begin{bmatrix} a & -b \\ b & a \end{bmatrix}$$

we can define:

$$e^{Bt} = \begin{bmatrix} e^{\lambda t} & 0 \\ 0 & e^{\mu t} \end{bmatrix}, e^{Bt} = e^{\lambda t} \begin{bmatrix} 1 & t \\ 0 & 1 \end{bmatrix} \text{ or } e^{Bt} = e^{at} \begin{bmatrix} \cos bt & -\sin bt \\ \sin bt & \cos bt \end{bmatrix}$$

respectively, and $e^{At} = Pe^{Bt}P^{-1}$.

15.6   THE FUNDAMENTAL THEOREM OF LINEAR SYSTEMS

For an $n \times n$ matrix A and a given $x_0 \in R^n$, the initial value problem can be written as

$$x = Ax$$

$$x(0) = x_0 \tag{15.21}$$

and has a unique solution, $x(t) = e^{At}x_0$.

Example 5
Solve the initial value problem

$$\dot{x} = Ax, \text{ with } x(0) = \begin{bmatrix} 1 \\ 0 \end{bmatrix} \text{ and the matrix } A = \begin{bmatrix} -2 & -1 \\ 1 & -2 \end{bmatrix}.$$

According to the fundamental theorem of linear system, the solution to the system of equation will be

$$x(t) = e^{At}x(0).$$

We now need to evaluate e^{AT}. We also know that e^{AT} with A on the form

$$A = \begin{bmatrix} a & -b \\ b & a \end{bmatrix}$$

is given by

$$e^{At} = e^{at} \begin{bmatrix} \cos bt & -\sin bt \\ \sin bt & \cos bt \end{bmatrix} \tag{15.22}$$

In this case, $a = -2$ and $b = 1$. Therefore, $e^{At} = e^{-2t} \begin{bmatrix} \cos t & -\sin t \\ \sin t & \cos t \end{bmatrix}$

and the solution to the system is $\begin{bmatrix} x \\ y \end{bmatrix} = e^{-2t} \begin{bmatrix} \cos t & -\sin t \\ \sin t & \cos t \end{bmatrix} \begin{bmatrix} 1 \\ 0 \end{bmatrix}$

that is

$$\begin{cases} x = e^{-2t}\cos t \\ y = e^{-2t}\sin t \end{cases} \tag{15.23}$$

The phase portrait is given by the following Mathematica command (Figure 15.3):

$$\text{StreamPlot}\Big[\{-2x - y, x - 2y\}, \{x, -3, 3\}, \{y, -3, 3\}, \text{Axes} \rightarrow \text{True}, \text{AxesLabel} \rightarrow \{x, y\}\Big].$$

Example 6
Show the stream vector plot of the system $\begin{cases} \dot{x} = -x - y \\ \dot{y} = 2x + 3y \end{cases}$

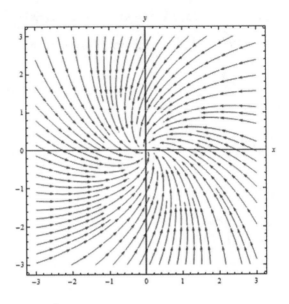

FIGURE 15.3 Phase portrait of $\begin{cases} \dot{x} = -2x - y \\ \dot{y} = x - 2y \end{cases}$.

Solution

The steam vector plot of the system

$$\begin{cases} \dot{x} = -x - y \\ \dot{y} = 2x + 3y \end{cases}$$

can be obtained using the Mathematica command (Figure 15.4):

$$\text{StreamPlot}\Big[\{-x - y, 2x + 3y\}, \{x, -3, 3\}, \{y, -3, 3\}, \text{AxesLabel} \to \{x, y\}, \text{Axes} \to \text{True}\Big]$$

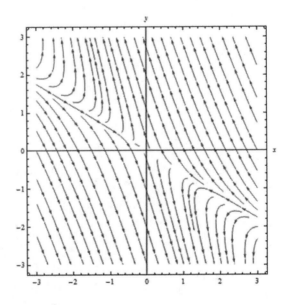

FIGURE 15.4 Phase portrait of $\begin{cases} \dot{x} = -x - y \\ \dot{y} = 2x + 3y \end{cases}$.

15.7 PHASE PORTRAIT AND SEPARATRICES

We will start this part with the following theorem

Theorem 3

Consider the linear system given by (15.1): $\dot{x} = Ax$.

If we assume $\delta = \det(A)$ and $\tau = \text{Trace}(A)$; Then, the following conditions are true:

a. If $\delta < 0$, then system (1) has a saddle at the origin.

b. If $\delta > 0, \tau^2 - 4\delta \geq 0$ and $\tau \neq 0$, then system (15.1) has a node at the origin. The node is stable if $\tau < 0$. It is unstable if $\tau > 0$.

c. If $\tau > 0$ and $\tau^2 - 4\delta < 0$ and $\tau \neq 0$, then system (15.1) has a focus at the origin. The node is stable if $\tau < 0$; it is unstable if $\tau > 0$; it is stable if $\tau < 0$; it is unstable if $\tau > 0$.

d. If $\tau > 0$ and $\tau = 0$, then, system (15.1) has a center at the origin.

Definition 1

A stable node or focus of the system (15.1) is called a sink of the linear system.

An unstable node or focus of the system (15.1) is called a source of the linear system.

The following graph summarizes the previous theorem and definition in a bifurcation diagram (Figure 15.5).

Example 7 (of a system of linear equations: the case of a simple harmonic motion)

An object of mass m is hung on a linear spring shown in Figure 15.6.

$x = l - l_o$ is the displacement from the equilibrium position. Using Newton's second law of motion, $\sum F = m\ddot{x}$. The only force F applying to the spring is the spring force $F = -kx$. The equation of motion becomes

$\ddot{x} + \dfrac{k}{m}x = 0$. Our goal is to analyze the system, to identify the state variable, and to graph the phase portrait. The state variables are the position x and the velocity v. So, the equation of motion can be split into

$$\begin{cases} \dot{x} = v \\ \ddot{x} = -\dfrac{k}{m}x \end{cases} \tag{15.24}$$

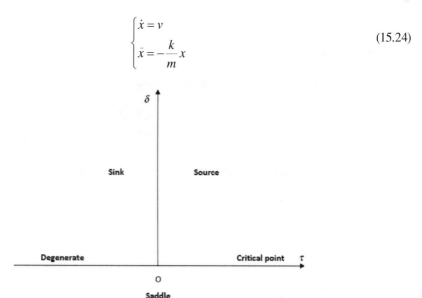

FIGURE 15.5 Bifurcation diagram of the linear system (15.1).

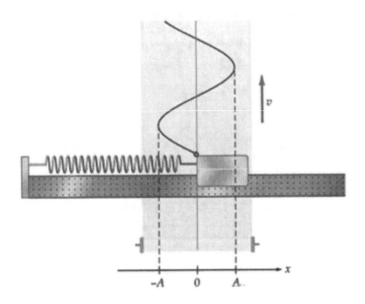

FIGURE 15.6 Simple harmonic motion.

We now consider $\dfrac{k}{m} = \omega^2$, and the equation becomes

$$\begin{cases} \dot{x} = 0x + 1v \\ \dot{v} = -\omega^2 x + 0v \end{cases} \tag{15.25}$$

with the matrix notation:

$$\begin{bmatrix} \dot{x} \\ \dot{v} \end{bmatrix} = \begin{bmatrix} 0 & 1 \\ -\omega^2 & 0 \end{bmatrix} \begin{bmatrix} x \\ v \end{bmatrix} \tag{15.26}$$

The fixed point occurs at $x = 0$, $y = 0$. It corresponds to a static equilibrium point of the system. The following command will help graph the phase portrait (Figure 15.7).

$$\text{StreamPlot}\Big[\{y, -0.5x\}, \{x, -10,10\}, \{y, -10,10\}, \text{Axes} \to \text{True}, \text{AxesLabel} \to \{x, v\}\Big]$$

The phase portrait shows orbits with the center, the fixed point (0,0). These orbits are closed; they represent the periodic motion of the system {mass, spring} or oscillations.

15.7.1 Fixed Point and Linearization

Let us consider the system of differential equations

$$\begin{cases} \dot{x} = f(x,y) \\ \dot{y} = g(x,y) \end{cases} \tag{15.27}$$

The fixed points (x^*, y^*) satisfy $f(x^*, y^*) = 0 = g(x^*, y^*)$. If $f(x,y)$ and $g(x,y)$ are nonlinear, they may not be easy to analyze. Even the phase portrait may not be that simple. In this case, it is interesting

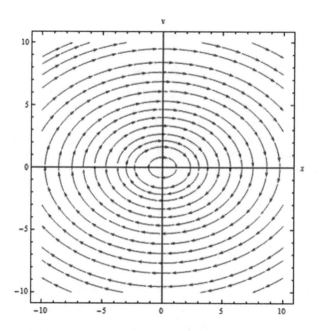

FIGURE 15.7   Phase portrait of the simple harmonic motion with $\omega^2 = 0.5 \, \text{Nm}^{-1} \, \text{kg}^{-1}$.

to introduce the displacements or disturbances $u = x - x^*$ and $v = y - y^*$, with respect to the fixed point. However, since x^* and y^* are constant values, the first derivatives of u and v will give $\dot{u} = \dot{x}$ and $\dot{v} = \dot{y}$. Therefore, $\dot{u} = \dot{x} = f(x, y) = f(u + x^*, v + y^*)$.

Taylor expansion to the first order gives:

$$\dot{u} = \dot{x} = f(x^*, y^*) + \frac{\partial f}{\partial u} u + \frac{\partial f}{\partial v} v + \theta(u^2, v^2, uv) + \cdots$$

The quadratic term and all the rest can be assumed very small so that they can be neglected. Also, notice that $\partial u = \partial x$ and $\partial v = \partial y$. Then,

$$\dot{u} = \dot{x} = f(x^*, y^*) + \frac{\partial f}{\partial x} u + \frac{\partial f}{\partial y} v + \theta(u^2, v^2, uv) + \cdots$$

We can have the same expression for v, that is:

$$\dot{v} = \dot{y} = g(x^*, y^*) + \frac{\partial g}{\partial x} u + \frac{\partial g}{\partial y} v + \theta(u^2, v^2, uv) + \cdots$$

If we remember that the point (x^*, y^*) is a fixed-point fulfilling $f(x^*, y^*) = 0 = g(x^*, y^*)$, we obtain:

$$\begin{cases} \dot{x} = \dfrac{\partial f}{\partial x} u + \dfrac{\partial f}{\partial y} v \\[3mm] \dot{y} = \dfrac{\partial g}{\partial x} u + \dfrac{\partial g}{\partial y} v \end{cases} \tag{15.28}$$

The previous system is now linear, the quadratic terms and others being neglected. The system can be written in matrix form:

$$\begin{bmatrix} \dot{u} \\ \dot{v} \end{bmatrix} = \begin{bmatrix} \dfrac{\partial f}{\partial x} & \dfrac{\partial f}{\partial y} \\ \dfrac{\partial g}{\partial x} & \dfrac{\partial g}{\partial y} \end{bmatrix} \begin{bmatrix} u \\ v \end{bmatrix} \qquad (15.29)$$

The matrix $J = \begin{bmatrix} \dfrac{\partial f}{\partial x} & \dfrac{\partial f}{\partial y} \\ \dfrac{\partial g}{\partial x} & \dfrac{\partial g}{\partial y} \end{bmatrix}$ is called the Jacobian matrix of the system in general and can be cal-

culated at the fixed point $\left(x^{*}, y^{*}\right)$.

Example 8

Find all the fixed points of the system:

$$\begin{cases} \dot{x} = -x^2 + x^3 \\ \dot{y} = -y \end{cases}$$

Linearize the system; compare both the nonlinear system and the linear system phase portraits.

First, the fixed points are obtained by setting,

$$\dot{x} = f(x, y) = 0 \Leftrightarrow x = 0, x = 1$$

$$\dot{y} = g(x, y) = 0 \Leftrightarrow y = 0$$

So, we have two fixed points (0,0) and (1,0).

Now, the Jacobian matrix is $J = \begin{bmatrix} -2x + 3x^2 & 0 \\ 0 & -1 \end{bmatrix}$. Therefore,

$$J(0,0) = \begin{bmatrix} 0 & 0 \\ 0 & -1 \end{bmatrix} \text{ and } J(1,0) = \begin{bmatrix} 1 & 0 \\ 0 & -1 \end{bmatrix}$$

So, we have two linear systems: $\begin{bmatrix} \dot{u} \\ \dot{v} \end{bmatrix} = \begin{bmatrix} 0 & 0 \\ 0 & -1 \end{bmatrix} \begin{bmatrix} u \\ v \end{bmatrix}$ and $\begin{bmatrix} \dot{u} \\ \dot{v} \end{bmatrix} = \begin{bmatrix} 0 & 0 \\ 0 & -1 \end{bmatrix} \begin{bmatrix} u \\ v \end{bmatrix}$.

StreamPlot[$\{0, -v\}, \{u, -10, 10\}, \{v, -10, 10\}$, Axes$->$True, AxesLabel$\rightarrow \{u, v\}$])

Phase portrait with degeneracy at the fixed point (0, 0) for the linearized system $\begin{bmatrix} \dot{u} \\ \dot{v} \end{bmatrix} = \begin{bmatrix} 0 & 0 \\ 0 & -1 \end{bmatrix} \begin{bmatrix} u \\ v \end{bmatrix}$..

StreamPlot[$\{u, -v\}, \{u, -10, 10\}, \{v, -10, 10\}$, Axes$->$True, AxesLabel$\rightarrow \{u, v\}$]

Phase portrait showing saddles points at the fixed point (1,0) for the linearized system $\begin{bmatrix} \dot{u} \\ \dot{v} \end{bmatrix} = \begin{bmatrix} 1 & 0 \\ 0 & -1 \end{bmatrix} \begin{bmatrix} u \\ v \end{bmatrix}$..

StreamPlot[$\{-x^2 + x^3, -y\}, \{x, -2, 2\}, \{y, -2, 2\}$, Axes$->$True, AxesLabel$\rightarrow \{x, y\}$]

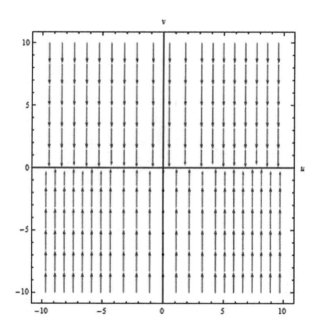

FIGURE 15.8   Phase portrait of the system $\begin{cases} \dot{x} = -x^2 + x^3 \\ \dot{y} = -y \end{cases}$.

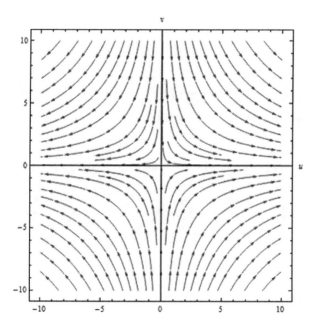

FIGURE 15.9   Phase portrait with degeneracy at the fixed point (0,0) for the linearized system $\begin{bmatrix} \dot{u} \\ \dot{v} \end{bmatrix} = \begin{bmatrix} 0 & 0 \\ 0 & -1 \end{bmatrix} \begin{bmatrix} u \\ v \end{bmatrix}$.

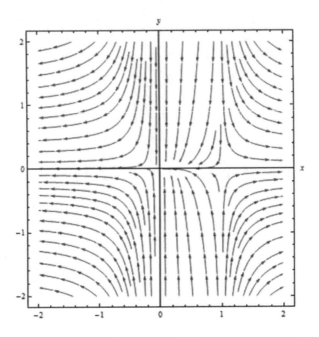

FIGURE 15.10 Phase portrait showing saddles points at (1,0) and a degeneracy at (0,0) for the nonlinearized

system, $\begin{bmatrix} 1 & 1 & 0 \\ 0 & 1 & 0 \\ 0 & 0 & 1 \end{bmatrix}$.

15.8   COMPLEX EIGENVALUES

If $\lambda_j = a_j + ib_j$ real matrix A has $2n$ distinct complex eigenvalues, $\lambda_j = a_j + ib_j$ and $\overline{\lambda}_j = a_j - ib_j$ and corresponding eigenvectors $w_j = u_j + iv_j$, with $j = 1, 2, 3, \ldots, n$, then $\{u_1, v_1, u_2, v_2, \ldots, u_n, v_n\}$ is a basis for $\mathbb{R}^{2n}$; the matrix $P = \begin{bmatrix} u_1 & v_1 & u_2 & v_2 \ldots, & u_n & v_n \end{bmatrix}$ is invertible and $P^{-1}AP = \operatorname{diag}\begin{bmatrix} a_j & -b_j \\ b_j & a_j \end{bmatrix}$ a real $2n \times 2n$ matrix with 2×2 blocks along the diagonal.

Corollary 5
Under these hypotheses, the solution of the initial value problem $\dot{x} = Ax$ and $x(0) = x_0$ is given by

$$x(t) = P\operatorname{diag}e^{a_jt}\begin{bmatrix} \cos b_j t & -\sin b_j t \\ \sin b_j t & \cos b_j t \end{bmatrix}P^{-1}x_0 \tag{15.30}$$

Remark 1
Note that the matrix $\begin{bmatrix} \cos b_j t & -\sin b_j t \\ \sin b_j t & \cos b_j t \end{bmatrix}$ represents a rotation through the angle $b_j t$ in radians.

Example 9
Solve the initial value problem for the matrix:

$$A = \begin{bmatrix} 1 & -1 & 0 & 0 \\ 1 & 1 & 0 & 0 \\ 0 & 0 & 3 & -2 \\ 0 & 0 & 1 & 1 \end{bmatrix}.$$

Solution

Matrix A has four complex eigenvalues, $\lambda_1 = 1+i, \overline{\lambda_1} = 1-i, \lambda_2 = 2+i$ and $\overline{\lambda_2} = 2-i$.

The corresponding eigenvectors are, respectively,

$$w_1 = u_1 + iv_1 = \begin{bmatrix} i \\ 1 \\ 0 \\ 0 \end{bmatrix} \text{ and } w_2 = u_2 + iv_2 = \begin{bmatrix} 0 \\ 0 \\ 1+i \\ 1 \end{bmatrix}. \text{ And the matrix } P \text{ is}$$

$$P = \begin{bmatrix} v_1 & u_1 & v_2 & u_2 \end{bmatrix} = \begin{bmatrix} 1 & 0 & 0 & 0 \\ 0 & 1 & 0 & 0 \\ 0 & 0 & 1 & 1 \\ 0 & 0 & 0 & 1 \end{bmatrix}$$

P is invertible with the inverse:

$$P^{-1} = \begin{bmatrix} v_1 & u_1 & v_2 & u_2 \end{bmatrix} = \begin{bmatrix} 1 & 0 & 0 & 0 \\ 0 & 1 & 0 & 0 \\ 0 & 0 & 1 & -1 \\ 0 & 0 & 0 & 1 \end{bmatrix}$$

and $P^{-1}AP = \begin{bmatrix} 1 & -1 & 0 & 0 \\ 1 & 1 & 0 & 0 \\ 0 & 0 & 2 & -1 \\ 0 & 0 & 1 & 2 \end{bmatrix}$

The solution to the initial value problem (15.21) is given by

$$x(t) = P \begin{bmatrix} e^t\cos t & -e^t\sin t & 0 & 0 \\ e^t\sin t & e^t\cos t & 0 & 0 \\ 0 & 0 & e^{2t}\cos t & -e^{2t}\sin t \\ 0 & 0 & e^{2t}\sin t & e^{2t}\cos t \end{bmatrix} P^{-1}x_0 \tag{15.31}$$

or

$$x(t) = \begin{bmatrix} e^t\cos t & -e^t\sin t & 0 & 0 \\ e^t\sin t & e^t\cos t & 0 & 0 \\ 0 & 0 & e^{2t}(\cos t + \sin t) & -2e^{2t}\sin t \\ 0 & 0 & e^{2t}\sin t & e^{2t}(\cos t - \sin t) \end{bmatrix} x_0 \tag{15.32}$$

15.9 JORDAN FORMS

Obtaining the Jordan canonical form may turn to be difficult for some matrices, but is useful in the theory of ordinary differential equations.

Theorem 4 [The Jordan canonical form]
Let A be a real matrix with real eigenvalues, $\lambda_j, j = 1,\ldots\ldots,k$, and complex eigenvalues, $\lambda_j = a_j + ib_j$ and $\overline{\lambda}_j = a_j - ib_j; j = k+1,\ldots,n$. There exists a basis $\{\vec{V}_1, \vec{V}_2,\ldots \vec{V}_k, \vec{V}_{k+1}, \vec{U}_{k+1}, \vec{V}_n, \vec{U}_n\}$ for R^{2n-k}, where $\vec{V}_j$, $j = 1,\ldots k$ and $\vec{w}_j, j = k+1,\ldots n$ are generalized eigenvectors of A, $\vec{U}_j = R_e(\vec{w}_j)$ and $\vec{V}_j = I_m(\vec{w}_j)$ for $j = k+1,\ldots,n$, such that the matrix $P = \vec{V}_1,\ldots \vec{V}_k \vec{V}_{k+1} \vec{U}_{k+1} \ldots \vec{V}_n \vec{U}_n$# is invertible and

$$P^{-1}AP = \begin{bmatrix} B_1 & & \\ & \ddots & \\ & & B_r \end{bmatrix} \tag{15.33}$$

where the elementary Jordan blocks $B = B_j, j = 1,\ldots,r$ are either of the form

$$B = \begin{bmatrix} \lambda & 1 & 0 & \cdots & 0 \\ 0 & \lambda & 1 & & \\ \cdots & \cdots & & \ddots & \\ 0 & \cdots & & \lambda & 1 \\ 0 & \cdots & & & \lambda \end{bmatrix} \tag{15.34}$$

for λ one of the real eigenvalues of A or of the form

$$B = \begin{bmatrix} D & I_2 & 0 & \cdots & 0 \\ 0 & D & I_2 & \cdots & 0 \\ \cdots & & & \ddots & \\ 0 & \cdots & & D & I_2 \\ 0 & & & 0 & D \end{bmatrix} \tag{15.35}$$

With $D = \begin{bmatrix} a & -b \\ b & a \end{bmatrix}, I_2 = \begin{bmatrix} 1 & 0 \\ 0 & 1 \end{bmatrix}$ and $O = \begin{bmatrix} 0 & 0 \\ 0 & 0 \end{bmatrix}$ the $2-2$ zero matrix for $\lambda = a + ib$, one of the complex eigenvalues of A.

The Jordan canonical form yields some information about the form of the solution of the initial value problem,

$$\begin{cases} \dot{X} = AX \\ X(0) = X_o \end{cases} \tag{15.36}$$

which gives, according to the fundamental theorem of linear systems, the solutions $x(t) = P\text{diag}\left|e^{B_j t}\right| P^{-1} X_o$.

If $B_j = B$ is an $m \times m$ matrix of the form (15.34) and λ is a real eigenvalue of A, then $B = \lambda I + N$ and

$$e^{Bt} = e^{\lambda t}e^{Nt} = e^{\lambda t}\begin{bmatrix} 1 & t & t^2/2! & \cdots & \cdots & \dfrac{t^{m-1}}{(m-1)!} \\ 0 & 1 & t & \cdots & \cdots & \dfrac{t^{m-2}}{(m-2)!} \\ 0 & 0 & 1 & & & \vdots \\ \vdots & \vdots & \vdots & & & \vdots \\ 0 & \vdots & \vdots & & & t \\ 0 & \vdots & \vdots & & & 1 \end{bmatrix}. \tag{15.37}$$

Similarly, if $B_j = B$ is a $2m \times 2m$ matrix of the form (15.35) and $\lambda = a + ib$ is a complex eigenvalue of A, then

$$e^{Bt} = e^{at}\begin{bmatrix} R & Rt & \dfrac{Rt^2}{2!} & \cdots & \cdots & \dfrac{Rt^{m-1}}{(m-1)!} \\ 0 & R & Rt & & & \dfrac{Rt^{m-2}}{(m-2)!} \\ \vdots & \vdots & \vdots & & & \dfrac{Rt^{m-3}}{(m-3)!} \\ \vdots & \vdots & \vdots & & & \vdots \\ 0 & \vdots & \vdots & & & Rt \\ 0 & \vdots & \vdots & & & R \end{bmatrix} \tag{15.38}$$

where the rotation matrix is $R = \begin{bmatrix} \cos bt & -\sin bt \\ \sin bt & \cos bt \end{bmatrix}$ since the $2m \times 2m$ matrix

$$B = \begin{bmatrix} 0 & I_2 & 0 & \cdots & 0 \\ 0 & 0 & I_2 & & 0 \\ \vdots & \vdots & \vdots & & \\ 0 & \vdots & \vdots & & I_2 \\ 0 & \vdots & \vdots & & 0 \end{bmatrix} \quad \text{is nilpotent of order } m \text{ and}$$

$$N^2 = \begin{bmatrix} 0 & 0 & I_2 & 0 & 0 \\ 0 & 0 & 0 & I_2 & 0 \\ \vdots & \vdots & \vdots & \vdots & \vdots \\ \vdots & \vdots & \vdots & \vdots & \vdots \\ 0 & \vdots & \vdots & \vdots & 0 \end{bmatrix}, \ldots N^{n-1} = \begin{bmatrix} 0 & 0 & \cdots & 0 & I_2 \\ 0 & \cdots & \cdots & \cdots & 0 \\ & & & & \\ & & \cdots & & \\ 0 & & & & 0 \end{bmatrix} \tag{15.39}$$

The solution of the initial value problem (15.21) leads to the following result.

Corollary 6
Each coordinate in the solution, $X(t)$, of the initial value problem (15.21) is a linear combination of functions of the form $t^k e^{at}\cos(bt)$ or $t^k e^{at}\sin(bt)$, where $\lambda = a + ib$ is an eigenvalue of matrix A and $0 \le k \le n-1$.

Definition 2

Let λ be an eigenvalue of the matrix A. The deficiency indices are given by $\delta_k = \dim\mathrm{Ker}(A - \lambda I)^k$.

Remember that the kernel of a linear operator $T\colon \mathbb{R}^n \to \mathbb{R}^n$ is given by the definition by

$$\mathrm{Ker}(T) = \left\{ X \in R^n \,\middle|\, T(X) = 0 \right\}.$$

δ_k can be found by using the Gauss–Jordan method of elimination. δ_k is the number of rows of zeros in the reduced row echelon form of $(A - \lambda I)^k$.

Example 10 (of finding a basis for $\mathbb{R}^3$)

Consider the matrix $A = \begin{bmatrix} 2 & 1 & 0 \\ 0 & 2 & 0 \\ 0 & -1 & 2 \end{bmatrix}$ and let us find a basis for R^3 that reduces it to its Jordan

canonical form. Now, let us find the eigenvalue: $|A - \lambda I| = O \Leftrightarrow \begin{vmatrix} 2-\lambda & 1 & 0 \\ 0 & 2-\lambda & 0 \\ 0 & -1 & 2-\lambda \end{vmatrix} = 0$

$\Leftrightarrow (2-\lambda)(2-\lambda)^2 = 0 \Leftrightarrow (2-\lambda)^3 = 0$;   so, $\lambda = 2$ is   eigenvalue   of   multiplicity   3;   also

$$A - \lambda I = \begin{bmatrix} 0 & 1 & 0 \\ 0 & 0 & 0 \\ 0 & -1 & 0 \end{bmatrix}.$$

Thus, $\delta_1 = 2$ and the eigenvector $\vec{v} \begin{pmatrix} X \\ Y \\ Z \end{pmatrix}$ is such that

$$(A - \lambda I)\vec{V} = \vec{O} \Rightarrow \begin{bmatrix} 0 & 1 & 0 \\ 0 & 0 & 0 \\ 0 & -1 & 0 \end{bmatrix}\begin{bmatrix} X \\ Y \\ Z \end{bmatrix} = \begin{bmatrix} 0 \\ 0 \\ 0 \end{bmatrix} \Leftrightarrow \begin{vmatrix} Y = 0 \\ -Y = 0 \end{vmatrix} \Rightarrow Y = 0$$

$$\begin{pmatrix} X \\ Y \\ Z \end{pmatrix} = \begin{pmatrix} X \\ 0 \\ Z \end{pmatrix} = Z\begin{pmatrix} 0 \\ 0 \\ 1 \end{pmatrix} + X\begin{pmatrix} 1 \\ 0 \\ 0 \end{pmatrix}$$

We can choose $V_1^{(1)} = \begin{bmatrix} 1 \\ 0 \\ 0 \end{bmatrix}$ and $V_2^{(1)} = \begin{bmatrix} 0 \\ 0 \\ 1 \end{bmatrix}$ as a basis for

$\mathrm{Ker}(A - \lambda I)$.

Consider the linear combination $V = C_1 V_1^{(1)} + C_2 V_2^{(2)}$. Now of these two vectors, let us solve $(A - \lambda I)\vec{V} = C_1 \vec{V}_1^{(1)} + C_2 \vec{V}_2^{(1)}$.

$$\begin{bmatrix} 0 & 1 & 0 \\ 0 & 0 & 0 \\ 0 & -1 & 0 \end{bmatrix}\begin{bmatrix} X \\ Y \\ Z \end{bmatrix} = \begin{bmatrix} C_1 \\ 0 \\ C_2 \end{bmatrix} \Leftrightarrow \begin{cases} Y = C_1 \\ -Y = C_2 \Rightarrow c_2 = -C_1 \end{cases}$$

we can choose $c_1 = 1 \Rightarrow c_2 = -1$ and the vector is $\begin{bmatrix} 1 \\ 0 \\ -1 \end{bmatrix}$; we can now put together the three vectors:

$$\overrightarrow{V_1^{(1)}} = \begin{bmatrix} 1 \\ 0 \\ 0 \end{bmatrix}; \ \vec{V}_2^{(1)} = \begin{bmatrix} 0 \\ 0 \\ 1 \end{bmatrix}; \ V_1^{(2)} = \begin{bmatrix} 1 \\ 0 \\ -1 \end{bmatrix}$$

If we choose $V_1^{(1)} = \begin{bmatrix} 1 \\ 0 \\ -1 \end{bmatrix}$, then we need to find $V_1^{(2)}$ such that $(A - \lambda I) v_j^{(2)} = V_1^{(1)}$

or

$$\begin{bmatrix} 0 & 1 & 0 \\ 0 & 0 & 0 \\ 0 & -1 & 0 \end{bmatrix} \begin{bmatrix} X \\ Y \\ Z \end{bmatrix} = \begin{bmatrix} 1 \\ 0 \\ -1 \end{bmatrix} \Rightarrow \begin{cases} Y = 1 \\ -Y = -1 \end{cases} \Rightarrow Y = 1 \Rightarrow \begin{bmatrix} X \\ Y \\ Z \end{bmatrix} = \begin{bmatrix} X \\ 1 \\ Z \end{bmatrix}$$

we can choose $X = Z = 0$ and $v_1^{(2)} = \begin{bmatrix} 0 \\ 1 \\ 0 \end{bmatrix} v$ and $V_1^{(2)} = \begin{bmatrix} 1 \\ 0 \\ 0 \end{bmatrix}$.

The three vectors $\begin{bmatrix} 1 \\ 0 \\ -1 \end{bmatrix}, \begin{bmatrix} 0 \\ 1 \\ 0 \end{bmatrix}$ and $\begin{bmatrix} 1 \\ 0 \\ 0 \end{bmatrix}$ help to write the matrix $\underline{P}$:

$$\underline{P} = \begin{bmatrix} 1 & 0 & 1 \\ 0 & 1 & 0 \\ -1 & 0 & 0 \end{bmatrix} \text{ and } P^{-1} = \begin{bmatrix} 0 & 0 & -1 \\ 0 & 1 & 0 \\ 1 & 0 & 1 \end{bmatrix}$$

It can be proved that

$$P^{-1}AP = \begin{bmatrix} 2 & 1 & 0 \\ 0 & 2 & 0 \\ 0 & 0 & 2 \end{bmatrix} \tag{15.40}$$

Now, we can give an algorithm for finding a basis of generalized eigenvectors of a $n \times n$ matrix with a real eigenvalue λ of multiplicity n.

1. Find a basis $\left\{ \overrightarrow{v_j^{(1)}} \right\}_{j=1}^{\delta_1}$ for $\text{Ker}(A - \lambda I)$.

2. If $\delta_2 > \delta_1$, choose a basis $\left\{ \overrightarrow{v_j^{(1)}} \right\}_{j=1}^{\delta_1}$ for $\text{Ker}(A - \lambda I)$ such that $(A - \lambda I) v_j^{(2)} = V_j^{(1)}$ has $\delta_2 > \delta_1$

   linearly independent solutions $v_j^{(2)}, j = 1, 2, \dots \delta_2 - \delta_1$. Then $\left\{ \overrightarrow{v_j^{(2)}} \right\}_{j=1}^{\delta_2} = \left\{ V_j^{(1)} \right\}_{j=1}^{\delta_1} \cup \left\{ v_j^{(2)} \right\}_{j=1}^{\delta_2 - \delta_1}$

   is a basis for $\text{Ker}(A - \lambda I)^2$.

3. If $\delta_3 > \delta_2$, choose a basis $\left\{V_j^{(2)}\right\}_{j=1}^{\delta_2}$ for $\mathrm{Ker}(A-\lambda I)^2$ with $V_j^{(2)} \in \mathrm{span}\left\{\vec{v}_j^{(2)}\right\}_{j=1}^{\delta_2-\delta_1}$ for

$j=1,\ldots,\delta_2-\delta_1$ such that $\left(A-\lambda I\right)\vec{v}_j^{(3)} = \vec{V}_j^{(2)}$ has $\delta_3-\delta_2$ linearly independent solutions

$v_j^{(3)}, j=1,\ldots.\delta_3-\delta_2$. If for $j=1,\ldots,\delta_2-\delta_1, \overline{V_j^{(2)}} = \Sigma_{i=1}^{\delta_2-\delta_1} c_i v_i^{(2)}$ let $\overline{V_j^{(1)}} = \displaystyle\sum_{i=1}^{\overline{\delta_2-\delta_1}} c_i V_i^{(1)}$ and

$\tilde{V}_j^{(1)} = V_j^{(1)}$ for $j=\delta_2-\delta_1+1,\ldots,\delta_1$. Then $\left\{\overrightarrow{V_j^{(1)}}\right\}_{j=1}^{\delta_3} = \left\{\overrightarrow{V_j^{(1)}}\right\}_{j=1}^{\delta_1} \cup \left\{\overrightarrow{V_j^{(2)}}\right\}_{j=1}^{\delta_2-\delta_1} \cup \left\{\overrightarrow{v_j^{(3)}}\right\}_{j=1}^{\delta_j-\delta_2}$ is a

basis for $\mathrm{Ker}(A-\lambda I)^3$.

4. Continue this process up to $\delta_k = n$ to obtain a basis $B = \left\{\overrightarrow{v_j^{(k)}}\right\}_{j=1}^n$ for R^n. Matrix A will then

assume its Jordan canonical form with respect to this basis.

15.10   STABLE AND UNSTABLE SPACE

Consider the linear system: $\dot{X} = AX$; assume $\lambda_i = a_i + ib_j$ the eigenvalues of A corresponding to the eigenvectors. We can define the following subspaces:

$$E^s = \mathrm{span}\left\{\overrightarrow{u_j}, \overrightarrow{v_j} \mid a_j < 0\right\}$$

$$E^s = \mathrm{span}\left\{\overrightarrow{u_j}, \overrightarrow{v_j} \mid a_j = 0\right\}$$

$$E^s = \mathrm{span}\left\{\overrightarrow{u_j}, \overrightarrow{v_j} \mid a_j > 0\right\}$$

As subspaces of $\mathbb{R}^n$

Example: For the matrix $\begin{bmatrix} -2 & -1 & 0 \\ 1 & -2 & 0 \\ 0 & 0 & 3 \end{bmatrix}$ we have $\lambda_1 = -2+i$, and $\lambda_2 = 3$.

- For $\lambda_1 = -2+i$, the corresponding eigenvector is $W_1 = U_1 + iV_1 = \begin{bmatrix} 0 \\ 1 \\ 0 \end{bmatrix} + i\begin{bmatrix} 1 \\ 0 \\ 0 \end{bmatrix}$

- For $\lambda_2 = 3$, $U_2 = \begin{bmatrix} 0 \\ 0 \\ 1 \end{bmatrix}$

The subspace E^s is stable and is the XY-plane and the unstable subspace E^u is the Z-axis.
    The phase portrait of the linear system $\dot{X} = AX$ is given in Figure 15.11.

15.10.1   Flow

The solution to the initial value problem is given by $\dot{X}(t) = AX(t)$.
    The set of mappings $e^{At} : \mathbb{R}^n \to \mathbb{R}^n$ is called the flow of the linear system $\dot{X}(t) = AX(t)$ with $X(0) = AX_0$.

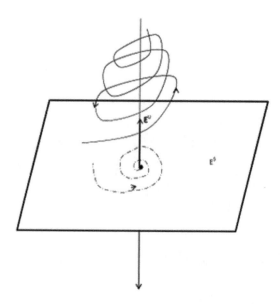

FIGURE 15.11   Stable and unstable subspaces of the system.

15.10.2   HYPERBOLIC FLOW

For the $n \times n$ matrix A, if all the eigenvalues have nonzero real parts, then the flow $e^{At} : \mathbb{R}^n \to \mathbb{R}^n$ is called a hyperbolic flow and the system (15.1) is called a hyperbolic linear system.

15.10.3   INVARIANT SUBSPACE

A subspace $E \subset \mathbb{R}^n$ is called invariant with respect to the flow $e^{At} : \mathbb{R}^n \to \mathbb{R}^n$ if $e^{At} E \subset E$ for all $t \in \mathbb{R}^n$.

Theorem 5a

Let A be a real $n \times n$ matrix. Then, $\mathbb{R}^n = E^s \oplus E^u \oplus E^c$ where E^s, E^u and E^c are the stable, unstable and center subspaces of (15.1), respectively. Furthermore, E^s, E^u and E^c are invariant with respect to the flow e^{At} of (15.1), respectively.

Definition 3

If all of the eigenvalues of A have negative (positive) real parts, the origin is called a sink (source) for the linear system (15.1).

Example 11

Consider the system (15.1) in which matrix A is given by

$$A = \begin{bmatrix} 1 & 0 & 0 \\ 2 & 1 & -1 \\ 3 & -1 & 1 \end{bmatrix}$$

Let us find the eigenvalues of A.

$$AV = \lambda V$$

$$A - \lambda I = \begin{bmatrix} 1-\lambda & 0 & 0 \\ 2 & 1-\lambda & -1 \\ 3 & -1 & 1-\lambda \end{bmatrix}$$

$$A - \lambda I = (1-\lambda)[(1-\lambda)^2 - 1]$$
$$= (1-\lambda)(1-\lambda-1)(1-\lambda+1)$$
$$= -\lambda(1-\lambda)(2-\lambda)$$

$$|A - \lambda I| = 0 \Leftrightarrow \lambda = 0 \quad \lambda = 1 \quad \lambda = 2$$

All these eigenvalues are real and positive for $\lambda = 0$, and we have the center subspace. Let us find the system that spans $AV = \lambda V$ for $\lambda = 0$ and $V = \begin{bmatrix} X \\ Y \\ Z \end{bmatrix}$.

For $\lambda = 0$, the center subspace E^c is spanned by $\begin{bmatrix} 0 \\ 1 \\ 1 \end{bmatrix}$.

For $\lambda = 1$

$$\begin{bmatrix} 1 & 0 & 0 \\ 2 & 1 & -1 \\ 3 & -1 & 1 \end{bmatrix}\begin{bmatrix} X \\ Y \\ Z \end{bmatrix} = \begin{bmatrix} X \\ Y \\ Z \end{bmatrix} \Leftrightarrow \begin{cases} X = X \\ 2X + Y - Z = Y \\ 3X - Y + Z = Z \end{cases}$$

$$\Leftrightarrow \begin{cases} X \in R \\ 2X - Z = 0 \quad Z = 2X \\ 3X - Y = 0 \quad Y = 3X \end{cases}.$$

For $\lambda = 2$

$$\begin{bmatrix} 1 & 0 & 0 \\ 2 & 1 & -1 \\ 3 & -1 & 1 \end{bmatrix}\begin{bmatrix} X \\ Y \\ Z \end{bmatrix} = 2\begin{bmatrix} X \\ Y \\ Z \end{bmatrix}$$

$$\begin{cases} X = 2X \\ 2X + Y - Z = 2Y \\ 3X - Y + Z = 2Z \end{cases}$$

$$\begin{cases} X = 0 \\ Y = -Z \\ Z = -Y \end{cases}$$

$$\text{So,}\begin{bmatrix} X \\ Y \\ Z \end{bmatrix} = \begin{bmatrix} 0 \\ Y \\ -Y \end{bmatrix} = \begin{bmatrix} 0 \\ 1 \\ -1 \end{bmatrix}$$

We can try to normalize the three previous eigenvectors: $\begin{bmatrix} 0 \\ 1 \\ 1 \end{bmatrix}$, $\begin{bmatrix} 1 \\ 3 \\ 2 \end{bmatrix}$ and $\begin{bmatrix} 0 \\ 1 \\ -1 \end{bmatrix}$.

$$\begin{bmatrix} 1 \\ 3 \\ 2 \end{bmatrix} = \sqrt{14}\begin{bmatrix} 1/\sqrt{14} \\ 3/\sqrt{14} \\ 2/\sqrt{14} \end{bmatrix} \Rightarrow \mu_2 \begin{bmatrix} 1/\sqrt{14} \\ 3/\sqrt{14} \\ 2/\sqrt{14} \end{bmatrix}$$

$$\sqrt{1^2 + 3^2 + 2^2} = \sqrt{1+9+4} = \sqrt{14}$$

$$\begin{bmatrix} 0 \\ 1 \\ -1 \end{bmatrix} = \sqrt{2}\begin{bmatrix} 0 \\ 1/\sqrt{2} \\ -1/\sqrt{2} \end{bmatrix} \Rightarrow \mu_3 = \begin{bmatrix} o \\ 1/\sqrt{2} \\ -1/\sqrt{2} \end{bmatrix}$$

μ_2 and μ_2 span the unstable subspace.

15.11   FUNDAMENTAL MATRIX

Definition 4

The $n \times n$ matrix $\psi(t) = \left[X^1(t), X^2(t), \dots, X^n(t) \right]$ each column of which is an independent solution of $\dot{X} = AX$,

with

$$A = \begin{bmatrix} 7 & -4 \\ 3 & -1 \end{bmatrix}, X_0 = X(0) = \begin{bmatrix} 4 \\ 8 \end{bmatrix}.$$

The characteristic equation is given by:

$$|A - \lambda I| = \begin{vmatrix} 7-\lambda & -4 \\ 3 & -1-\lambda \end{vmatrix} = (7-\lambda)(-1-\lambda) + 12$$

$$= \lambda^2 - 6\lambda + 5$$

$$|A - \lambda I| = (\lambda - 1)(\lambda - 5)$$

We have two eigenvalues, $\lambda_1 = 1$ and $\lambda_2 = 5$. Let us now find the eigenvectors.

$$\lambda_2 = 5 \Rightarrow \begin{cases} 2x - 4y = 0 \\ 3x - 6y = 0 \end{cases} \Rightarrow X = 2Y$$

$$\Rightarrow \begin{pmatrix} x \\ y \end{pmatrix} = \begin{pmatrix} 2y \\ y \end{pmatrix} = y\begin{pmatrix} 2 \\ 1 \end{pmatrix}$$

$$v_2 = \begin{pmatrix} 2 \\ 1 \end{pmatrix}$$

$$\lambda_1 = 1 \Rightarrow \begin{cases} 6x - 4y = 0 \\ 3x - 2y = 0 \end{cases}$$

$$\Leftrightarrow \begin{cases} y = \dfrac{3}{2}x \\ y = \dfrac{3}{2}x \end{cases}$$

$$\begin{pmatrix} x \\ y \end{pmatrix} = \begin{pmatrix} 3^x \\ 2 \end{pmatrix} = \frac{1}{2}x\begin{pmatrix} 2 \\ 3 \end{pmatrix} \Rightarrow v_1 = \begin{pmatrix} 2 \\ 3 \end{pmatrix}$$

$$\underline{P} = \begin{pmatrix} 2 & 2 \\ 3 & 1 \end{pmatrix} \underline{P}^{-1} = \begin{pmatrix} -1/4 & 1/2 \\ 3/4 & -1/2 \end{pmatrix} \text{ and assume}$$

$$P^{-1}x_0 = \begin{pmatrix} 3 \\ -1 \end{pmatrix} \equiv \begin{pmatrix} c_1 \\ c_2 \end{pmatrix}. \text{ Then,}$$

$$x(t) = P \text{diag}\left[e^{\lambda it} \right] P^{-1}x_o = \begin{pmatrix} 2 & 2 \\ 3 & 1 \end{pmatrix} \begin{pmatrix} e^t & 0 \\ 0 & e^{5t} \end{pmatrix} \begin{pmatrix} 3 \\ -1 \end{pmatrix}$$

$$= c_1 \begin{pmatrix} 2 \\ 3 \end{pmatrix} e^t + c_2 \begin{pmatrix} 2 \\ 1 \end{pmatrix} e^{5t} = c_1 v_1 e^{\lambda_1 t} + c_2 v_2 e^{\lambda_2 t}$$

15.12   STABILITY

15.12.1   ASYMPTOTIC STABILITY

Definition 5. Assume the system of ODE $\dot{x}(t) = Ax$. The equilibrium position is given by the condition $x(t) = 0$.

Definition 6. The equilibrium position $x(t) = 0$ for the system of ODE $\dot{x}(t) = Ax$ is said to be locally asymptotically stable if for all $x(t)$ starting at $x(0)$, sufficiently close to 0, $\lim_{t \to \infty} x(t) = 0$.

Remark 2

If x^* in the solution $x(t) = e^{At} \cdot x_0$, then the system of ODE $\dot{x}(t) = Ax(t)$ is stable.

Definition 7 (Global Stability. Liapunov's Second Method)

Assume x^* the equilibrium points of $x(t)$; any differentiable weighted distance function $V(x) \geq 0$ is called a Liapunov function.

Theorem 5b

The system $\dot{x} = Ax$ is globally stable if there exists a function $V(x) = x'Bx$ where B is positive definite such that $W \equiv A'B + BA$ is negative definite.

Qualitative solution of $\dot{X} = f(x)$

Consider the linear equation $\dot{X} = aX \Rightarrow \dot{X} = f(x) = aX$; we have two cases: $a > 0$ and $a > 0$ or $a < 0$ and $a > 0$.

So, for $a > 0$, $f'(x) > 0$ as X increases, $\dot{X} = f(x)$ also increases and as X decreases, $\dot{X} = f(x)$ also decreases. For $a < 0 f'(x) < 0 f$ is a decreasing function. In other words, as X increases in the positive values, $\dot{X} = f(x)$ decreases. The velocity field is negative.

$f(x) = 0 \Leftrightarrow x = 0$ is an equilibrium of fixed point. For $a < 0, f'(x) < 0$ and the stable point $\dot{X} = 0$ is called attractor. For $a > 0, f'(x) > 0$ and the unstable point $\dot{X} = 0$ is called a repeller.

Example 12

We can give the example of the logistic curve.

$\dot{x} = x(a - bx)$ with $a \neq 0$ and $b \neq 0$ $a, b > 0$

$$f(x) = x(a - bx)$$

$$f'(x) = a - bx + x(-b) = a - 2bx$$

The fixed points are given by $f'(x) = 0 \Leftrightarrow x = 0$ or $x = \dfrac{a}{b}$

$$f'(x) = 0 \Leftrightarrow x = \frac{a}{2b}$$

$$f\left(\frac{a}{2b}\right) = \frac{a}{2b}\left(a - \frac{ba}{2b}\right)$$

$$f\left(\frac{a}{2b}\right) = \frac{a^2}{2b} - \frac{a^2}{4b}$$

$$f\left(\frac{a}{2b}\right) = \frac{a^2}{4b}$$

At $x = 0$, $x > 0$, $f'(x) > 0$, $\dot{x} = f(x)$ is increasing; the velocity is positive the system is moving away from the fixed point $x = 0$.

For $x = 0$ and $x < 0, f'(x) > 0 \Rightarrow \dot{x} < 0 \Rightarrow x$ decreases over time. The system moves away from the point. So, the fixed point $x = 0$ is a Repeller.

$$\lim_{x \to \left(\frac{a}{b}\right)^+} f(x) < 0$$

with $x > 0$.

$$x > \frac{a}{b} \Leftrightarrow bx > a \Leftrightarrow a - bx < o \Leftrightarrow x(a - bx) < 0 \Leftrightarrow f(x) < 0$$

So, x decreases over time.

$$\lim_{x \to \left(\frac{a}{b}\right)^-} f(x) < 0$$

with $x > 0$

$$x\left(\frac{a}{b} \Leftrightarrow a - bx\right) o \Leftrightarrow x(a - bx) > 0 \Leftrightarrow f(x) > 0$$

$$\frac{dx}{dt} > 0 \Rightarrow x$$

increases over time.

So, the system moves toward the fixed point, $x < 0, f'(x) > 0 \Rightarrow \dot{x} < 0 \Rightarrow x$, which is then called an attractor.

Example 12

Consider the equation $\dot{x} = x^2 - 1$; the fixed points occur for $f(x) = x^2 - 1 = 0 \Leftrightarrow x = -1$ or $x = +1$. So, the system has two fixed points. Also $f'(x) = 2x$.

$x = -1$ is positive outside the roots $x = -1$ and $x = +1$ and is negative inside. In other words, at $x = -1$, on the right $f(x) < 0 \Rightarrow x$ decreases over time; therefore, the arrow of the velocity field $\dot{x}$ is going toward the fixed point $x = -1$ (Figure 15.12).

On the left of $x = -1$, $f(x) > 0 \Rightarrow x$ increases over time $\Rightarrow$ the velocity field $\dot{x}$ moves toward the fixed point $x = -1$. As a result, the fixed point $x = -1$ represents an attractor.

At $x = +1$, on the right side, $f(x) > 0$, this means that x increases over time and the system runs away from that fixed point. But on the left of $x = +1 f(x) < 0$ and x decreases over time: the system runs away from $x = +1$ on the left. As a result, the fixed point $x = +1$ represents a Repeller.

Example 13
Case of $\dot{x} = 2x^2$

$$f(x) = 2x^2 f'(x) = 4x$$

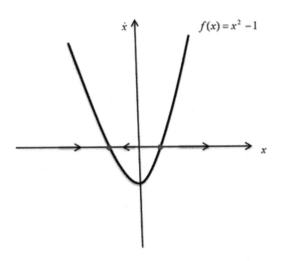

FIGURE 15.12   Fixed pour of $\dot{x} = x^2 - 1$.

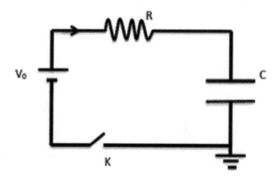

FIGURE 15.13 An *RC* circuit.

The fixed points are given by $f(x) = 0 \Leftrightarrow x = 0$
  $\forall x \neq 0 \, f(x) > 0$; so, x always increases with time.
  The fixed point $x = 0$ is a shunt.

Case of $\dot{x} = -2x^2$

$$f(x) = -2x^2 \Leftrightarrow f'(x) = -4x$$

Fixed point $f(x) = 0 \Leftrightarrow x = 0$
  $\forall x \neq 0 \, f(x) < 0 \Leftrightarrow \dot{x} < 0 \Leftrightarrow x$ decreases over time. The fixed-point $x = 0$ is shunt.

Example 14 (in a one-dimensional dynamical system): the *RC* circuit
Let us consider the following *RC* circuit (Figure 15.13).
  At the initial time $t = 0$ s, the switch is closed. There's no charge on the capacitor. For $t \geq 0$, using the Kirchhoff rule of the loop we have:

$$-V_0 + Ri + V_C = 0$$

Where V_0 is the voltage across the generator, i is the current intensity flowing throughout the circuit, V_c is the voltage across the capacitor C and R is the resistance of the resistor inserted in the circuit.
  We know that $i = \dfrac{dq}{dt}$, and the above equation becomes:

$$-V_0 + R\frac{dq}{dt} + \frac{q}{c} = 0$$

$$\text{or } \frac{dq}{dt} = \frac{V_0}{R} - \frac{1}{Rc}q$$

So, the *RC* circuit is modeled by the differential equation, $\dot{q} = \dfrac{dq}{dt} = f(q) = \dfrac{V_0}{R} - \dfrac{1}{Rc}q$.

For the fixed point, it occurs at, $q^* = CV_0$. Let us now graph $\dot{q} = f(q)$ (Figure 15.14).
For $q > q^*, \dot{q} < 0$, and the flow is to the left
For $q < q^*, \dot{q} > 0$, and the flow is to the right

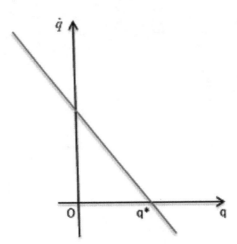

FIGURE 15.14   Phase portrait of the RC circuit.

Because the flow is always toward $i = \dfrac{dq}{dt}$, we conclude that the fixed point q^* is stable. It is in fact globally stable for it is approached from all initial conditions.

15.12.2   GRAPH OF $Q(T)$

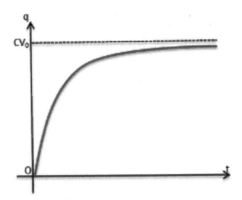

FIGURE 15.15   Graph of the charge of a capacitor.

15.12.3   APPLICATION

Solve the following autonomous dynamical system of nonlinear differential equations:

$$(S)\begin{cases} \dfrac{dx}{dt} = y \\ \dfrac{dy}{dt} = ay - by^2 - cxy \end{cases}. \qquad (15.41)$$

The Jacobian is needed now in order to study theoretically the behavior of the system. It can be written as

$$J(x,y) = \begin{bmatrix} 0 & y \\ -cy & a - 2by - cx \end{bmatrix}. \qquad (15.42)$$

The equilibrium points are given by $y = 0$ and $(a - by - cx) = 0$.

As a result, every point $B = (x_0, 0)$ is an equilibrium point to the system. The Jacobian then becomes

$$J(x_o, 0) = \begin{bmatrix} 0 & 0 \\ 0 & a - cx_o \end{bmatrix} \tag{15.43}$$

Let τ be the trace and δ the determinant of the Jacobian, of the system S at $B = (x_0, 0)$. $\tau = a - cx_o$ and $\delta = 0$. However, because $\tau^2 - 4\delta = (a - cx_o)^2 > 0$, we identify three cases as follows:

If $\tau > 0$ $x_0 < \dfrac{a}{c}$ we have a Source.

If $\tau < 0$ we have a Sink.

If StreamPlot$\left[\{ y, y(1 - y - x) \}, \{ x, -5, 5 \}, \{ y, -5, 5 \} \right]$ for all x.

Let us sketch the phase portrait of the following dynamical system S (Figure 15.16):

$$\begin{cases} \dfrac{dx}{dt} = y \\ \dfrac{dy}{dt} = y(a - by - cx) \end{cases}$$

$$\text{StreamPlot}\left[\{ y, y(1 - y - x) \}, \{ x, -5, 5 \}, \{ y, -5, 5 \} \right]$$

$$\text{StreamPlot}\left[\{ y, y(2 - y - 3x) \}, \{ x, -5, 5 \}, \{ y, -5, 5 \} \right]$$

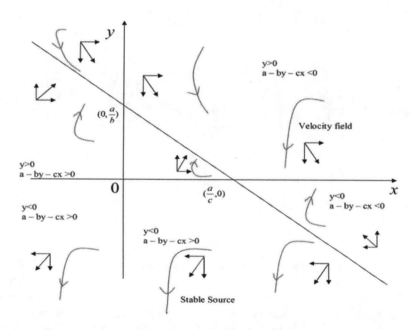

FIGURE 15.16   General phase portrait of the dynamical system (S).

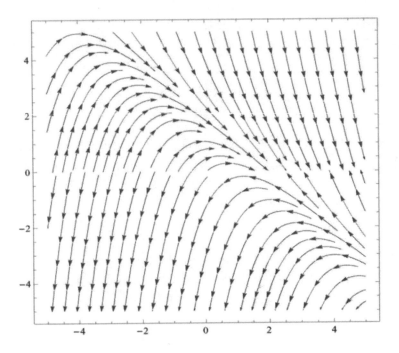

FIGURE 15.17   Phase portrait of the dynamical system for $a = 1, b = 1, c = 1$.

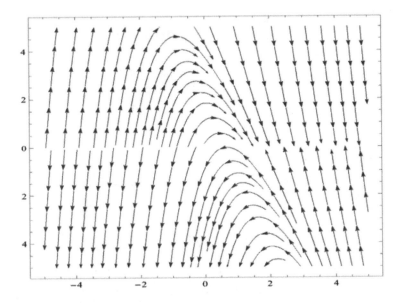

FIGURE 15.18   Phase portrait of the dynamical system for $a = 2, b = 1, c = 3$.

15.13   INTRODUCTION TO CHAOS

Chaos theory is the science of surprise, of nonlinear deterministic phenomenon. It describes the phenomenon that changes due to small changes in the initial conditions. Edward Lorenz is the pioneer of chaos who worked intensively on non-linear systems and later applied his mathematical knowledge to Meteorology where systematically he proved how atmospheric phenomena are such unpredictable. Chaos theory also known as the Butterfly effect is used in phenomena like turbulence, weather, the stock market, our brain states, biological phenomena and atmospheric phenomena. These phenomena are often described by fractal mathematics, which captures the infinite complexity of nature. Recognizing the chaotic, fractal nature of our world can give us new insight, power and wisdom.

15.13.1 Lorenz's Equations

Lorenz has done an unprecedented work in the chaos. He was an American mathematician and meteorologist who discovered the strange attractor also called the butterfly effect embedded in his equations:

$$\begin{cases} \dot{x} = \sigma\left(y - x\right) \\ \dot{y} = rx - y - xz \\ \dot{z} = xy - bz \end{cases} \tag{15.44}$$

Lorentz derived these equations in 1963 from a simplified model of convection rolls in the atmosphere. These equations are also used in lasers and dynamos models. The equations are deterministic and the solutions oscillate irregularly in the phase space. The coefficients or parameters σ, r and b are positive, σ is called the Prandtl number, r is the Rayleigh number, but b is a constant number. Lorenz proved that in a given range of parameters, there could be no stable fixed points, no stable limit cycles and all trajectories remain confined to a bounded region and are eventually attracted to a set of zero volume. These trajectories look very complicated and are called strange attractors. The strange attractor is a fractal with a fractional dimension between 2 and 3.

15.13.2 Mechanical Model of Lorenz Equation: The Chaotic Waterwheel

The chaotic waterwheel was invented in the 1970s by Willem Malkus and Lou Howard. It consists of a wheel with some leaky cups in its rim. Water poured into one of the cups triggers the motion of the wheel. The flow of water is steady so that all cups will get water. The cups will not be too full. This will help avoid friction. The wheel can rotate in any direction, clockwise and counterclockwise. When the flow rate increases, the rotation can be destabilized. We then have a chaotic motion. Newton's second law of motion applies to the wheel in motion: $F = ma$ and the equation of motion of the wheel is given by

$$I\dot{\omega} = \text{damping torque}\left(\tau_d\right) + \text{gravitational torque}\left(\tau_g\right)$$

I is the moment of inertia of the wheel. The damping torque is given by $\tau_d = -v\omega$, the rotational damping rate, $v > 0$. The negative sign indicates the fact that the damping is against the motion of the wheel. The gravitational torque can be calculated considering a portion of the mass of the wheel rotating as a pendulum. Consider that an infinitesimal mass dm is chosen around point A as in Figure 15.19. The weight of that mass is dmg, and the torque of the mass around the axis of rotation (OB) is given by

$$d\tau^2{}_d = dmgr\sin\theta d\theta \tag{15.45}$$

with r the radius of the wheel and $d\theta$ the angle between (OA) and (OB). The notation τ_d^2 is used because of the two differentials dm and $d\theta$ at the second side of the equation. By integrating to get rid of one differential, we obtain:

$$d\tau_d = mgr\sin\theta d\theta \tag{15.46}$$

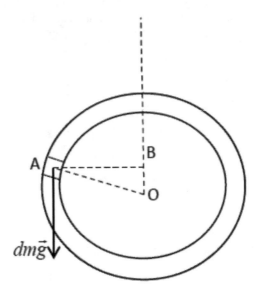

FIGURE 15.19   Schematic of the rotating wheel.

and finally,

$$\tau_d = gr \int_0^{2\pi} m(\theta,t)\sin\theta d\theta \tag{15.47}$$

The mass m depends on θ and t; that's why we used $m = m(\theta,t)$ in equation (15.47).

15.13.2.1   Equation of Motion of the Wheel

The equation of motion of the wheel will be given by

$$I\dot{\omega} = -v\omega + gr \int^{2\pi} m(\theta,t)\sin\theta d\theta \tag{15.48}$$

Equation (15.48) is an integro-differential equation.

15.13.2.2   Equation of Continuity

We will give here without proof the equation of continuity. The proof can be seen in several books and documents*. It is given by

$$\frac{\partial m}{\partial t} = Q - Km - \omega \frac{\partial m}{\partial \theta} \tag{15.49}$$

where K is the leakage rate of the wheel as it rotates. $Q = Q(\theta)$ is the inflow that is the rate at which water flows into the nozzles at the position θ.

15.13.3   PROPERTIES OF THE LORENTZ EQUATIONS

In his original paper, Lorenz used the case $\sigma = 10, b = 8/3, r = 28$. For $r < 1$, the origin $(0,0,0)$ is globally attracting; all trajectories tend to the origin.

For $r > 1$, there are three stationary or equilibrium points: $x = y = r - 1, z = \dfrac{(r-1)^2}{b}$.

15.13.3.1 Lorenz Equation Is Symmetric

By changing x into $-x$, and y into $-y$, Lorenz's equation doesn't change. This generates two types of images that are symmetric to each other. The z-axis is invariant under the flow.

15.13.3.2 Nonlinearity

Lorenz's equation is nonlinear. They are two nonlinear terms in $\dot{y}$ and $\dot{z}$.

15.13.3.3 Fixed Points

At the origin, and by neglecting the nonlinear terms, Lorenz's equation becomes

$$\begin{cases} \dot{x} = \sigma(y-x) \\ \dot{y} = rx - y \\ \dot{z} = -bz \end{cases} \tag{15.50}$$

We can see that z decreases exponentially. Considering the first equation, we have the system:

$$\begin{pmatrix} x \\ y \end{pmatrix} = \begin{pmatrix} -\sigma & \sigma \\ r & -1 \end{pmatrix} \begin{pmatrix} x \\ y \end{pmatrix} \tag{15.51}$$

The matrix associated with the system is $A = \begin{pmatrix} -\sigma & \sigma \\ r & -1 \end{pmatrix}$. The trace of A is $\tau = -\sigma - 1 < 0$, and the determinant is $\delta = \sigma - r\sigma$.

For $r > 1$, the origin is a saddle point.

For $r < 1$, the origin is a sink; it is a stable node.

In general, for $r > 1$, there are three stationary points, which are: the origin, unstable and two other points $C^{\pm} = \left\{ \pm b\sqrt{r-1}, \pm b\sqrt{r-1}, r-1 \right\}$ (Figures 15.20 and 15.21).

15.13.3.4 Global Stability

For $r < 1$ and $t \to \infty$, the origin is globally stable. Therefore, there is no limit cycles or chaos.

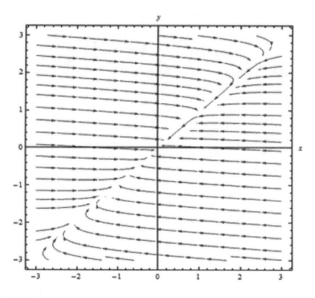

FIGURE 15.20 Graph of the system represented by (15.51) with $\sigma = 10$ and $r = 0.5 < 1$.

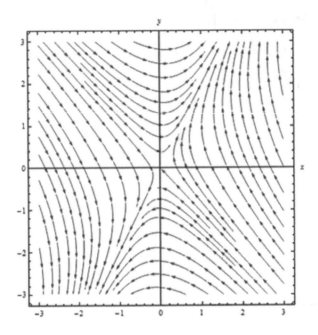

FIGURE 15.21   Graph of the system represented by (15.51) with $\sigma = 10$ and $r = 21$.

PROBLEM SET 15

15.1 Study the quantitative solutions of the differential equations; graph the phase-portrait

$$\dot{x} = -x^2 - 4 \qquad \dot{x} = -4x^2 \qquad \dot{x} = x^2$$
$$\dot{x} = x^2 - x - 2 \qquad \dot{x} = x + 2x - 15$$
$$\dot{x} = 2x^3 \qquad \dot{x} = -5x^3$$

15.2 Use the stability analysis to study the qualitative behavior of the system determined by
$\dot{x} = \sin(x)$

15.3 Analyze the following equation qualitatively and graphically.

$$\dot{x} = e^{-2x}\sin x; \quad \dot{x} = e^{-x}\cos x; \quad \dot{x} = 2 + 3\cos x$$

15.4 Find the fixed points of the following systems. Graph the phase-portrait and analyze the stability.

$$\dot{x} = x(x-3)(x+2); \ \dot{x} = (x-1)(x+1); \ \dot{x} = x^3(2-x)$$
$$\dot{x} = \ln x; \ \dot{x} = 1 - e^{-x^2}$$

15.5 Find the fixed points and analyze the stability of the systems.

$$\dot{x} = \frac{1}{x^2+1}; \ \dot{x} = \sqrt{4-x^2}; \ \dot{x} = \ln(x-5)$$

15.6 Analyze $\dot{x} = 2x^2$, show the fixed point and draw the stability diagram.

15.7 Analyze $\dot{x} = -2x^2$, show the fixed point and draw the stability diagram.

15.8 a. Use the geometric series to compute the exponential of the following matrix

$$T = \begin{bmatrix} 2 & 0 \\ 0 & -1 \end{bmatrix}, T = \begin{bmatrix} 1 & 3 \\ 0 & 2 \end{bmatrix}, T = \begin{bmatrix} 1 & 0 \\ 4 & 1 \end{bmatrix}$$

b. For each matrix find the eigenvalues of e^T

c. Show that if $\vec{u}$ is an eigenvector of T corresponding to the eigenvalue, λ, then $\vec{u}$ is also eigenvector of e^T corresponding to the eigenvalue e^λ.

15.9 Consider the matrix $A = \begin{bmatrix} a & -b \\ b & a \end{bmatrix}$; prove that $e^A = e^a \begin{bmatrix} \cos b & -\sin b \\ \sin b & \cos b \end{bmatrix}$

15.10 Consider the matrix $A = \begin{bmatrix} a & b \\ 0 & a \end{bmatrix}$; prove that $e^A = e^a \begin{bmatrix} 1 & b \\ 0 & 1 \end{bmatrix}$.

15.11 Find e^{At} and solve the near system $x = Ax$ for the following matrix A

$$A = \begin{bmatrix} 1 & 5 \\ 5 & 1 \end{bmatrix}; A = \begin{bmatrix} 4 & -3 \\ -3 & 4 \end{bmatrix}; A = \begin{bmatrix} 3 & 1 & -1 \\ 1 & 3 & 1 \\ 2 & 1 & -1 \end{bmatrix}$$

15.12 Solve the initial value problem (15.1) with

a. $A = \begin{bmatrix} 3 & -2 \\ 1 & 1 \end{bmatrix}$ b. $A = \begin{bmatrix} 1 & 0 & 0 \\ 0 & 2 & -3 \\ 1 & 3 & 2 \end{bmatrix}$

15.13 Find the Jordan canonical forms of the matrices:

(1) $A = \begin{bmatrix} 1 & 0 \\ 0 & 1 \end{bmatrix}$

(2) $A = \begin{bmatrix} 0 & 1 \\ 1 & -1 \end{bmatrix}$

(3) $A = \begin{bmatrix} 1 & 1 & 0 \\ 0 & 1 & 0 \\ 0 & 0 & 1 \end{bmatrix}$

(4) $A = \begin{bmatrix} 1 & 1 & 2 \\ 0 & 2 & 1 \\ 0 & 0 & 2 \end{bmatrix}$.

16 Variational Methods

The search for truth is in one way hard and in another easy. For it is evident that no one can master it fully nor miss it wholly. But each adds a little to our knowledge of nature and from all the facts assembled these arises a certain grandeur.

–Aristotle

INTRODUCTION

This chapter gives the background of variational methods. Examples have been given in classical and quantum mechanics and in optics.

16.1   BACKGROUND

In this chapter, we will consider the variable $x = x(t)$ and the function $f\left(t, x, \dfrac{dx}{dt}\right)$. Consider the integral:

$$S = \int_0^t f\left(t, x, \frac{dx}{dt}\right) dt. \tag{16.1}$$

In the theory of variations, a basic question is: for what function x, is the integral S stationary for small variations of x? In these conditions, assume $p = \dfrac{dx}{dt}$, then $f(t, x, p)$. Assume now that we consider x, and x', varying very slightly by $x'(t) = x(t) + \delta x(t)$ where α is a constant value and g a function of t which is differentiable and has continuous second derivatives. We can write

$$\begin{aligned} x'(t) &= x(t) + \delta x(t) \\ \delta x(t) &= \alpha g(t) \end{aligned} \tag{16.2}$$

It is interesting to note that $x(t)$ makes the integral (22.1) stationary if

$$\frac{\partial S}{\partial \alpha} = 0 \tag{16.3}$$

that is:

$$\frac{\partial}{\partial \alpha} \left[\int_0^t f'\left(t, x', \frac{dx'}{dt}\right) dt \right] = 0 \tag{16.4}$$

For $\alpha = 0$,

$$\begin{aligned} &\frac{\partial}{\partial \alpha} \left[\int_{t_0}^{t_1} f\left(p + \alpha \frac{dg}{dt}, x + \alpha g(t), t\right) dt \right] \\ &= \left[\int_{t_0}^{t_1} \frac{\partial}{\partial \alpha} \left[f\left(p + \alpha \frac{dg}{dt}, x + \alpha g(t), t\right) \right] dt \right] \\ &= \int_{t_0}^{t_1} \left(\frac{dg}{dt} \frac{\partial f}{\partial p}, x + g(t) \frac{\partial f}{\partial x} \right) dt \end{aligned} \tag{16.5}$$

DOI: 10.1201/9781003478812-16

Integration by parts gives

$$\left[g \frac{\partial f}{\partial p} \right]_{t_0}^{t_1} + \int_0^t g \left(\frac{\partial f}{\partial x} - \frac{d}{\partial t} \frac{\partial f}{\partial p} \right) dt \tag{16.6}$$

We can choose g in such a way that the expression $\delta y = 0$ and

$$L = \frac{\partial f}{\partial x} - \frac{d}{\partial t} \left(\frac{\partial f}{\partial p} \right) = 0 \tag{16.7}$$

An important case to look at occurs when f is not an explicit function of t. Also, by multiplying the differential equation by p we have:

$$\frac{\partial f}{\partial x} \frac{dx}{\partial t} - p \frac{d}{\partial t} \left(\frac{\partial f}{\partial p} \right) = 0 \tag{16.8}$$

However, because

$$\frac{df}{dx} = \frac{\partial f}{\partial x} \frac{dx}{\partial t} + \frac{\partial f}{\partial p} \frac{dp}{\partial t} = \frac{dx}{\partial t} \left(\frac{\partial f}{\partial p} \right) + p \frac{d}{\partial t} \left(\frac{\partial f}{\partial p} \right) \text{ and } \frac{\partial f}{\partial t} = 0, \tag{16.9}$$

we have

$$\frac{d}{\partial t} \left(p \frac{\partial f}{\partial p} - f \right) = p \frac{d}{\partial t} \frac{\partial f}{\partial p} - \frac{\partial f}{\partial x} \frac{dx}{\partial t} = 0 \tag{16.10}$$

As a result,

$$p \frac{\partial f}{\partial p} - f = \text{constant} \tag{16.11}$$

Also, considering $p = \frac{\partial y}{\partial t}$ and $f = \left(p^2 + 1 \right)^{1/2}$, we have $\frac{\partial f}{\partial p} = p \left(p^2 + 1 \right)^{-1/2}$

$$p \frac{\partial f}{\partial p} - f = \frac{p^2}{\left(p^2 + 1 \right)^{1/2}} - \left(p^2 + 1 \right)^{1/2} = \frac{-1}{\left(p^2 + 1 \right)^{1/2}} \tag{16.12}$$

For $p \dfrac{\partial f}{\partial p} - f$ to be constant, p must be constant, that is $\dfrac{\partial y}{\partial x} = \text{Constant} = c$, or $y = cx + y_0$, or y is a straight line with respect to x.

16.2   VARIATIONAL METHOD IN QUANTUM MECHANICS

Variational methods help to solve many problems in physics. One example occurs when physicists try to solve the eigenvalue equation:

$$H\psi_n = E_n \psi_n \tag{16.13}$$

where H is the Hamiltonian of an arbitrary physical system. Assume that H is a time-dependent Hamiltonian and E_n the different energy states of the quantum system. Assume that the energy of the ground state E_0 is known and that the corresponding wave function of the ground state, ψ_0, is also known. Consider the functional form:

$$\ell = \langle \phi | H | \psi \rangle - \lambda \langle \phi \# \psi \rangle \tag{16.14}$$

where ϕ and ψ are variable functions; λ is a complex number. We look for the conditions for which λ will be stationary, for very small changes in the functions ϕ and ψ. The variational derivatives of λ with respect to ϕ and ψ gives

$$\frac{\partial \lambda}{\partial \phi} = 0, \text{ implies} \left(H - \lambda \right) \# \psi >= 0;$$

$$\frac{\partial \lambda}{\partial \psi} = 0, \text{ implies} \left(H^+ - \lambda^* \right) \# \phi >= 0.$$

So, the condition of λ being stationary implies that ψ must be the eigenvector of H with the corresponding eigenvalue λ and that ϕ must be the eigenvector of H with the corresponding eigenvalue λ^*. If H is a Hermitian operator, that is $H^+ = H$, then, $\lambda = \lambda^*$, and the functions ϕ and ψ need to be proportional, specially we can choose $|\psi \rangle = | \phi \rangle$.

Sometimes, it is not easy to obtain stationary conditions. L may not have a maximum or a minimum. But it can have an inflexion point or a saddle point. In such conditions, the variational theorem appears to be useful.

Theorem 1 (Variational theorem)

If $H = H^+$ and the following inequality holds for every, ψ:

$$E_0 \leq \frac{< \psi | H | \psi >}{\langle \psi | \psi >} \tag{16.15}$$

Proof

To proof inequality (16.15), let us consider the operator $|\psi_n >< \psi_n|$ and the expansion of $| \psi >$:

$$| \psi > = \sum_n |\psi_n >< \psi_n | H | \psi >$$

then,

$$\langle \psi | H | \psi \rangle = \sum_n \langle \psi | H | \psi_n \rangle < \psi_n|$$

But $H | \psi_n >= E_n | \psi_n >$; therefore, the previous expression becomes $< \psi | H | \psi >= \sum_n E_n \langle \psi | \psi_n > < \psi_n | \psi \rangle$.

16.2.1 HAMILTON'S PRINCIPLE

Consider a system of particles of mass m_r located at the position x_{ri}. The force F_{ri} acting upon these particles is given by

$$F_{ri} = m_r \ddot{x}_{ri} \left(r = 1, 2, \ldots, n; \quad i = 1, 2, 3, \ldots \right) \tag{16.16}$$

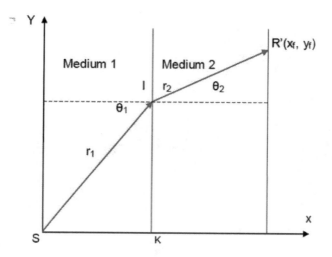

FIGURE 16.1   Schematic of the Fermat principle.

16.3   THE LAW OF REFRACTION AND FERMAT'S PRINCIPLE

Consider light traveling from medium 1 to medium 2 according to Figure 16.1.

The impact point of the incident light is $I = (x, y)$; $K = (x, 0)$ is the intercept of the surface of separation and the x-axis:

$R' = (x_f, y_f)$ is the impact point of the light in medium 2.

The path of the light is SIR' and the length of the path is

$(SR') = SI + IR' = r_1(y) + r_2(y)$. The optical path is defined as $(SR') = n_1 r_1(y) + n_2 r_2(y)$.

The time for light to propagate from S to I is $t_1 = \dfrac{r_1(y)}{v_1}$.

The time light takes to travel from I to R' is $t_2 = \dfrac{r_2(y)}{v_2}$.

The total time light takes to travel from S to R' is given by

$$t = \frac{r_1(y)}{v_1} + \frac{r_2(y)}{v_2}.$$ (16.17)

By using Pythagoras theorem, we have $r_1(y) = \sqrt{x^2 + y^2}$ and $r_2(y) = \sqrt{(x_f - x)^2 + (y_f - y)^2}$.

The total time the path of light from point S to point R' can be written as $t = \dfrac{\sqrt{x^2 + y^2}}{v_1} + \dfrac{\sqrt{(x_f - x)^2 + (y_f - y)^2}}{v_2}$. The optimum condition requires that the first derivative of the total time with respect to the specific variable y be zero: $\dfrac{dt}{dy} = 0$ or

$$\frac{2y}{2v_1\sqrt{x^2+y^2}} + \frac{-2(y_f-y)}{2v_2\sqrt{(x_f-x)^2+(y_f-y)^2}} = 0.$$

that is,

$$\frac{y}{v_1 r_1} = \frac{(y_f-y)}{v_2 r_2} = 0$$

But we know that $\sin\theta_1 = \dfrac{y}{r_1} = 0$ and $\sin\theta_2 = \dfrac{(y_f-y)}{r_2}$.

As a result, we have $\dfrac{\sin\theta_1}{v_1} = \dfrac{\sin\theta_2}{v_2}$.

Using the relation $v = \dfrac{c}{n}$ for each medium, we have the law of refraction:

$$n_1\sin\theta_1 = n_2\sin\theta_2 \tag{16.18}$$

16.4   MINIMUM PRINCIPLES AND APPLICATIONS

Consider Newton's second law of motion expressed with an arbitrary potential $V(\vec{r})$:

$$m\ddot{\vec{r}} = \vec{F} = -\vec{\nabla}V(\vec{r}) \tag{16.19}$$

These types of equations are mostly nonlinear differential equations and difficult to solve; that's what explains the use sometimes of some special potential which makes it easy to solve. In this sub-chapter, we will consider the mapping $V \to V$ such that a vector x in V is connected to Ax in V, with A, a linear operator. Let us consider the inner product quantity $I = (x, Ax)$ of x and Ax. Consider that vector x is continuous at some parameter, ε; this means that $x = x(\varepsilon)$. We can then rewrite $I(\varepsilon) = (x(\varepsilon), Ax(\varepsilon))$. Now, let us take the first derivative of $I(\varepsilon)$ with respect to ε; we get

$$\frac{dI(\varepsilon)}{d\varepsilon} = \left[\frac{dx(\varepsilon)}{d\varepsilon}, Ax(\varepsilon)\right] + \left(x(\varepsilon), A\frac{dx(\varepsilon)}{d\varepsilon}\right)$$

The intent is, in fact, to extremize the quantity, $I(\varepsilon)$; so, operator A needs to be Hermitian. Moreover, for an arbitrary vector x, the only stationary value of I is $I = 0$. However, to make this problem simple, let's add the constraint, $\|I(\varepsilon)\| = (x, x) = J = 1..$ Let us also consider that x is a function of two parameters ε_1 and ε_2. x could be function of many more parameters. But two are sufficient for the purpose of this section. So, let us consider $\bar{x} = \bar{x}(\varepsilon_1, \varepsilon_2)$ such that the following conditions are satisfied:

1. $\bar{x}(\varepsilon_1, \varepsilon_2) \in V$, for all ε_1 and ε_2.

2. $\bar{x}(0,0) = x$ is the extremizing vector.

3. $\dfrac{\partial \overline{x}}{\partial \varepsilon_1}$ and $\dfrac{\partial \overline{x}}{\partial \varepsilon_2}$ exist and belong to V.

Now, we can write

$$I\left(\varepsilon_1,\varepsilon_2\right)=\left(\overline{x}\left(\varepsilon_1,\varepsilon_2\right),A\overline{x}\left(\varepsilon_1,\varepsilon_2\right)\right) \tag{16.20}$$

and

$$J\left(\varepsilon_1,\varepsilon_2\right)=\left(\overline{x}\left(\varepsilon_1,\varepsilon_2\right),\overline{x}\left(\varepsilon_1,\varepsilon_2\right)\right)=1 \tag{16.21}$$

Let us now consider the linear combination of the quantities $I\left(\varepsilon_1,\varepsilon_2\right)$ and $J\left(\varepsilon_1,\varepsilon_2\right)$ as follows:

$$K\left(\varepsilon_1,\varepsilon_2\right)=I\left(\varepsilon_1,\varepsilon_2\right)-\lambda J\left(\varepsilon_1,\varepsilon_2\right) \tag{16.22}$$

where λ is the Lagrange multiplier.

The condition to obtain an extremum is given by

$$\left(\frac{\partial K}{\partial \varepsilon_1}\right)_{\varepsilon_1,\varepsilon_2=0}=0=\left(\frac{\partial K}{\partial \varepsilon_2}\right)_{\varepsilon_1,\varepsilon_2=0} \tag{16.23}$$

Also, at the extremum point $\overline{x}\left(\varepsilon_1,\varepsilon_2\right)$, we can write

$$K\left(\varepsilon_1,\varepsilon_2\right)=\left(\overline{x}\left(\varepsilon_1,\varepsilon_2\right),A\overline{x}\left(\varepsilon_1,\varepsilon_2\right)\right)-\lambda\left(\overline{x}\left(\varepsilon_1,\varepsilon_2\right),\overline{x}\left(\varepsilon_1,\varepsilon_2\right)\right)$$

$$K\left(\varepsilon_1,\varepsilon_2\right)=\left(\overline{x}\left(\varepsilon_1,\varepsilon_2\right),\left[A-\lambda I\right]\overline{x}\left(\varepsilon_1,\varepsilon_2\right)\right) \tag{16.24}$$

Since operator A is Hermitian and $I\left(\varepsilon_1,\varepsilon_2\right)$ is real, then λ is real. We can now calculate the partial derivatives of K with respect to ε_1 and ε_2, respectively.

$$\left(\frac{\partial K}{\partial \varepsilon_1}\right)=\left(\frac{\partial \overline{x}}{\partial \varepsilon_1},\left[A-\lambda I\right]\overline{x}\right)+\left(\overline{x},\left[A-\lambda I\right]\frac{\partial \overline{x}}{\partial \varepsilon_1}\right) \tag{16.25}$$

$$\left(\frac{\partial K}{\partial \varepsilon_2}\right)=\left(\frac{\partial \overline{x}}{\partial \varepsilon_2},\left[A-\lambda I\right]\overline{x}\right)+\left(\overline{x},\left[A-\lambda I\right]\frac{\partial \overline{x}}{\partial \varepsilon_2}\right) \tag{16.26}$$

For $\varepsilon_1=\varepsilon_2=0$, we have

$$\left(x_{\varepsilon_1},\left[A-\lambda I\right]x\right)+\left(x,\left[A-\lambda I\right]x_{\varepsilon_1}\right)=0 \tag{16.27}$$

and

$$\left(x_{\varepsilon_2},\left[A-\lambda I\right]x\right)+\left(x,\left[A-\lambda I\right]x_{\varepsilon_2}\right)=0 \tag{16.28}$$

with $x_{\varepsilon_i} = \left(\dfrac{\overline{\partial x}}{\partial \varepsilon_i} \right)_{\varepsilon_i = 0}$, $i = 1,2$

In case x is function of several variables, such as, ε_1, ε_2, ..., ε_n, we can generalize the two expressions,

$$\left(x_{\varepsilon_i}, \left[A - \lambda I \right] x \right) + \left(x, \left[A - \lambda I \right] x_{\varepsilon_i} \right) = 0, \quad i = 1,2 \tag{16.29}$$

Assume $x_{\varepsilon_i} = \left[A - \lambda I \right] x$, we obtain:

$$2 \left(\left[A - \lambda I \right] x, \left[A - \lambda I \right] x \right) = 0,$$

which gives $A - \lambda I = 0$ or $Au = \lambda I u$ for a given vector u, that is the eigenvalue problem in the calculus of variations. Now, we can state the following theorems.

Theorem 2a
Consider A, a linear self-adjoint operator. The eigenvalue problem, $Ax = \lambda x$, holds if and only if the inner product (x, Ax) is extremized with respect to the constraint $(x, x) = 1$. The extremized value of I is λ, the Lagrange multiplier.

Theorem 2b
Consider A, a linear self-adjoint operator in an n-dimensional inner space. Assume λ_1 the least eigenvalue of A and λ_n, the largest eigenvalue of A, with n a non-zero integer. The minimum and maximum values of (x, Ax) are λ_1 and λ_n, respectively, for all vectors $x \in V$ such that the constraint $(x, x) = 1$ holds.

In physics and engineering, it is intuitively assumed that for a Hermitian operator A, there always exists the lowest eigenvalue of A for a given eigenvector x such that (x, Ax) is minimum.

Example 1
The Hamiltonian of the harmonic oscillator in 1-D is given by

$$H = \frac{p^2}{2m} + \frac{1}{2} m\omega^2 x^2,$$

which gives the operator

$$H = -\frac{\hbar^2}{2m} \frac{d^2}{dx^2} + \frac{1}{2} m\omega^2 x^2.$$

Let us use the trial vector ϕ and consider $I = (\phi, H\phi) = \int_{\infty}^{+\infty} \phi^*(x) H\phi(x) dx$.

In physics, the eigenfunctions such as ϕ are chosen such that they converge to zero at infinity. As an example, let us consider the eigenfunctions, $\phi(x) = C \exp\left(-\frac{\alpha}{2}|x| \right)$ with $\alpha > 0$ and C a constant real value. The constraint condition would be $(\phi, \phi) = 1$, which means $C^2 \int_{\infty}^{+\infty} \exp(-\alpha|x|) dx = 1$. The constant C will be then found to be $c = \left(\dfrac{\alpha}{2} \right)^{1/2}$, and the eigenfunctions can be written as

$$\phi(x) = \left(\frac{\alpha}{2}\right)^{1/2} \exp\left(-\frac{\alpha}{2}|x|\right), \quad \alpha > 0 \tag{16.30}$$

$$I = \frac{\hbar^2\alpha^2}{8m} + \frac{m\omega^2}{\alpha^2} \tag{16.31}$$

The previous quantity I is extremum if $\dfrac{dI}{d\alpha} = 0$, which gives $\alpha = \alpha_0 = \left(\dfrac{8m^2\omega^2}{\hbar^2}\right)^{1/4}$. In this case the

extremum of I is a minimum. At the minimum point, I takes the value $I(\alpha_0) = \dfrac{\hbar\omega}{\sqrt{2}}$. This value of

I is greater than the eigenvalue, $\lambda = \dfrac{1}{2}\hbar\omega$, by 40%. However, a better trial eigenfunction could be

$$\phi(x) = C\exp\left(-\frac{\alpha x^2}{2}\right).$$

PROBLEM SET 16

16.1   If $\int ds$ is stationary for variations of a path with fixed termini,
       where $ds^2 = g_{ik}dx_i dx_k, (i,k = 1,2,3,4)$

   Prove that $\dfrac{d}{ds}\left(g_{ik}\dfrac{dx_k}{ds}\right) = \dfrac{\partial g_{km}}{\partial x_i}\dfrac{\partial x_k}{\partial s}\dfrac{\partial x_m}{\partial s}$,

   three of these equations being independent.

16.2   If in question 1

$$ds^2 = c^2\left(1 - \frac{2fM}{c^2 r}\right)dt^2 - \left(1 - \frac{2fM}{c^2 r}\right)^{-1} dr^2 - r^2\left(d\theta^2 + \sin^2\theta d\lambda^2\right),$$

   find three first integrals of the equation of motion; and if a particle moves nearly in a
   circle in the plane $\theta = \pi/2$, find the apsidal angle.

16.3   Use the trial function $\phi(x) = C\exp\left(-\dfrac{\alpha x^2}{2}\right)$ to find an upper bound for the lowest eigen-

   value of the quantum harmonic oscillator.

16.4   Given a linear transformation A on a finite-dimensional vector space, prove that in each
       coordinate system, the adjoint matrix always exists.

17 Introduction to Banach and Hilbert's Spaces

The eternal mystery of the world is its comprehensibility.

–Albert Einstein

INTRODUCTION

This chapter has importance in quantum mechanics, where physicists use Hilbert space. Banach space and Hilbert's space are related to each other.

17.1 LINEAR SPACE

Definition

A set $\mathbb{R}$ of elements $f, g, h, \ldots$ is a linear space if there are defined,

(i) an operation called addition denoted as "+" with respect to $\mathbb{R}$. The elements $f, g, h, \ldots$ are called points or vectors.

(ii) an operation called multiplication denoted as "×" or "·" of elements of $\mathbb{R}$ by real or complex numbers $\alpha, \beta, \gamma, \ldots$, such that

$$\alpha \cdot (f + g) = \alpha \cdot f + \alpha \cdot g \tag{17.1}$$

$$(\alpha + \beta) \cdot f = \alpha \cdot f + \beta \cdot g \tag{17.2}$$

$$(\alpha \beta) \cdot f = \alpha \cdot (\beta f) \tag{17.3}$$

$$1 \cdot f = f \tag{17.4}$$

$$0 \cdot f = 0 \tag{17.5}$$

17.2 METRIZABLE LINEAR SPACE

A linear space is metrizable if for $f, g, h, \ldots$, there is a real or complex number (f,g) called the product of f and g which satisfies the conditions:

a. $(f,g)^* = (g,f)$ $\tag{17.6}$

b. $(\alpha f_1 + \beta f_2, g) = \alpha (f_1, g) + \beta (f_2, g)$ $\tag{17.7}$

c. $(f,f) \geq 0$ $\tag{17.8}$

d. $(f,f) = 0,$ if and only if $f = 0$ $\tag{17.9}$

DOI: 10.1201/9781003478812-17

17.3 FUNDAMENTAL SEQUENCE

A sequence $\{f_n\}_{n=1}^{\infty}$ is called fundamental (or Cauchy) if and only if

$$\forall \varepsilon > 0, \exists N(\varepsilon) / \forall n > N(\varepsilon), m > N(\varepsilon), f_m, f_n < \varepsilon \qquad (17.10)$$

We can write

$$\lim_{x \to \infty} \left\| f_m, f \right\|_n = 0 \qquad (17.11)$$

17.4 BANACH SPACE

Definition
A set B of elements $u, v, w, \ldots$ is called a Banach space if and only if the following properties are satisfied:

i. **Property 1**

   B is a linear vector space over the field of real numbers or over the field of complex numbers

ii. **Property 2**

$$\forall u, v \in B, \left\| u + v \right\| \le \left\| u \right\| + \left\| v \right\| \qquad (17.12)$$

$$\forall u \in B, \forall a \in R, \left\| \alpha u \right\| = 0 \le \left| \alpha \right| \left\| u \right\| \qquad (17.13)$$

$$\forall u \in B, \left\| u \right\| = 0 \Leftrightarrow u = 0 \qquad (17.14)$$

iii. **Property 3**

For every sequence, $u_1, u_2, \ldots, u_n \in B$, such that $\lim_{n,m \to \infty} \left\| u_n - u_m \right\| = 0$, there exists only one element u, such that $\lim_{n \to \infty} \left\| u_n - u \right\| = 0$.

The limit of u_n as n tends to infinity is given by $\lim_{n \to \infty} u_n = u$.

We would need to state the situation where each element u of B is assigned to an element v of B through an operator A, that is, we write
$V = Au$.

Lemma 1
Let A be an operator in Banach space B, and α a real number such that $0 \le \alpha < 1$, so that for all $u, v \in B, \left\| Au - Av \right\| \le a \left\| u - v \right\|$, then, there exists one z element of B such that $z = Az$.

Proof
Let $u_0 \in B; u_0$ is arbitrary. Consider the sequence

$$
\begin{aligned}
u_1 &= Au_0 \\
u_2 &= Au_1 \\
u_3 &= Au_2 \\
&\vdots \\
u_n &= Au_{n-1}
\end{aligned}
\qquad (17.15)
$$

So,

$$\begin{aligned}
\|u_n - u_{n+1}\| = \|Au_{n-1} - Au_n\| &\le a\|u_{n-1} - u_n\| \\
&\le a^2 \|u_{n-2} - u_{n-1}\| \\
&\le a^n \|u_0 - u_1\|
\end{aligned} \tag{17.16}$$

Let us now prove by induction the proposition P_n: for all natural numbers n,

$$\|u_n - u_{n+1}\| \le a^n \|u_0 - u_1\|. \tag{17.17}$$

For $n = 0$, the first member is $\|u_0 - u_1\|$ and the second member is $a^0\|u_0 - u_1\|$. The inequality $\|u_n - u_1\| \le a^0\|u_0 - u_1\|$ is true. So the proposition is true at the zeroth order, that is, for $n = 0$.

Assume now that the proposition P_n is true at the order n, that is, $\forall n \in N, \|u_n - u_{n+1}\| \le a^n\|u_0 - u_1\|$ and let us prove that it is true at the next immediate order, that is, at the order $n + 1$.

$$\forall n \in N, \|u_{n+1} - u_{n+1+1}\| \quad \begin{aligned} &\le \|Au_n - Au_{n+1}\| \\ &\le a\|u_n - u_{n+1}\| \end{aligned} \tag{17.18}$$

Because $\forall n \in N, \|u_n - u_{n+1}\| \le a^n\|u_0 - u_1\|$, which infers that $\forall n \in N, \|u_{n+1} - u_{n+1+1}\| \le a^{n+1}\|u_0 - u_1\|$.

This means that the proposition is true at the order $n + 1$. Therefore, it is true for all n and we can write

$$\forall n \in N, \|u_n - u_{n+1}\| \le a^n\|u_0 - u_1\|. \tag{17.19}$$

Now, we need to prove that

$$\exists z \in B \,/\, z = Az.$$

To prove this, let us consider $\|u_n - u_m\|$ by adding and subtracting the successive sequence terms, $u_{n+1}, u_{n+2}, u_{n+3}, \ldots, u_m$. In this situation, $m > n$ and $m = n + p$, with p being a non-null natural number.

$$\begin{aligned}
\|u_n - u_m\| &= \|u_n - u_{n+p}\| \\
&= \|u_n - u_{n+1} + u_{n+1} - u_{n+2} + u_{n+2} + \cdots + u_{n+p}\| \\
&\le \|u_n - u_{n+1}\| + \|u_{n+1} - u_{n+2}\| + \cdots + \|u_{n+p-2} - u_{n+p-1}\| + \|u_{n+p-1} - u_{n+p}\| \\
&\le a^n\|u_0 - u_1\| + a^{n+1}\|u_0 - u_1\| + \cdots + a^{n+p-2}\|u_0 - u_1\| + a^{n+p-1}\|u_0 - u_1\| \\
&\le \left(a^n + a^{n+1} + a^{n+2} + \cdots + a^{n+p-1}\right)\|u_0 - u_1\|
\end{aligned} \tag{17.20}$$

The sum $S_n = a^n + a^{n+1} + \ldots + a^{n+p-2} + a^{n+p-1}$ is a geometric series of common ratio a and converges for $|a| < 1$, to, $S_n = \dfrac{a^n}{1-a}$.

So,

$$\|u_n - u_m\| \le \left(\frac{a^n}{1-a}\right)\|u_0 - u_1\| \tag{17.21}$$

$\lim\limits_{n,m\to\infty} \|u_n - u_m\| = 0$, which implies that there exists at least $z \in B$ / $\forall \varepsilon > 0, n > N(\varepsilon)$, we have

$\|z - u_n\| < \dfrac{\varepsilon}{2}$. We then have

$$\begin{aligned}
\|z - Az\| = \|z - u_n + u_n - Az\| &\le \|z - u_n\| + \|u_n - Az\| \\
&\le \|z - u_n\| + \|Au_{n-1} - Az\| \\
&\le \|z - u_n\| + a\|u_{n-1} - z\| \\
&\le \frac{\varepsilon}{2} + \frac{\varepsilon}{2} \\
&\le \varepsilon
\end{aligned} \tag{17.22}$$

So,

$$\|z - Az\| \to 0 \Rightarrow z = Az \tag{17.23}$$

We now need to prove that z is unique. To do so, suppose that there exists a z' different from z, such that $z' = Az'$.

Therefore,

$$\|z' - z\| = \|Az' - Az\| \le a\|z' - z\| \tag{17.24}$$

which implies that $1 \le a$, which is a contradiction. In conclusion, $z' = z$ and z is then unique.

17.5 HILBERT'S SPACE

Definition
A set H of elements $u, v, w, \ldots$ is called a real Hilbert's space, if H satisfies properties P_1, P_3 and the following property, P_4.

Property P_4
To every pair of elements, u, v there's assigned the scalar product (u, v) satisfying the following conditions:

$$\begin{aligned}
&(u, v) = (v, u)^* \\
&(u, u) \ge 0 \\
&(u, u) = 0 \Leftrightarrow u = 0 \\
&\forall \alpha \in R, (\alpha u, v) = \alpha(u, v) \\
&(u + v, w) = (u, w) + (v, w)
\end{aligned} \tag{17.25}$$

Other Properties
If we consider u, an element of the Hilbert's space H and denoted by

$$\|u\| = \sqrt{(u, u)} \tag{17.26}$$

Then, the following inequities are satisfied:

$$\left|(u,v)\right| \le \|u\|.\|v\| \quad \left(\text{Schwarz's inequality}\right) \tag{17.27}$$

$$\|u+v\| \le \|u\| + \|v\| \quad \left(\text{Minkowsky's inequality}\right) \tag{17.28}$$

$$\|u-v\| \le \|u-w\| + \|w-v\| \quad \left(\text{triangle inequality}\right) \tag{17.29}$$

As we can see, Hilbert's space contains all Banach space properties. As a result, every Hilbert's space is also a Banach space. But there are plenty of Banach spaces that are not Hilbert's space.

17.5.1   Continuity

An operator A in Hilbert's space is completely continuous if every infinite sequence $\{u_j\} \equiv u_1, u_2, \ldots, u_n \in H$, with $\|u_j\|$, finite, contains a subsequence $\{u_{ji}\}$, for which $\lim\limits_{u \in H}\|Au_{ji} - u\| = 0$, for an appropriate u element of H. We can also say that Au_{ji} converges to u or write $\lim\limits_{i \to \infty} Au_{ji} = u$.

17.5.1.1   Example of Hilbert's Space

The space $L^2(a,b)$ is the space of Lebesgue measurable functions:

$f:(a,b) \to C$, which are square-integrable, that is:

$$\int_a^b |f(t)|^2\, dt < \infty \tag{17.30}$$

The inner product of two functions f and g is defined as

$$(f,g) = \int_a^b f(t) \cdot g(t)^* dt \tag{17.31}$$

Theorem 1

For f, g element of H,

$$\|f\| - \|g\| \le \|f-g\| \tag{17.32}$$

Proof

$$\|f\| = \|(f-g)+g\| \le \|f-g\| + \|g\|$$

$$\|g\| = \|(g-f)+f\| \le \|f-g\| + \|f\|$$

Combining both inequalities, one has

$\|f\| - \|g\| \le \|f-g\|$ and $\|g\| - \|f\| \le \|f-g\|$, which give together,

$$-\|f\| - \|g\| \le \|f-g\| \le \|f\| - \|g\|$$

which can also be written as

$$\|f\| - \|g\| \le \|f - g\|. \tag{17.33}$$

17.5.2 Completeness

To illustrate the completeness of a space, consider the linear space $\mathbb{R}^2$ composed of vectors with two components (a, b). Recall that a sequence $\{a_n\}_{n=1}^{\infty}$ in $\mathbb{R}^2$ converges to an element of $\mathbb{R}^2$ if the distance $\|a_n - a\|$ converges to zero as n tends to ∞. Also it is important to mention that the following statements are equivalent:

$$\lim_{n \to \infty} a_n = a \tag{17.34}$$

$$\lim_{n \to \infty} \|a_n - a\| = 0 \tag{17.35}$$

For every real number, $\varepsilon > 0$, there exists an index, $n(\varepsilon)$, such that

$$\|a_n - a\| < \varepsilon \quad \text{for all} \quad n \ge n(\varepsilon) \tag{17.36}$$

17.5.3 Fundamental Sequence

A sequence, $\{a_n\}_{n=1}^{\infty}$, is called fundamental (or Cauchy) if the distances, $\|a_n - a\|$, become very small for very large values of n and m. This is the same as to say that for every real number, $\varepsilon > 0$, there exists an index, $n(\varepsilon)$, such that, $\|a_m - a_n\| < \varepsilon$ for all $m \ge n(\varepsilon)$ and for all $n \ge n(\varepsilon)$.

We can also write for the fundamental sequence the following limit:

$$\lim_{m,n \to \infty} \|a_m - a_n\| = 0 \tag{17.37}$$

We finally need to know that every fundamental sequence in $\mathbb{R}^2$ converges to a vector in $\mathbb{R}^2$, which means that $\mathbb{R}^2$ is complete.

17.5.3.1 Normed Linear Space

A linear space L over F is called normed linear space over F if with every vector $f \in L$, there is associated a unique number $\|f\| \in \mathbb{R}$, called the norm of f, in such a way that the following conditions are satisfied:

a. $\|f\| \ge 0$;

b. $\|f\| = 0 \Leftrightarrow f = 0$;

c. $\|\lambda f\| = |\lambda| \|f\|$;

d. $\|f + g\| \le \|f\| + \|g\|$.

17.5.3.2   Definition of an Open Sphere

Consider a vector $f_0 \in L$; for any $\in > 0$, the set $\left\{ f : \| f - f \|_0 < \varepsilon \right\}$ is called the open sphere of radius ε or simply the open ε-sphere about f_0. It is also called ε-neighborhood of f_0 or open ball of radius ε centered at f_0.

A subset $N \subset L$ is called open if for every $f \in N$, there is some $\in > 0$ such that the open ε-sphere about f is contained in N.

17.5.4   ACCUMULATION POINT

A vector f_1 is called the accumulation point of a subset $M \subset L$ if every open sphere about f_1 contains at least one vector $f \in M$ different from f_1. A subset $M \subset L$ is said to be closed if it contains all its accumulation points.

17.5.5   COMPLETENESS OF A NORMED LINEAR SPACE

A normed linear space L is complete if every fundamental sequence in L converges to some vector in L. A complete normed linear space over F is called a Banach space. F is either $\mathbb{R}$ (the field of real numbers) or $\mathbb{C}$ (the field of complex numbers). If $F = \mathbb{R}$, L is referred to as a real Banach space. If $F = \mathbb{C}$, then it is a complex Banach space.

17.6   EXAMPLE OF HILBERT'S SPACE: THE SPACE OF ALMOST PERIODIC FUNCTIONS

Consider the set of all functions of the form $\exp(i\lambda t)$, with $t \in (-\infty, +\infty)$, and λ a real number. Let us call L the linear envelope of this set, that is, the set of all polynomials of the form $\sum_k a_k \exp(i\lambda_k t)$.

If we also consider the limit of sequences of functions of L that are uniformly convergent in $\mathbb{R}$, we obtain a set of continuous functions. A continuous function, defined on the real axis, is the element of the collection B if and only if it is almost periodic, i.e., if for each, $\varepsilon > 0, \exists l = l(\varepsilon) \in \mathbb{R}$ such that for every length, l, there exists at least one number τ, for which $\left| f(t+\tau) - f(t) \right| < \varepsilon$, with $t \in (-\infty, +\infty)$.

Let us now explain how we can metrize L. We need to consider two polynomials,

$$f(t) = \sum_{r=1}^{m} a_r \exp(i\lambda_r t) \text{ and } g(t) = \sum_{s=1}^{n} b_s \exp(i\mu_s t), \text{ and define the scalar product:}$$

$$(f, g) = \lim_{T \to \infty} \frac{1}{T} \int_{-T}^{T} f(t) \overline{g(t)} dt = \lim_{T \to \infty} \sum_{r,s}^{m,n} \delta(\lambda_r, \mu_s) a_r b_s \qquad (17.38)$$

where

$$\delta(\lambda_r, \mu_s) = \begin{cases} 0 & \lambda \neq \mu \\ 1 & \lambda = \mu \end{cases} \qquad (17.39)$$

If L is close with this scalar product, then we obtain a complete Hilbert space B^2 which contains B as a linear manifold. The space B^2 is not separable.

17.7 LINEAR MANIFOLDS

Let's call L the set of all finite linear combinations, $\sum_{k=1}^{n} \alpha_k f_k$, with the f_k, elements of a subset M of $\mathbb{R}$. Subsets of $\mathbb{R}$ or of H are most of the time called linear manifolds if for f and g elements of L and for any arbitrary real numbers α and β, $\alpha f + \beta g$ is also an element of L. If $\mathbb{R}$ is a metric space, then the closure of the linear envelope of a set M is called the closed linear envelope of M.

PROBLEM SET 17

17.1 Consider the function $f(u) = u^a$. For which value of a does the appropriate restriction of f belongs to $L^2(0,1)$?

17.2 Prove that a closed subspace of a complete metric space is complete with respect to the induced metric. If C_0 represents the subspace of ℓ^∞, comprising all sequences (x_n) which tend to zero as $n \to \infty$, deduce that $\left(C_0, \|\cdot\|_\infty\right)$ is a Banach space.

17.3 Let F be a linear functional on a normed space, $\left(E, \|\cdot\|_\infty\right)$. Prove that the following statements are equivalent:

(i) F is continuous

(ii) F is continuous at 0

(iii) $\operatorname{Sup}\left\{|F(x)| : x \in E, \|x\| \le 1\right\} < \infty$.

17.4 Prove that for any continuously differentiable function f on the interval $[-\pi, \pi]$,

$$\left| \int_\pi^\pi \left[f(t)\cos t - f'(t)\sin t \right] dt \right| \le \sqrt{2\pi} \left\{ \int_\pi^\pi \left[|f(t)|^2 + |f'(t)|^2 \right] dt \right\}^{1/2}.$$

18 Probability and Statistics

Our passions, our prejudices, and dominating opinions, by exaggerating the probabilities which are favorable to them and by attenuating the contrary probabilities, are the abundant sources of dangerous illusions.

A Philosophical Essay on Probabilities, p. 160.

–Pierre Simon, Marquis De Laplace

INTRODUCTION

Probability and statistics play a key role in mathematical methods in physics and engineering. In this chapter, we will contend ourselves with a review of some of the most important topics in probability and statistics.

18.1 COMBINATORY

18.1.1 BINOMIAL THEOREM

The binomial theorem gives a useful formula to evaluate the expansion of $(x + a)^n$ for any positive integer n. Let us try to expand $(x + a)^n$ for the first five values of n:

For $n = 0, (x+a)^0 = 1$
For $n = 1, (x+a)^1 = x + a$
For $n = 2, (x+a)^2 = x^2 + 2ax + a^2$
For $n = 3, (x+a)^3 = x^3 + 3x^2a + 3xa^2 + a^3$
For $n = 4, (x+a)^4 = x^4 + 4ax^3 + 6a^2x^2 + 4a^3x + a^4$

Evaluation of $\begin{pmatrix} n \\ j \end{pmatrix}$

In probability, the symbol $\begin{pmatrix} n \\ j \end{pmatrix}$ is read "n taken j at a time" or "n taken j by j."

Definition 1

If j and n are positive integers, such that $0 \leq j \leq n$, then the symbol $\begin{pmatrix} n \\ j \end{pmatrix}$ is defined as

$$\begin{pmatrix} n \\ j \end{pmatrix} = \frac{n!}{j!(n-j)!}.$$

Example 1

Evaluate $\begin{pmatrix} n \\ 0 \end{pmatrix}, \begin{pmatrix} n \\ 1 \end{pmatrix}, \begin{pmatrix} n \\ n \end{pmatrix}, \begin{pmatrix} n \\ n-1 \end{pmatrix}.$

Solutions

$$\begin{pmatrix} n \\ 0 \end{pmatrix} = \frac{n!}{0!n!} = 1$$

DOI: 10.1201/9781003478812-18

$$\binom{n}{1} = \frac{n!}{1!(n-1)!} = \frac{n(n-1)!}{(n-1)!} = n$$

$$\binom{n}{0} = \frac{n!}{0!n!} = 1$$

$$\binom{n}{n-1} = \frac{n!}{(n-1)!1!} = n$$

18.1.2 PASCAL TRIANGLE

Suppose that we arrange the values of the symbol $\binom{n}{j}$ in a triangular display such as

$$\binom{0}{0} n = 0 \; j = 0$$

$$\binom{1}{0}\binom{1}{1} \quad n = 1 \; j = 0,1$$

$$\binom{2}{0}\binom{2}{1}\binom{2}{2} \quad n = 2 \; j = 0,1,2$$

$$\binom{3}{0}\binom{3}{1}\binom{3}{2}\binom{3}{3} \quad n = 3 \; j = 0,1,2,3$$

$$\binom{4}{0}\binom{4}{1}\binom{4}{2}\binom{4}{3}\binom{4}{4} \quad n = 4 \; j = 0,1,2,3,4$$

or

1					$n = 0$
1	1				$n = 1$
1	2	1			$n = 2$
1	3	3	1		$n = 3$
1	4	6	4	1	$n = 4$

We can see the coefficients of a and b in the expansion of $(a+b)^n$; the maximum value of j being n.
For example,

$$(a+b)^4 = 1a^4b^0 + 4a^3b^1 + 6a^2b^2 + 4a^1b^3 + 1a^0b^4.$$

The coefficients are given for $n = 4$ by 1, 4, 6, 4, 1.

Starting from $n = 4$, while the exponent of a decreases to 0, the one of b increases from 0 to maximum 4. The coefficients can be found by the following set.

1				
1	1			
1	2	1		
1	3	3	1	
1	4	6	4	1

The above set looks like a triangle. It is called Pascal's triangle. Pascal's triangle helps to find the coefficients in the expansion of the $(a+b)^n$.

18.1.3 BINOMIAL THEOREM

Given j and n positive integers such that $0 \le j \le n$, the expression of $(a+b)^n$ can be expanded as

$$(a+b)^n = \binom{n}{0}a^n b^0 + \binom{n}{1}a^{n-1}b^1 + \binom{n}{2}a^{n-2}b^2 +$$

$$\ldots + \binom{n}{n}a^{n-n}b^n = \sum_{j=0}^{n}\binom{n}{j}a^{n-j}b^j \qquad (18.1)$$

The coefficients $\binom{n}{j}$ are called binomial coefficients.

Remark

In Pascal's triangle, it is easy to notice as indicated by the arrow in the schematic below that $1 + 1 = 2$, $1 + 2 = 3$, etc. In fact, this can help to easily construct Pascal's triangle.

$$
\begin{array}{ccccccccc}
1 & & & & & & & & \\
1 & + & 1 & & & & & & \\
  &   & \downarrow & & & & & & \\
1 & + & 2 & + & 1 & & & & \\
  &   & \downarrow & & \downarrow & & & & \\
1 & + & 3 & + & 3+ & & 1 & & \\
  &   & \downarrow & & \downarrow & & \downarrow & & \\
1 & & 4 & & 6 & & 4 & & 1
\end{array}
$$

Example 2

Expand $(x+2)^5$ in a polynomial form.

Solution

To find the binomial coefficients here we need to go one step further. That is, up to $n = 5$. The Pascal triangle will then give us

$$
\begin{array}{cccccc}
1 & & & & & \qquad n=0 \\
1 & 1 & & & & \qquad n=1 \\
1 & 2 & 1 & & & \qquad n=2 \\
1 & 3 & 3 & 1 & & \qquad n=3 \\
1 & 4 & 6 & 4 & 1 & \qquad n=4 \\
1 & 5 & 10 & 10 & 5 & 1 \quad n=5
\end{array}
$$

The coefficients in the binomial law are 1   5   10   10   5   1 for the exponent $n = 5$. The expansion will be

$$(x+2)^5 = 1x^5 2^0 + 5x^4 2^1 + 10x^3 2^2 + 10x^2 2^3 + 5x^1 2^4 + 1x^0 2^5$$
$$= x^5 + 10x^4 + 40x^3 + 80x^2 + 80x + 32$$

18.1.4   COUNTING

18.1.4.1   Subsets of a Set

Let us explain the notion of the subset by using an example. A set is a collection of objects identified by names. These objects are called elements of the set. As an example, we can assume P is the set of all the planets of the solar system.

$$P = \{venus, mercury, earth, mars, jupiter, saturn, uranus, neptune, pluto\}.$$

P has 2^8 subsets. In general, if n is the number of elements in the set, the number of subsets of the set will be 2^n.

Example 3

How many subsets can be formed with the set $\{a,b,c\}$. Express these subsets.

Solution

The set $\{a,b,c\}$ has $2^3 = 8$ subsets, which are: $\{\}$ $\{a\}$, $\{b\}$, $\{c\}$, $\{a,b\}$, $\{a,c\}$, $\{b,c\}$, $\{a,b,c\}$. Note the empty set is always a subset.

18.1.4.2   The Fundamental Counting Principle

We first need to define an event in counting as we are going to use this word more frequently in this chapter. Indeed, in some experiments, we may be interested in the occurrence of certain events rather than in the outcome of a specific element in the sample space. For example, we might be interested in event A in such a way that when a die is tossed, the outcome is an even natural number. In this case, event A will be the subset $A = \{2,4,6\}$ of the sample space $\{1,2,3,4,5,6\}$. So, an event can be defined as a subset of a sample space.

Consider now two events E_1 and E_2. If event E_1 can occur in n_1 different ways and event E_2 can occur in n_2 different ways, then the number of ways that the two events E_1 and E_2 can occur is $n_1 \cdot n_2$.

Example 4

How many pairs of letters can be formed with the English alphabet?

Solution

The English alphabet has 26 letters. To form a pair of letters, one must choose from two events E_1 and E_2. Each of these events has 26 alphabets. The number of different possible pairs is 26 x 26=676.

18.1.4.3   Permutation

Permutation is a selection of objects in which the order of the objects matters. Example: The permutations of the letters in the set $\{a, b, c\}$ are as follows:
    abc, acb, bac, bca, cab and cba. This gives six permutations.

Example 5

How many permutations can we have with the first three letters of the English alphabet?

Solution

The first three letters of the English alphabet are abc. Here are the possibilities of ordering them:
    abc, acb, bac, bca, cab and cba. If one letter is chosen, we can permute the remaining two. This leads to the number $3! = 6$ of possibilities.

18.1.4.4   Number of Permutations

The number of permutations of n elements is given by

$$n! = n \cdot (n-1) \cdot (n-2) \cdot 4 \cdot 3 \cdot 2 \cdot 1. \tag{18.2}$$

18.1.4.5   Permutation of n Elements Taken r at a Time

The number of possible permutations of r objects from a set of n objects is given by

$$P(n,r) = \frac{n!}{(n-r)!} = n \cdot (n-1) \cdot (n-2) \cdots (n-r+1). \tag{18.3}$$

18.1.4.6   Distinguishable Permutations

Consider n distinguishable objects that have n_1 of one kind, n_2 of a second kind, n_3 of a third kind, ..., n_k of a kth kind. The number of distinguishable permutations of these n objects is

$$\frac{n!}{n_1! n_2! n_3! n_4! \ldots n_{k-1}! n_k!} \tag{18.4}$$

Example 6

In how many distinguishable ways can the letters in ANA be written?

Solution

The word ANA has two of the A's and one of the N's. Then the number of distinguishable ways the letters in ANA can be written is $\dfrac{3!}{2!!!} = 3$

These three different distinguishable permutations are
AAN, NAA, ANA

Example 7

In how many ways can a chairperson, a vice-chairperson and a secretary be selected from a committee of ten people?

Solution

These leaders must be selected so that the chairperson comes first, the vice-chairperson the second and the secretary the third. We must order 1, 2 and 3 people in a total number of 10 people which means a permutation:

$$P(10,3) = \frac{10!}{(10-3)!} = \frac{10!}{7!} = 10.9.8 = 720$$

18.1.4.7   Combination

Combination of n elements taken r at a time is the number of subsets of r elements contained in a set of n elements. Order does not matter in combination.

18.1.4.8   Number of Combinations

The number of combinations of n elements taken r at a time – that is taken r by r is

$$C(n,r) = \frac{n!}{r!(n-r)!}. \tag{18.5}$$

We can notice that a combination of n elements taken r at a time, $C(n,r)$, can be written by the symbol used for binomial coefficients, $\begin{pmatrix} n \\ r \end{pmatrix}$. Also, it can be proved that $C(n,r) = \dfrac{P(n,r)}{r!}$.

Example 8

In a classroom of 10 students, find the number of combinations of students taken 5 at a time.

Solution

$$C(10,5) = \frac{10!}{5!(10-5)!} = \frac{10!}{5!(5)!} = \frac{10 \cdot 9 \cdot 8 \cdot 7 \cdot 6}{5 \cdot 4 \cdot 3 \cdot 2 \cdot 1} = 252.$$

Example 9

A coin is flipped with possible outcomes, H for head and T for tail. Then, a single die is rolled with the possible outcome 1, 2, 3, 4, 5, 6. How many combined outcomes are there?

Solution

There's no order here so we will have to use the combination. When the coin is flipped, there are two possible outcomes: $C(2,1) = 2$. When the die is rolled, there are six possibilities: $C(6,1) = 6$.

The combined outcomes will be $C(2,1) \cdot C(6,1) = 2 \cdot 6 = 12$.

18.2 PROBABILITY

18.2.1 EXPERIMENTS

When we performed an experiment in science we expect a result. Most of the time, a scientist will perform, under certain conditions, the same experiments several times to assure the result. These types of experiments depending on some conditions are called deterministic. There are other types of experiments called random experiments. They do always not yield the same results. Example: the experiment consisting of rolling a die can give any of the results, 1, 2, 3, 4, 5 or 6. It is a random experiment.

18.2.1.1 Sample Space

The sample space is the set of the possible outcomes of an experiment. A sample space can be finite or infinite.

Example 10

If an experiment is to roll a single die, then the sample space will be $S = \{1, 2, 3, 4, 5, 6\}$.

18.2.1.2 Event

Consider the sample space S of an experiment. An event is a subset of S.

Example 11

Consider that you toss two coins. The sample space will be

$$S = \{HH, HT, TH, TT\}.$$

The statement "Two heads are up" is the event $E = \{HH\}$. The statement "Exactly one tail is up" is the event $Ex = \{HT, TH\}$.

18.2.2 Probability of a Simple Event

Definition 2

Consider a sample space $S = \{e_1, e_2, \ldots, e_n\}$. To each event e_i, it can be assigned a real number $P\{e_i\}, i = 1, 2, \ldots, n$, such that the following conditions are satisfied:

1. For every $e_i, 0 \leq P\{e_i\} \leq 1$, that is, probability of an event is always between 0 and 1.

2. $\sum_{i=1}^{n} P\{e_i\} = 1$; the sum of the probabilities of all the events of the sample space is 1.

Also, one can regard the probability of an event is the chance of that event to occur.

Example 12

Flipping a coin would give either H or T. The space sample is $S = \{H, T\}$. There is a 50% of chance of H to occur and 50% of chance of T to occur. Therefore, $P\{H\} = P\{T\} = 0.5$. The two probabilities are between 0 and 1 and their sum gives 1.

18.2.2.1 Consequence

Consider $P\{E\}$ the probability of an arbitrary event E of a sample space S.
If E is an empty set, then $P\{E\} = P\{\phi\} = 0$.
If $E = A \cup B$ and $A \cap B = \phi$, then $P\{E\} = P\{A\} + P\{B\}$.

$$\text{If } E = S, \text{ then } P\{E\} = P\{S\} = 1.$$

Example 13

A child is playing with a nickel and a dime by tossing them.

a. Determine the sample space.
b. What is the probability of getting one head (or one tail)?
c. What is the probability of getting at least one head?
d. What is the probability of getting three heads?

Solution

We assume that H and T have the same chance of occurring.

a. When the child tosses the nickel and the dime together, the possible outcomes are HH, HT, TH and TT. So, the sample space is the set: $S = \{HH, HT, TH, TT\}$
b. If we call A the event "getting one head," then,
$A = \{HT, TH\}$ and the probability of A would be $P\{A\} = P\{HT\} + P\{TH\} = 1/4 + 1/4 = 1/2$.
c. If B is the event "getting at least one head," this means that the child can have HH, HT, TH and the probability of B would be: $P\{B\} = 1/4+1/4+1/4 = 3/4$.
d. Consider C the event "having three heads;" C is impossible since the child only has two head coins. As a result, $P\{C\} = 0$.

We can summarize the previous calculations in Table. 18.1.

TABLE 18.1
Table of Probability Distribution

Events	HH	HT	TH	TT
Probability $P\{e_i\}$	¼	¼	¼	¼

18.2.2.2 Equiprobability

Assume the sample $S = \{e_1, e_2, \ldots, e_n\}$ with n possible events. Consider that each simple event is as likely to occur as any other event. This means that they have the same probability or that we have equiprobability: $P\{e_1\} = P\{e_2\} = \ldots P\{e_n\}$.

Since $\sum_{i=1}^{n} P\{e_i\} = 1$, we have $nP\{e_i\} = 1$; so $P\{e_i\} = \dfrac{1}{n}$ is the probability of each event.

18.2.3 PROBABILITY OF AN ARBITRARY EVENT

Definition 3

Under the assumption equiprobability, the probability of an event E obtained from a sample space, S, is given by

$$P\{E\} = \frac{\text{number of element in } E}{\text{number of element in } S} = \frac{n(E)}{n(S)}. \tag{18.6}$$

Example 14

In a game of 52 cards, a player is supposed to draw 5 cards without replacement. What is the probability of getting 5 spades?

Solution

Assume the event E: "set of 5 cards out of 13 spades." Since there is no order, when the 5 cards are drawn, the number of ways the players can draw 5 cards out of 13 spades is $C(13, 5)$; and by doing so, the player has drawn these 5 cards out of the whole 52 cards, the number of ways of which is $C(52, 5)$. The probability of event E to occur is

$$P\{E\} = \frac{n(E)}{n(S)} = \frac{C(13,5)}{C(52,5)} = \frac{\dfrac{13!}{5!(13-5)!}}{\dfrac{52!}{5!(52-5)!}} = 0.0005.$$

18.2.4 DEFINITION OF EMPIRICAL PROBABILITY

Empirical probability is concerned mostly with many elements of a sample space and the frequency of occurrence of the events. If the experiment is conducted n times and the event occurs with a frequency $f(E)$, then the empirical probability of occurrence of E will be defined by

$$P\{E\} = \frac{\text{frequency of occurence of } E}{\text{total number of trials}} = \frac{f(E)}{n} \tag{18.7}$$

Example 15: In an experiment, two coins are tossed 2,000 times with the following frequencies:

2 heads occur 400 times.
1 head occurs 1120 times.
0 head occur 480 times.

1. What is the approximate empirical probability for each type of outcome?
2. Compare with the corresponding theoretical probabilities.

Solution

1. **The approximate empirical probabilities are**

$$P\{2\ heads\} = \frac{f(E)}{n} = \frac{400}{2000} = 0.2$$

$$P\{1\ head\} = \frac{f(E)}{n} = \frac{1120}{2000} = 0.56$$

$$P\{0\ head\} = \frac{f(E)}{n} = \frac{480}{2000} = 0.24$$

We can notice that the sum of all these probabilities is $0.2 + 0.56 + 0.24 = 1$.

2. **Theoretical Probability**

Remember that the sample space of tossing two coins is
   $S = \{HH, HT, TH, TT\}$. So, the probability of having two heads is $P\{2\ heads\} = 1/4 = 0.25 >$ 0.2 for the approximate empirical probability counterpart. The probability of having 1 head is $P\{1\ head\} = 1/4 + 1/4 = 0.50 < 0.56$ for the approximate empirical probability counterpart. The probability of having 0 head is the same as the probability of having two tails, which is $P\{0\ head\} = 1/4 = 0.25 > 0.24$ for the approximate empirical probability counterpart.

Conclusion: Probabilities can be approached in two different ways: the theoretical probability and the approximate empirical probability.

18.2.5   PROBABILITY OF THE UNION OF TWO EVENTS

Definition 4
Consider two subsets A and B of a simple event E of a sample space S such that $E = A \cup B$. The probability of $A \cup B$ is given by

$$P\{A \cup B\} = P\{A\} + P\{B\} - \{P\{A \cap B\}$$ (18.8)

If $A \cap B = \phi$ then, A and B are independent which means, $P\{A \cap B\} = P\{\phi\} = 0$ and $P\{A \cup B\} = P\{A\} + P\{B\}$.

Example 16
A student is rolling a single fair die. Assume equiprobability.

   a. What is the sample space?
   b. What is the probability of rolling an even number or the number 3?

Solution
   a. The sample space is $S = \{1, 2, 3, 4, 5, 6\}$.
   b. Assume A the event "rolling an even number" and B the event "rolling the number 3."

$A = \{2, 4, 6\}$ and $B = \{3\}$. The event $E = A \cup B$ is "rolling an even number or the number 3."
   $A \cap B = \phi$, then, A and B are independent, so $P\{A \cap B\} = P\{\phi\} = 0$ and

$$P\{A \cup B\} = P\{A\}_+ P\{B\} = 3/6 + 1/6 = 2/3.$$

18.2.6   PROBABILITY OF THE COMPLEMENTS

Consider two events E and Ex. These events are called complement to each other if

$$\begin{cases} E \cap E' = \phi \\ EUE' = S \end{cases} \tag{18.9}$$

We customary write $C_E^S = E'$ and $C_{E'}^S = E$.

So, $P\{S\} = 1 = P\{E\} + P\{Ex\} x P\{E \cap Ex\}$

Since $E \cap E' = \phi, P\{E \cap E'\} = 0$ and $P\{Ex\} = 1 \, x \, P\{E\}$.

Example 17

The probability of rain on a day of Saturday is 80%. What is the probability that it would not rain the same day?

Solution

Consider E the event "there's a chance of rain" and Ex the event "there is no chance of rain."

$$P\{E\} = 0.80 \text{ and } P\{Ex\} = 1 \, x \, P\{E\} = 1 \, x \, 0.8 = 0.20.$$

So, there's a 20% chance that it would not rain.

Example 18

If there's a 52% of chance for a baby to be a girl, what is the probability that a baby would be a boy?

Solution

$P\{Ex\} = 1 \, x \, P\{E\} = 1 \, x \, 0.52 = 0.48$. So, there's a 48% chance that a baby would be a boy.

18.2.7   CONDITIONAL PROBABILITY

Definition 5

In the equiprobability situation, the conditional probability of event A given event B is given by

$$P\{A/B\} = \frac{P\{A \cap B\}}{P\{B\}} \text{ with } P\{B\} \neq 0 \tag{18.10}$$

Example 19

A game consists of tossing three coins.

  a. What is the sample space?
  b. Determine the probability that at least two coins turn heads.
  c. Find the probability that at least two coins turn heads, given that the first coin turns up tail.

Solution

When three coins are tossed together in disorder, the sample space would be:

$$S = \{HHH,\ HHT,\ HTH,\ THH,\ HTT,\ THT,\ TTH,\ TTT\}.$$

If A is the event "at least two coins turn heads" and B is the event "the first coin turns up tail," then

$$P\{A\,/\,B\} = \frac{P\{A \cap B\}}{P\{B\}} = \frac{1/8}{4/8} = 0.25.$$

18.2.8 INDEPENDENT EVENTS

Definition 6

Two events A and B are independent is $P\{A\,/\,B\} == P\{A\}$; otherwise, the two events are said to be dependent.

Consequences

If two events A and B are independent,

$$P\{A\,/\,B\} = \frac{P\{A \cap B\}}{P\{B\}} = P\{A\},$$

which implies that $P\{A \cap B = P\{A\} \cdot P\{B\}.$

18.2.9 BAYES' RULES

The term "mutually exclusive" is also used in probability. Thus, two events A and B are mutually exclusive if and only if $A \cap B = \phi$, and so $P\{A \cup B\} = P\{A\} + P\{B\}.$

General Case

In general, if the events A_1, A_2, ..., A_n are mutually exclusive, then

$$P\{A_1 \cup A_2 \cup ... \cup A_n\} = P\{A_1\} + P\{A_2\} + ... + P\{A_n\}. \tag{18.11}$$

Corollary

If the events A_1, A_2, ..., A_n represent a partition of the sample space S, then

$$P\{A_1 \cup A_2 \cup ... \cup A_n\} = P\{A_1\} + P\{A_2\} + ... + P\{A_n\} = P\{S\} = 1. \tag{18.12}$$

18.3   RANDOM VARIABLES AND PROBABILITY DISTRIBUTIONS

18.3.1 RANDOM VARIABLES

Definition 7

Consider a sample space S. A random variable is a function that associates to each element of the sample space a real number.

Example 20

Two balls are drawn successively without replacement from an urn containing four red balls and three black balls. What are the possible outcomes and the values y of the random variable Y associated with the number of red balls drawn?

Solution

RR, RB, BR and BB are the events of the sample space. We can summarize the number of possibilities in Table 18.2.

TABLE 18.2

Table of Sample Space and Outcome

Sample space	Y
RR	2
RB	1
BR	1
BB	0

18.3.2 DISCRETE SAMPLE SPACE

A sample space is said to be discrete if it contains a finite number of possibilities. A random variable is called a discrete random variable if the set of its possible outcomes is countable.

18.3.3 CONTINUOUS SAMPLE SPACE

A sample space is said to be continuous if it contains an infinite number of possibilities. A random variable is called a continuous random variable number if its set of possible outcomes is not countable or can take continuous values.

Example 21

Examples of continuous random variables can be represented in measured data which gives, for example, the heights, weight, life expectancy, etc. Conversely, examples on discrete random variables can be represented by count data such as the Y of the number of red balls in the previous example.

18.3.4 DISCRETE PROBABILITY DISTRIBUTIONS

Consider the experiment of tossing a coin three times. Assume that the number of heads is represented by the variable X. Then, $P\{X = 2\} = 3/8$ since 3 of the 8 equally likely sample points result in two heads and one tail.

Definition 8

The probability function also called probability mass function or probability distribution of the discrete variable X is the set of ordered pairs $(x, f(x))$ where x is a possible outcome such that

1. $f(x) \geq 0$

2. $\sum_{x} f(x) = 1$ 　　　　　　　　　　　　　　　　　　(18.13)

3. $P\{X = x\} = f(x)$

Example 22

A shipment of eight similar microcomputers to a retail outlet contains three that are defective. If a school makes a random purchase of two of these computers, find the probability distribution for the number of defectives.

Solution

Assume X the random variable whose values x are the possible numbers of defective computers purchased by the school; x can take the values 0, 1 or 2.

$$P\{X=0\}=f(0)=\frac{C(5,2)C(3,0)}{C(8,2)}=\frac{10}{28}$$

$$P\{X=1\}=f(1)=\frac{C(3,1)C(5,1)}{C(8,2)}=\frac{15}{28}$$

$$P\{X=2\}=f(2)=\frac{C(3,2)C(5,0)}{C(8,2)}=\frac{3}{28}$$

and $f(0)+f(1)+f(2)=1=\sum_{x}f(x)$ is satisfied.

The probability distribution will be given in the following probability distribution table (Table 18.3).

18.3.5 CUMULATIVE DISTRIBUTION

The cumulative distribution $F(x)$ of a discrete random variable X with a probability distribution $f(x)$ is given by

$$F(x)=P\{X\le x\}=\sum_{t\le x}f(t)\ \ for\ \ x\in(-\infty,+\infty).\qquad(18.14)$$

For the previous example, the cumulative distribution would be

$$F(x)=\begin{cases}0 & x<0\\ 10/28 & 0\le x<1\\ 25/28 & 1\le x<2\\ 1 & x\ge 2\end{cases}$$

18.3.6 BAR CHART

To construct the bar chart, we need to join point x and $f(x)$ by dash or solid lines. The bar chart of our previous example about the shipment of microcomputers is shown in Figures 18.1–18.3.

The graph of $(x, F(x))$ in Figure 18.3 is called the cumulative distribution graph for the case of Example 22.

18.4 CONTINUOUS PROBABILITY DENSITY

We would give here two important definitions which are going to be useful in future calculations.

TABLE 18.3

Table of Probability Function of a Random Variable

x	0	1	2
$f(x)=P\{X=x\}$	10/28	15/28	3/28

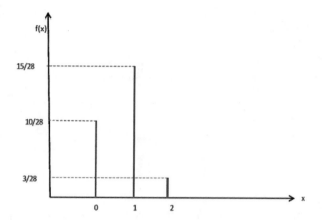

FIGURE 18.1   Probability bar chart.

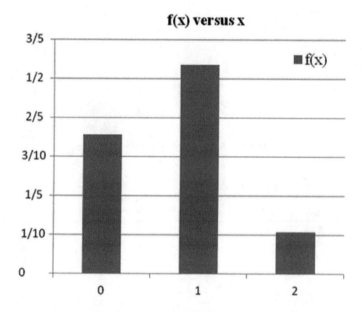

FIGURE 18.2   Probability histogram chart.

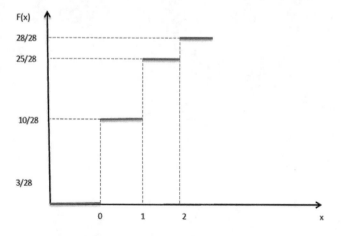

FIGURE 18.3   Cumulative distribution graph.

18.4.1 Definitions

Definition 9
If x is a continuous random variable defined over the set $\mathbb{R}$ of real numbers, the function $f(x)$ is called a probability density function (p.d.f.) if the following three conditions are satisfied:

1. $f(x) \geq 0$ for all real number x

2. $\int_{-\infty}^{+\infty} f(x)\,dx = 1$ (18.15)

3. $P\{a < X < b\} = \int_a^b f(x)\,dx$

Definition 10
The cumulative distribution $F(x)$ of a continuous random variable X with density function $f(x)$ is given by

$$F(x) = P\{X \leq x\} = \int_{-\infty}^x f(t)\,dt$$ (18.16)

for all real number x.

18.4.2 Consequence

A direct consequence is that the derivative of $F(x)$ is $f(x)$, that is:

$$f(x) = \frac{dF(x)}{dx} \quad \text{and} \quad P\{a < X < b\} = F(b) - F(a)$$ (18.17)

Example 23
The error in the reaction temperature, in °C, for a controlled laboratory experiment is a continuous random variable X with the probability density function defined as follows:

$$f(x) = \begin{cases} \dfrac{x^2}{3} & -1 < x < 2 \\ 0 & \text{elsewhere} \end{cases}$$

a. Prove that $\int_{-\infty}^{+\infty} f(x)\,dx = 1$.

b. Evaluate $F(x)$ and calculate $P\{0 < X < 1\}$.

Solution

a. $\int_{-\infty}^{+\infty} f(x)\,dx = \int_{-\infty}^{-1} f(t)\,dt + \int_{-1}^{2} f(t)\,dt + \int_{2}^{+\infty} f(t)\,dt$

$= 0 + \int_{-1}^{2} \dfrac{t^2}{3}\,dt + \int_{2}^{+\infty} 0\,dt$

$\left[\dfrac{x^3}{9} \right]_{-1}^{2} = 1$

b. $\int_{-\infty}^{x} f(t)\,dt = \int_{-1}^{t} \dfrac{t^2}{3}\,dt = \left[\dfrac{x^3}{9} \right]_{-1}^{x} = \dfrac{x^3}{9} + \dfrac{1}{9}$

So, the expression of $F(x)$ will be given by

$$F(x) = \begin{cases} 0 & x < -1 \\ \dfrac{x^3 + 1}{9} & -1 < x < 2 \\ 1 & x \geq 2 \end{cases}$$

18.5   MATHEMATICAL EXPECTATION

Definition 11

We consider X a random variable and $f(x)$ the probability distribution function. The distribution function is $f(x) = P(X = x)$ and $f(x) = F'(x)$. We also consider $u(X)$, a function of X, such that $\int_{-\infty}^{+\infty} f(x)\,dx$ in the continuous case or $\sum_x u(x) f(x)$ in the discrete case exists, then

$E[u(X)] = \int_{-\infty}^{+\infty} u(x) f(x)\,dxv$ is referred to as the mathematical expectation or the expected value of $u(x)$, if X is continuous. The expression will change to $E[u(X)] = \sum_x u(x) f(x)$ if the random variable X is discrete.

Assume $Y = u(X)$, a random variable with its own distribution of probability. If $g(y)$ is the probability density distribution of Y, then the mathematical expectation is $E(Y) = \int_{-\infty}^{+\infty} y g(y)\,dy$ or $E(Y) = \sum_x y g(y)$ for Y a continuous or discrete random variable.

18.5.2   GENERAL CASE

Let $X_1, X_2, \ldots, X_n$ be random variables with probability density function $f(x_1, x_2, \ldots, x_n)$. Consider $u = u(X_1, X_2, \ldots, X_n)$ a function of these variables. The n-fold integral

$$\int_{-\infty}^{\infty} \ldots \int_{-\infty}^{+\infty} u(x_1, x_2, \ldots, x_n) f(x_1, x_2, \ldots, x_n)\,dx_1 dx_2 \ldots dx_n,$$

if it exists, represents the mathematical expectation of u for the continuous case of the variables $X_1, X_2, \ldots, X_n$.

Moreover, $\sum_{x_n} \ldots \sum_{x_1} u(x_1, x_2, \ldots, x_n) f(x_1, x_2, \ldots, x_n)$, if it exists, represents the mathematical expectation of u when the random variables $X_1, X_2, \ldots, X_n$ are discrete.

18.6   PROPERTIES OF THE MATHEMATICAL EXPECTATION

Here are some basic properties of the mathematical expectation:

a. If k is constant, $E(k) = k$.
b. If k is a constant and v a function, then $E(kv) = kE(v)$.
c. If k_1 and k_2 are constants and v_1 and v_2 are functions, then $E(k_1 v_1 + k_2 v_2) = k_1 E(v_1) + k_2 E(v_2)$.
d. In general, for $k_1, \ldots, k_n$ and functions $v_1, \ldots, v_n$, we have $E(k_1 v_1 + k_2 v_2 + \ldots k_n v_n) = k_1 E(v_1) + k_2 E(v_2) + \ldots k_n E(v_n)$. This last property shows that E is a linear operator.

Example 24

Let X have the probability density function f such that

$$f(x) = \begin{cases} 2(1-x) & 0 < x < 1 \\ 0 & \text{elsewhere} \end{cases}$$

Calculate $E(X)$, $E(X^2)$ and $E(6X + 3X^2)$.

Solution

$$E(X) = \int_{-\infty}^{+\infty} xf(x)dx = \frac{1}{3};$$

$$E(X^2) = \int_{-\infty}^{+\infty} x^2 f(x)dx = \frac{1}{6};$$

$$E(6X + 3X^2) = \frac{5}{2}.$$

18.7   MEAN VALUE AND VARIANCE OF X

Assume that $u(X) = X$ a random discrete variable having a p.d.f. $f(x)$, then the mathematical expectation is $E[X] = \Sigma_x xf(x)$.

If $a_1, a_1, ..., a_n$ are discrete points in the space of positive probability density, then:

$$E(X) = a_1 f(a_1) + a_2 f(a_2) + ... + a_n f(a_n) \tag{18.18}$$

The expression (18.18) shows that $E(X)$ represents a weighted average of the values $a_1, a_1, ..., a_n$, the weight being $f(a_1), f(a_2), ..., f(a_n)$. We can therefore call the expression $\mu = E(X)$, the arithmetic mean of the discrete values of X or simply the mean value of X. Let us now consider the expectation value of the expression $(X - \mu)^2$; That is:

$$E\left[(X-\mu)^2\right] = \sum_x (x-\mu)^2 f(x) \tag{18.19}$$
$$= (a_1 - \mu)^2 f(a_1) + (a_2 - \mu)^2 f(a_2) + ... + (a_n - \mu)^2 f(a_n)$$

which is the mean value of the square of the deviation of X from its mean μ. It is called the variance of X and is expressed by

$$\sigma^2 = E\left[(X-\mu)^2\right] = E\left[X^2 - 2\mu X + \mu^2\right] \tag{18.20}$$
$$= E\left[X^2\right] - \mu^2$$

The number $\sigma = \sqrt{E\left[(X-\mu)^2\right]}$ is called the standard deviation of X or the standard deviation of the distribution. Recall that σ is interpreted as a measure of the dispersion of the points of the space relative to the mean value μ. So, if the space contains only one point x for which $f(x) > 0$, then $\sigma = 0$.

18.8   MOMENT–GENERATING FUNCTION

For a random variable X, if e^{tX} exists with $-h < t < h$ and $h > 0$, the mathematical expectation of e^{tX} is called the moment-generating function of X and is expressed by

$$M(t) = E\left[e^{tX}\right] = \int_{\infty}^{+\infty} e^{tx} f(x) dx \qquad (18.21)$$

for X continuous and by:

$$M(t) = E\left[e^{tX}\right] = \sum_{x} e^{tx} f(x) \qquad (18.22)$$

for X discrete.

If $t = 0$, then $e^{tx} = e^0 = 1$, and $M(0) = E(1) = 1$.

Example 25
Let the random variable X have the p.d.f. $f(x)$ such that

$$f(x) = \begin{cases} \dfrac{1}{2}(x+1)^2 & -1 < x < 1 \\ 0 & \text{elsewhere} \end{cases}$$

   a. Calculate the mean value of X.
   b. What is the variance of X?

Solution
The mean value of X is

a. $\mu = \int_{-\infty}^{+\infty} xf(x) dx = \int_{-1}^{1} x\dfrac{(x+1)}{2} dx = \dfrac{1}{3}$

b. The variance of X is $\mu = \int_{-\infty}^{+\infty} x^2 f(x) dx - \mu^2 = \dfrac{2}{3}$

18.9   GAUSS' DISTRIBUTION

The most important continuous probability distribution is the normal distribution whose graph is called the normal curve and is represented in Figure 18.4. The normal distribution graph often called the Gauss distribution describes many physical phenomena in nature. Its equation was first derived by Abraham De Moivre in 1733.

Definition 12
Consider X the continuous random variable which has the shape distribution in Figure 18.4. X is called the normal random variable. The probability distribution of the normal variable n depends on its mean μ and its standard deviation σ. It is represented by the density function:

$$n(x,\mu,\sigma) = \dfrac{1}{\sqrt{2\pi}\sigma} e^{-\frac{1}{2}\left[\frac{(x-\mu)}{\sigma}\right]^2},$$

where $\pi = 3.14159\ldots$ and $e = 2.71828\ldots$
   n is completely known once μ and σ are known.

Examples 26

For $\mu = 0, \sigma = 2$, the Mathematica Command gives:

Plot[Evaluate@Table[PDF[NormalDistribution[0,σ],x],{σ,{2}}]],{x,x6,6},Filling $\rightarrow$ Axis, AxesLabel $\rightarrow$ {x,n}]]

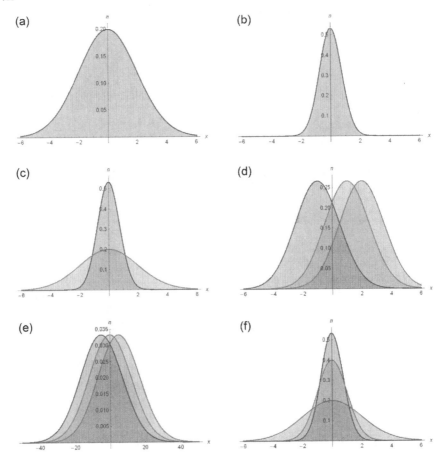

FIGURE 18.4 (a) Normal Distribution for $\mu = 0$ and $\sigma = 2$. (b) Normal Distribution for $\mu = 0$ and $\sigma = 0.75$. (c) Normal Distribution for $\mu = 0$, $\sigma = 0.75$ and $\sigma = 2$ in the same graph. (d) Normal Distribution for $\sigma = 1.5$ and $\mu = -1, 1, 2$ in the same graph. (e) Normal Distribution for $\sigma = 12$ and $\mu = -5, 0, 5$ in the same graph. (f) Normal Distribution for $\sigma = 0.75, 1, 2$ and $\mu = 0$ in the same graph.

For $\mu = 0, \sigma = 0.75$, the Mathematica Command:

Plot[Evaluate@Table[PDF[NormalDistribution[0,σ],x],{σ,{.75}}]],{x,x6,6},Filling$\rightarrow$Axis, AxesLabel $\rightarrow$ {x,n}] gives the following graph:

Note that the area under the curve is smaller than in the case of $\sigma = 2$, the Mathematica command gives:

For $\mu = 0$, $\sigma = 0.75$ and $\sigma = 2$ in the same graph, the Mathematica Command gives:

Plot[Evaluate@Table[PDF[NormalDistribution[0,σ],x],{σ,{.75,2}}]],{x,x6,6},Filling $\rightarrow$Axis, AxesLabel $\rightarrow$ {x,n}]

For $\sigma = 1.5$ and $\mu = -1, 1, 2$ in the same graph, the Mathematica command:

Plot[Evaluate@Table[PDF[NormalDistribution[μ,1.5],x],{μ,{x1,1,2}}]],{x,x6,6},Filling $\rightarrow$ Axis, AxesLabel $\rightarrow$ {x,n}], gives the following graph:

For $\sigma = 12$ and $\mu = -5, 0, 5$ in the same graph, the Mathematica command:

Plot[Evaluate@Table[PDF[NormalDistribution[μ,12],x],{μ,{x5,0,5}}],{x,x50,50},Filling   →Axis, AxesLabel →{x,n}], gives the following graph:

For $\sigma = 0.75, 1, 2$ and $\mu = 0$ in the same graph, the Mathematica Command gives:

Plot[Evaluate@Table[PDF[NormalDistribution[0,σ],x],{σ,{.75,1,2}}],{x,x6,6},Filling   →   Axis, AxesLabel →{x,n}]

18.9.1   PROPERTIES OF THE NORMAL CURVE

From the normal curve we can identify the following properties:

1. The maximum occurs at $x = \mu$.
2. The curve is symmetric about the vertical axis passing through the mean $x = \mu$.
3. The curve has two points of inflection: $x = \mu \pm \sigma$.
4. The curve is concave downward for $\mu - \sigma < X < \mu + \sigma$ and is concave upward otherwise.
5. $\lim_{x \to \pm \infty} n(x, \mu, \sigma) = 0$; so, the x-axis is an asymptote to the curve as $x \to \pm \infty$.
6. The area under the curve is 1: $\int_{-\infty}^{+\infty} n(x, \mu, \sigma) dx = 1$.

Let us now show that the parameters μ and σ^2 are, respectively, the mean and the variance of the normal distribution. To evaluate the mean, let us consider

$$E(X) = \frac{1}{\sqrt{2\pi}\sigma} \int_{-\infty}^{+\infty} e^{-\frac{1}{2}\left[\frac{(x-\mu)}{\sigma}\right]^2} dx \text{ and suppose } z = \frac{x-\mu}{\sigma}; dx = \sigma dz; \text{ then, we have}$$

$$E(X) = \frac{1}{\sqrt{2\pi}} \int_{-\infty}^{+\infty} (\mu + \sigma z) e^{-z^2/2} dz$$

$$E(X) = \frac{1}{\sqrt{2\pi}} \int_{-\infty}^{\infty} (\mu + \sigma z) e^{-z^2/2} dz$$

The first term gives μ. The second term will give zero after integration by parts. Therefore, $E(X) = \mu$.

Now, let us consider the variance of the normal distribution:

$$E\left[(X-\mu)^2\right] = \frac{1}{\sqrt{2\pi}\sigma} \int_{-\infty}^{+\infty} (x-\mu)^2 e^{-\frac{1}{2}\left[\frac{(x-\mu)}{\sigma}\right]^2} dx;$$

with the substitution $z = \frac{x-\mu}{\sigma}; dx = \sigma dz$, we obtain

$$E\left[(X-\mu)^2\right] = \frac{\sigma^2}{\sqrt{2\pi}} \int_{-\infty}^{\infty} z^2 e^{-z^2/2} dz.$$

After integration by part $u = z$ and $dv = ze^{-z^2/2}$, we obtain:

$$E\left[(X-\mu)^2\right] = \sigma^2 (0+1) = \sigma^2.$$

PROBLEM SET 18

18.1 A student is tossing a six-sided die four times. What is the probability of getting a 5 on each roll?

18.2 What is the probability of having at least one tail when you toss a coin five times?

18.3 Use the Pascal triangle to evaluate $(2x + 3y)^{10}$.

18.4 Eight people are going on a ride in a boat that seats eight people. One person will drive and only three of the remaining people are willing to ride in the two bow seats. How many seating arrangements are possible?

18.5 A single fair die is rolled twice. What is the probability of getting a 5 on the second roll given that a 5 turned up on the first roll?

18.6 A coin is tossed twice. What is the probability of getting a head on the second toss given that a head turned up on the first toss?

18.7 If the space sample $C = A \cup B$ with $P(A) = 0.9$ and $P(B) = 0.4$, what is $P(A \cap B)$?

18.8 If A and B are subsets of the sample space C, show that

$$P(A \cap B) \leq P(A) \leq P(A \cup B) \leq P(A) + P(B).$$

18.9 A pair of dice is thrown. If it is known that one die shows a 4, what is the probability that

a. the second die shows a 5
b. the total of both dice is greater than 7

18.10 Let X have a p.d.f. $f(x)$ such that

$$f(x) > 0 \text{ at } x = -1, 0, 1$$

$$f(x) = 0 \text{ elsewhere}$$

a. If $f(0) = \dfrac{1}{2}$ find $E(X^2)$

b. If $f(0) = \dfrac{1}{2}$ and $E(X) = \dfrac{1}{6}$ determine $f(-1)$ and $f(1)$

18.11 If X has the p.d.f.,

$$f(x) = \frac{1}{x^2} \text{ for } 1 < x < \infty$$

$f(x) = 0$ elsewhere
What is the mean value of X?

18.12 If X has the p.d.f.,

$$f(x) = \frac{1}{2} \text{ for } -1 < x < 2$$

$f(x) = 0$ elsewhere
then, show that the moment-generating function of the random variable X is

$$M(t) = \frac{e^{2t} - e^{-t}}{3t} \text{ for } t \neq 0$$

$M(t) = 1$ for $t = 0$.

18.13 Find the mean value μ and the variance σ^2 of the distribution function $F(x)$ such that

$$F(x) = \begin{cases} 0 & x < 0 \\ \dfrac{x}{8} & 0 \le x < 2 \\ \dfrac{x^2}{16} & 2 \le x < 4 \\ 1 & 4 \le x \end{cases}.$$

18.14 Find the mean and variance, if they exist, of the distribution $f(x) = \dfrac{2}{x^3}$ $\quad 1 < x < \infty$

$$= 0 \quad \text{elsewhere}$$

Appendix A
Basic Laws of Electromagnetism

We will review here some basic laws in electromagnetism. We first need to introduce Coulomb's law in electrostatics.

A.1 COULOMB'S LAW

Coulomb's law gives the force exerted between two electric points' charges at a given distance from each other. It states that the electrostatic force between the two-point charges is directly proportional to the absolute value of each individual charge and is inversely proportional to the square of the separation between the two-point charges.

Assume two charges q_1 and q_2 separated by a distance r_{12}. The electrostatic forces exerted between these two charges have a common magnitude:

$$F_{12} = F_{21} = k \frac{|q_1||q_2|}{r_{12}^2} \tag{A1}$$

$\vec{F}_{12}$ is the force exerted by q_1 on q_2 and $\vec{F}_{21}$ is the force exerted by q_2 on q_1.

The constant k is called the Coulomb's constant: $k = \dfrac{1}{4\pi\varepsilon_0} \approx 9\times10^9\,\mathrm{Nm^2/C^2}$

where ε_0 represents the permittivity of free space: $\varepsilon_0 = 8.8542\times10^{-12}\,\mathrm{C^2\,N^{-1}\,m^{-2}}$.

The SI unit of the forces is Newton (N) and the SI unit of the charges is Coulomb (C); r_{12} is in meter (m).

It is understood that like charges repel and opposite charges attract. In the case of a system of n charges $q_1, q_2, q_3, \ldots, q_n$ the force of the ith charge is given by

$$\vec{F}_i = q_i \sum_{j \neq i}^{n} \frac{q_j}{4\pi\varepsilon_0} \frac{\vec{r}_{ij}}{r_{ij}^3} \tag{A2}$$

where $\vec{r}_{ij} = \vec{r}_i - \vec{r}_j$ is the vector distance between the ith charge and the jth charge.

Equation (A2) represents the superposition principle for forces.

A.1.1 CONTINUOUS DISTRIBUTION OF CHARGE

For a continuous distribution of charge we can define:

The volume charge density, $\rho = \dfrac{dq}{dV}$, where dV is the volume element containing the charge element dq; the surface charge density $\sigma = \dfrac{dq}{dS}$, where dSV is the surface element containing the charge element dq; and the line charge density $\lambda = \dfrac{dq}{dl}$, where dl is the line element containing the charge element dq.

A.1.2 Electric Field

Assume that we have a charge q_B located at the position $\vec{r}_B$ (source point) and create some field condition in the environment surrounding it such that a force will be exerted at distance $\vec{r}_A$ (field point) where there is a test charge q_A. The electric field at any point is defined as the force per unit of charge on a small test charge held stationary at that point.

The test charge is assumed to be small to ensure that it does not cause any motion of the source charges responsible for the field. The electric field magnitude is denoted by E and its MKS unit is Newton per Coulomb (N/C). The latter is equivalent to volt per meter.

For the static case (or time-independent electric field) we can define:

$$\vec{E}\left(\vec{r}_A,\vec{r}_B,q_B\right) = \lim_{q_A \to 0} \frac{F_{A,B}}{q_A} = kq_B \frac{\vec{r}_A - \vec{r}_B}{\left|\vec{r}_A - \vec{r}_B\right|^3} \tag{A3}$$

A.2 MAXWELL EQUATIONS

Maxwell's equations are a series of four important equations of electromagnetic field wave properties published in a Maxwell 1861 paper on physical lines of force. They express how electric charges produce electric fields (Gauss' law in electricity), the experimental absence of magnetic monopoles (Gauss' law in magnetism), how electric current and change in electric fields can produce magnetic fields (Ampere's circuital law) and how magnetic field and change in magnetic fields can produce electric fields (Faraday's law of induction).

For convenience, we recapitulate them here in the MKSA unit system (Meter for length, Kilogram for mass, Second for time and Ampere for current intensity), which is adopted in Système International (SI).

A.2.1 Maxwell's Equations in Inhomogeneous Medium: Differential Form

Maxwell's equations were put together by the Scottish physicist James Clerk Maxwell (1831–1879). It is a set of four equations encompassing the following:

A.2.1.1 Gauss' Law for Electricity

Gauss' law for electricity states that the electric flux of the electric field $\vec{E}$ throughout a closed surface is equal to the total charge within the surface divided by the electric permittivity of free space, ε_0.

$$div\vec{D} = \rho \quad \text{or} \quad \vec{\nabla} \cdot \vec{D} = \rho \tag{A4}$$

with $\vec{D} = \varepsilon\vec{E}$ the vector electric displacement and ρ the volume charge density.

A.2.1.2 Gauss' Law for Magnetism (Absence of Monopole)

Magnetic field lines form closed loops which will go from the north pole of the magnet and come back through the south pole of the magnet. They never begin and never end. Gauss' law for magnetism states that the magnetic flux through any closed surface is zero. There's no magnetic monopole.

$$div\vec{B} = 0 \text{ or } \vec{\nabla} \cdot \vec{B} = 0 \tag{A5}$$

A.2.1.3   Ampere's Circuital Law

Ampere's circuital law due to Ampere himself and improved later states that the line integral of the magnetic field over a closed line equals μ_0 times the current in the closed line. A more general Ampere's circuital law adds a second term and can be read as the curl of the magnetic field $\vec{H}$ is equal to the free current density $\vec{J}$ plus the time rate of change of the electric displacement $\vec{D}$.

$$\mathrm{curl}\vec{H} = \vec{J} + \frac{\partial \vec{D}}{\partial t} \quad \text{or} \quad \vec{\nabla} \times \vec{H} = \vec{J} + \frac{\partial \vec{D}}{\partial t} \tag{A6}$$

A.2.1.4   Faraday's Law of Induction

Faraday's law of induction explains that the curl of the electric field is proportional to the time rate of change of the magnetic field. Faraday's law implies the creation of an induced electromotive force due to the change in the magnetic flux throughout a loop.

$$curl\vec{E} = -\frac{\partial \vec{B}}{\partial t} \quad \text{or} \quad \vec{\nabla} \times \vec{E} = -\frac{\partial \vec{B}}{\partial t} \tag{A7}$$

A.2.2   Maxwell Equations in Inhomogeneous Medium: Integral Form

We will use here the integral form of the above laws but in an inhomogeneous medium.
   We will successively have the following:

(a) Gauss' law for electricity

$$\oint_S \vec{D} \cdot \vec{dA} = q \tag{A8}$$

(b) Gauss' law for magnetism (absence of monopole)

$$\oint_S \vec{B} \cdot \vec{dA} = 0 \tag{A9}$$

(c) Faraday's law of induction

$$\oint_C \vec{E} \cdot \vec{dl} = -\int_S \frac{\partial \vec{B}}{\partial t} \cdot \vec{dA} \tag{A10}$$

(d) Ampere's circuital law

$$\oint_C \vec{H} \cdot \vec{dl} = \int_S \vec{J} \cdot \vec{dA} + \int_S \frac{\partial \vec{D}}{\partial t} \cdot \vec{dA} \tag{A11}$$

$k = \dfrac{1}{4\pi\varepsilon_0}$ is coulomb's constant

$$c^2 = \frac{1}{\mu_0 \varepsilon_0} \tag{A12}$$

$\vec{H}$ is the magnetic field strength vector of the magnetic field in A/m

$\vec{E}$ is the electric field in V/m or N/C

$\vec{D}$ is the electric displacement field in C/m^2

$\vec{J}$ is the free current density in A/m^2

$d\vec{A}$ is the differential vector element of surface area $\vec{A}$ in m^2

$\vec{P}$ is the polarization density vector in C/m^2

$\vec{M}$ is the magnetization density vector in A/m

μ is the magnetic permeability in Henry/m or N/A^2

A.2.3 POLARIZATION AND MAGNETIZATION IN A LINEAR MATERIAL

The polarization density vector is proportional to the electric field and is written as

$$\vec{P} = \chi_e \varepsilon_0 \vec{E} \tag{A13}$$

and the magnetization density vector is proportional to the magnetic field and is given by:

$$\vec{M} = \chi_m \vec{H} \tag{A14}$$

The electric displacement field is written as

$$\vec{D} = \varepsilon_0 \vec{E} + \vec{P} = \left(1 + \chi_e\right)\varepsilon_0 \vec{E} = \varepsilon \vec{E} \tag{A15}$$

and the magnetic induction

$$\vec{B} = \mu_0 \left(\vec{H} + \vec{M}\right) = \left(1 + \chi_m\right)\mu_0 \vec{H} = \mu \vec{H} \tag{A16}$$

with the following definitions:

χ_e is the electric susceptibility of the material
χ_m is the magnetic susceptibility of the material
ε is the electric permittivity of the material
μ is the magnetic permeability of the material
In the vacuum, $\varepsilon = \varepsilon_0$ and $\mu = \mu_0$:

$$\varepsilon_0 = 8.85419 \times 10^{-12} \ \text{Farads / m}$$

$$\mu_0 = 4\pi \times 10^{-7} \ \text{H / m} \tag{A17}$$

A.2.4 EQUATION OF CONSERVATION OF CHARGE

Let us take the divergence of the curl of $\vec{B}$

$$\vec{\nabla}\left(\text{curl } \vec{B}\right) = 0 = \vec{\nabla} \cdot \left(\vec{\nabla} \times \vec{B}\right) = \vec{\nabla} \cdot \left(\mu \vec{J} + \mu \frac{\partial \vec{D}}{\partial t}\right) \tag{A18}$$

which gives the following equations:

$$\vec{\nabla}\vec{J} + \frac{\partial\left(\vec{\nabla}\cdot\vec{D}\right)}{\partial t} = 0 \quad \text{or} \quad \vec{\nabla}\cdot\vec{J} + \frac{\partial\rho}{\partial t} = 0 \tag{A19}$$

A.3 ELECTROMAGNETIC ENERGY

A.3.1 THE ELECTROSTATICS ENERGY

The electrostatics energy in linear dielectrics of constant of permittivity ε can be written as

$$U_E = \frac{1}{2}\int_{\text{Volume}}\vec{D}\cdot\vec{E}dv \tag{A20}$$

The integration is over the volume of the system external to the conductor. It can be of course extended to all space included inside the conductor where the electric field $\vec{E}$ is zero. The electrostatic energy is stored inside the electric field itself. The electrostatic energy density or electrostatic energy per unit of volume can then be defined as

$$u_E = \frac{dU_E}{dv} = \frac{1}{2}\vec{D}\cdot\vec{E} = \frac{1}{2}\frac{\vec{D}^2}{\varepsilon} \tag{A21}$$

A.3.2 THE MAGNETIC ENERGY

The formulation of the magnetic energy in terms of the vectors
$\vec{B}$ and $\vec{H}$ is of considerable interest because it provides picture of how the magnetic energy is stored in the magnetic field itself. The magnetic energy can be written as

$$U_M = \frac{1}{2}\int_{\text{volume}}\vec{H}\cdot\vec{B}dv \tag{A22}$$

The previous expression is analogous to the one obtained for the electrostatic energy. It is important to note that it is restricted to systems containing linear magnetic media.

Similarly, the magnetic energy density or magnetic energy per unit of volume in isotropic, linear and magnetic materials can be written as follows:

$$u_M = \frac{dU_M}{dv} = \frac{1}{2}\vec{H}\cdot\vec{B} = \frac{1}{2}\frac{\vec{B}^2}{\mu} \tag{A23}$$

A.3.3 THE POYNTING VECTOR

The Poynting vector $\vec{S}$ can be defined as the local energy flow per unit time per unit area. It is named after John Henry Poynting (1852–1914) who discovered it and represents the energy flux per unit of area in (W/m²) of an electromagnetic field.

$$\vec{S} = \vec{E}\times\vec{H} \tag{A24}$$

A.3.4   THE TOTAL ENERGY IN A LINEAR MEDIUM

The total energy in a linear medium is the sum of the electrostatic and magnetic energies:

$$U = U_E + U_M \tag{A25}$$

The total electromagnetic energy density can be written as follows:

$$u = u_E + u_M = \frac{1}{2}\left(\vec{E} \cdot \vec{D} + \vec{H} \cdot \vec{B}\right) \tag{A26}$$

A.3.5   ENERGY CONSERVATION

The expression that shows the electromagnetic energy conservation law is given by

$$\vec{\nabla} \cdot \vec{S} + \frac{\partial u}{\partial t} = -\vec{J} \cdot \vec{E}. \tag{A27}$$

In this expression the term $\vec{J} \cdot \vec{E}$ represents the work done by the local field on charged particles per unit volume. The energy conservation relation can be proved from Maxwell's equations.

A.4   THE ELECTROMAGNETIC WAVE EQUATIONS

A.4.1   FOR THE MAGNETIC FIELD

From Maxwell's equation in vacuum, we know

$$\vec{\nabla} \times \vec{B} = \varepsilon_0 \mu_0 \frac{\partial \vec{E}}{\partial t} \tag{A28}$$

By taking the rotational or the curl of both sides, we have

$$\vec{\nabla} \times \left(\vec{\nabla} \times \vec{B}\right) = \varepsilon_0 \mu_0 \frac{\partial\left(\vec{\nabla} \times \vec{E}\right)}{\partial t} \tag{A29}$$

$$\text{or} \quad \vec{\nabla}\vec{\nabla} \cdot \vec{B} - \vec{\nabla} \cdot \vec{\nabla}\vec{B} = \varepsilon_0 \mu_0 \frac{\partial\left(\vec{\nabla} \times \vec{E}\right)}{\partial t} \tag{A30}$$

but one of the Maxwell equation properties is that

$$\vec{\nabla} \cdot \vec{B} = 0 \ \text{ and that } \vec{\nabla} \times \vec{E} = -\frac{\partial \vec{B}}{\partial t}$$

Then equation (A30) becomes

$$\vec{\nabla} \cdot \vec{\nabla}\vec{B} = \varepsilon_0 \mu_0 \frac{\partial^2 \vec{B}}{\partial t^2} \tag{A31}$$

or

$$\vec{\nabla}\cdot\vec{\nabla}\vec{B} = \frac{1}{c^2}\frac{\partial^2\vec{B}}{\partial t^2} \tag{A32}$$

The previous equation is the wave equation for magnetic field $\vec{B}$ in a vacuum. It involves the scalar Laplacian $\Delta = \vec{\nabla}\cdot\vec{\nabla}$ of the magnetic field $\vec{B}$.

A.4.2 FOR THE ELECTRIC FIELD

From Maxwell's equation in a vacuum, we know

$$\vec{\nabla}\times\vec{B} = \varepsilon_0\mu_0\frac{\partial\vec{E}}{\partial t} \tag{A33}$$

Let's take now time differentiation of both sides of the previous equality. We then have

$$\vec{\nabla}\times\frac{\partial\vec{B}}{\partial t} = \varepsilon_0\mu_0\frac{\partial^2\vec{E}}{\partial t^2} \tag{A34}$$

and using $\vec{\nabla}\times\vec{E} = -\dfrac{\partial\vec{B}}{\partial t}$,

we obtain:

$$\vec{\nabla}\times\left(\vec{\nabla}\times\vec{E}\right) = -\varepsilon_0\mu_0\frac{\partial^2\vec{E}}{\partial t^2} \tag{A35}$$

or

$$\vec{\nabla}\vec{\nabla}\cdot\vec{E} - \vec{\nabla}\cdot\vec{\nabla}\vec{E} = -\varepsilon_0\mu_0\frac{\partial^2\vec{E}}{\partial t^2} \tag{A36}$$

But one of Maxwell's equation properties in a vacuum is that $\vec{\nabla}\cdot\vec{E} = 0$; therefore, the wave equation for the electric field $\vec{E}$ in a vacuum becomes

$$\vec{\nabla}\cdot\vec{\nabla}\vec{E} = \varepsilon_0\mu_0\frac{\partial^2\vec{E}}{\partial t^2} \tag{A37}$$

or

$$\vec{\nabla}\cdot\vec{\nabla}\vec{E} = \frac{1}{c^2}\frac{\partial^2\vec{E}}{\partial t^2} \tag{A38}$$

This equation involves the scalar Laplacian $\Delta = \vec{\nabla}\cdot\vec{\nabla}$ of the electric field, $\vec{E}$.

A.4.3 ELECTROMAGNETIC WAVE EQUATIONS IN AN INHOMOGENEOUS MEDIUM

Consider both the scalar potential φ and the vector potential $\vec{A}$; the vector electric field $\vec{E}$ can be written as:

$$\vec{E} = -\vec{\nabla}\varphi - \frac{\partial \vec{A}}{\partial t} \tag{A39}$$

With $\vec{B} = \vec{\nabla} \times \vec{A}$

Using Gauss' law $\vec{\nabla} \cdot \vec{E} = \frac{\rho}{\varepsilon_0}$ and

Oersted's,

$$\vec{\nabla} \times \vec{B} - \frac{1}{c^2}\frac{\partial \vec{E}}{\partial t} = \mu_0 \vec{J} \tag{A40}$$

equation (A39) leads to the wave equation for the potentials φ and $\vec{A}$ if we have the constraint:

$$\frac{1}{c^2}\frac{\partial \varphi}{\partial t} + \vec{\nabla} \cdot \vec{A} = 0 \tag{A41}$$

The above constraint which consists of fixing the divergence of the vector potential is called the Lorentz gauge.

Substituting the electric field into Gauss' law yields

$$\frac{\rho}{\varepsilon_0} = \vec{\nabla} \cdot \vec{E} = -\vec{\nabla}^2\varphi - \frac{\partial}{\partial t}\left(\vec{\nabla} \cdot \vec{A}\right) = -\vec{\nabla}^2\varphi + \frac{1}{c^2}\frac{\partial^2 \varphi}{\partial t^2} \tag{A42}$$

which is the wave equation for electric potential. Also, substituting $\vec{B} = \vec{\nabla} \times \vec{A}$ into Oersted's law and using the Lorentz gauge condition, we can write the wave equation for the vector potential:

$$-\vec{\nabla}^2\vec{A} + \frac{1}{c^2}\frac{\partial^2 \varphi}{\partial t^2} = \mu_0 \vec{J} \tag{A43}$$

A.5   POISSON AND LAPLACE'S EQUATIONS

Let's consider the differentiation form of Gauss' law for electricity:

$\vec{\nabla} \cdot \vec{D} = \rho$ with the electric displacement field $\vec{D} = \varepsilon_0 \vec{E}$. Therefore, we have $\vec{\nabla} \cdot \vec{E} = \frac{\rho}{\varepsilon_0}$; also, the electrostatic field derived from a potential φ that is: $\vec{E} = -\vec{\nabla}\varphi$ (the electric field is the opposite of the gradient of a potential). Combining the two previous expressions, we have

$$\vec{\nabla}^2\varphi = -\frac{\rho}{\varepsilon_0} \tag{A44}$$

The scalar differential operator, $\vec{\nabla}^2$, can also be written as Δ and is called the Laplacian. Equation (A44) is called Poisson's equation. For $\rho = 0$, Poisson's equation is referred to as Laplace's equation and is very useful in physics in the computation of the electric potential and the electric field in a given region of space.

Laplace's equation can be used to find the electric field, $\vec{E}$, at a given point in the space between two infinitely large parallel plates separated by a distance d as shown in Figure A.1. The object is to first find the electric potential which satisfies Laplace's equation, $\vec{\nabla}^2\phi = 0$. Since the plates are

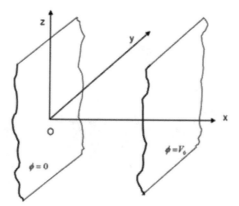

FIGURE A.1   Laplace equation and the electric potential.

perpendicular to the x-axis, we will only consider the x-component of the Laplacian and as a result, Laplace's equation reduces to

$$\vec{\nabla}_x^2 \phi = 0 \text{ or } \frac{d^2\phi}{dx^2} = 0 \qquad (A45)$$

The electric potential can be found to be $\phi = k_1 x + k_2$ where k_1 and k_2 are real numbers. Using now the boundary conditions, that is, $\phi = 0$ for $x = 0$ and $\phi = V_0$ for $x = d$, the electric potential is finally written as $\phi(x) = \dfrac{V_0}{d} x$ and the electric field $\vec{E} = -\vec{\nabla}\varphi$ will have the components:

$$E_x = -\frac{d\varphi}{dx} = -\frac{V_0}{d}; \qquad E_y = 0; \qquad E_z = 0$$

Example 1

Consider a spherical capacitor with two concentric spheres of radius R_1 and R_2. The inner sphere is at the potential V_0 and the outer at the potential zero. Find the electric potential and the electric field using the following boundary conditions:

$$\phi(r) = \begin{cases} V_0 & \text{at} \quad r = R_1 \\ 0 & \text{at} \quad r = R_2 \end{cases}$$

Because of the geometry of the problem, we will use the spherical coordinates (r, θ, φ) of the position vector $\vec{r}$ (see Figure A.2).

The Laplacian in the spherical coordinates is given by

$$\Delta\phi = \vec{\nabla}^2\phi = \frac{1}{r^2}\left[\frac{\partial}{\partial r}\left(r^2\frac{\partial\phi}{\partial r}\right) + \frac{1}{\sin\theta}\frac{\partial}{\partial\theta}\left(\sin\theta\frac{\partial\phi}{\partial\theta}\right) + \frac{1}{\sin^2\theta}\frac{\partial^2\phi}{\partial\varphi^2}\right] \qquad (A46)$$

It is not possible to distinguish a point (r, θ, φ) from another point (r, θ', φ') with the same radius and different polar and azimuthal angles. As a result, the position vector is only radial and ϕ depends only on r. We can then write

$$\frac{\partial\phi}{\partial\theta} = \frac{\partial\phi}{\partial\varphi} = 0 \qquad (A47)$$

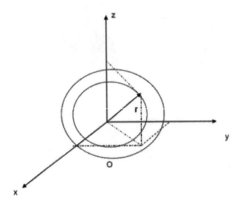

FIGURE A.2 A spherical capacitor.

And Laplace equation reduces to

$$\frac{1}{r^2}\frac{\partial}{\partial r}\left(r^2\frac{\partial\phi}{\partial r}\right)=0.$$ (A48)

Using the boundary condition, we can find the solution to be

$$\phi(r)=\frac{V_0 R_1}{R_1 - R_2}\left(1-\frac{R_2}{r}\right)$$ (A49)

For $R_1 < r < R_2$

The electric field components would therefore be

$$E_r = -\frac{d\phi}{dr} = \frac{V_0 R_1 R_2}{R_1 - R_2}\frac{1}{r^2}$$ (A50)

$E_\theta = E_\varphi = 0$ for $R_1 < r < R_2$.

Appendix B
Introduction to Lebesgue Theory

Before defining Lebesgue's theory, it is important to recall the set theory and more importantly some basic terminologies and notations in the set theory.

B.1 NOTATION AND TERMINOLOGY

B.1.1 DEFINITION

A set can be described by its members, that is, $\{x_1,...,x_n\}$ is the set whose members also called elements which are $x_1,...,x_n$. More often, sets are described properties. $\{x : P\}$, for example, is the set of all elements of x which have the property P. The empty set is represented by the symbol ϕ or $\{\}$.

If x is an element of or belongs to a set, A, then we write $x \in A$. Otherwise $x \notin A$.

Consider B a subset of A; we can write $(x \in B)$ implies $(x \in A)$, which means $B \subset A$.

Note that if $A \subset B$ and $B \subset A$, then $A = B$.

If $B \subset A$ and $A \neq B$, then B is a proper subset of A.

Note that $\phi \subset A$ for every set A.

$A \cup B$ represents the union of A and B and $A \cap B$ represents the intersection of A and B.

If $\{A_\alpha\}$ is a collection of sets, where α runs through some indices set I, we refer $\underset{\alpha \in I}{\cup} A_\alpha$ as the union of $\{A_\alpha\}$ and $\underset{\alpha \in I}{\cap} A_\alpha$ as the intersection of $\{A_\alpha\}$. We can also express $\underset{\alpha \in I}{\cup} A_\alpha$ and $\underset{\alpha \in I}{\cap} A_\alpha$ as follows:

$$\underset{\alpha \in I}{\cup} A_\alpha = \{x : x \in A_\alpha \text{ for at least one } \alpha \in I\}.$$

$$\underset{\alpha \in I}{\cap} A_\alpha = \{x : x \in A_\alpha \text{ for every } \alpha \in I\}.$$

B.1.2 DISJOINT SETS

If the sets $\{A_\alpha\}$ and $\{A_\beta\}$ have no element in common, they are said to be disjoint sets.

B.1.3 COMPLEMENT OF A

The complement of A in B represents the set of all elements in A that are not in B. The complement of A in B is denoted by $A - B$ and is expressed as

$$A - B = \{x : x \in A; x \notin B\}.$$

The complement of A in B can also be denoted by A^c.

B.1.4 CARTESIAN PRODUCT

The Cartesian product $A_1 \times A_2 \times \cdots \times A_n$ of the sets $A_1, A_2, \cdots; A_n$ is the set of all ordered n-tuples $(a_1, a_2, \cdots, a_n)$ where $a_i \in A_i$ for $i = 1, 2, ..., n$.

B.1.5   THE REAL LINE OR REAL NUMBER SYSTEM

The real line is denoted by $\mathbb{R}^l$ and $r^k = \mathbb{R}^l \times \mathbb{R}^l \times ... \times \mathbb{R}^l$ (k factors). The extended real number system is $\mathbb{R}^l$ with two symbols $-\infty$ and $+\infty$ with the ordering. If $-\infty \le a \le b \le +\infty$, the interval $[a,b]$ and the segments (a,b), $[a,b)$ and $(a,b]$ are defined, respectively, by

$$\left[a,b\right] = \left\{x : a \le x \le b\right\},$$

$$\left(a,b\right) = \{x : a < x < b\},$$

$$\left[a,b\right) = \{x : a \le x < b\}$$

and $\left(a,b\right] = \{x : a < x \le b\}$.

If $E \subset \left[-\infty,+\infty\right]$ and $E \ne \phi$, the least upper bound or supremum of E exists in $\left[-\infty,+\infty\right]$ and is denoted by $\mathrm{Sup}E$; the greatest lower bound or infimum of E exists in $\left[-\infty,+\infty\right]$ and is denoted by $InfE$.

The symbol in $f : X \to Y$ means that f is the function (or mapping or transformation) of the set X into the set Y, i.e., f assigns to each input x in X and output y in Y. The output which is the element of Y will be denoted by $y = f\left(x\right)$.

If $A \subset X$ and $B \subset Y$, the image of A and the inverse image (or pre-image) of B are defined, respectively, by

$$f\left(A\right) = \left\{y : y = f\left(x\right) \text{ for some } x \in A\right\}$$

and $f^{-1}\left(A\right) = \left\{x : f\left(x\right) \in B\right\}$.

The domain of f is the set of all elements x of X that have an image or output; the range of f is the set of all elements $f(x)$ of Y that have a pre-image or input in X. If $f(X) = Y$, f is said to map X into Y.

If $f^{-1}\left(y\right)$ consists of at most one point, for each $y \in Y$, then f is said to be one-to-one. If f is one-to-one, then f^{-1} is a function with domain $f(X)$ and range X.

If $f : X \to Y$ and $g : Y \to Z$, the composite function $gof : X \to Z$ is defined by the formula $gof\left(x\right) = g\left(f\left(x\right)\right)$ for all $x \in X$.

B.1.6   DEFINITION OF COVERING

The concept of coverings is essential to the topic of measure theory. A collection U of subsets of M covers $A \subset M$ if A is contained in the union of the sets belonging to U. The collection U is a covering of A.

If two collections U and V cover A and if $V \subset U$ in the sense that every set v of V belongs to U, then U reduces to V and V is a subcovering. If all the sets in U are open, U is called open covering of A. If every covering of A reduces to a finite subcovering then we say that A is covering compact. A covering U of A can also be called a cover of A. The members of U are called scraps.

B.2   THE MEASURABILITY CONCEPT

B.2.1   DEFINITION OF A TOPOLOGY

A collection τ of a subset of a set X is said to be a topology in X if τ has the following properties:

i. $\phi \in \tau$ and $X \in \tau$.

ii. If $V_i \in \tau$ for $i = 1, 2, 3, \ldots, n$,

then $V_1 \cap V_2 \cap \cdots \cap V_n \in \tau..$

iii. If $\{V_\alpha\}$ is an arbitrary collection of members of τ (finite, countable or uncountable) then

$\cup_\alpha V_\alpha \in \tau$.

Remarks

If τ is a topology in X, then X is called a topological space and the members of τ are called the open set in X.

B.2.2 TOPOLOGICAL SPACE

If τ is a topology in X, then X is called a topological space, and the members of τ are called the open sets in X.

Remark: Consider X and Y as two topological spaces. If f is a mapping of X into Y, then f is said to be continuous provided that the inverse $f^{-1}(V)$ is an open set in X for every open set V in Y.

B.3 INTRODUCTION TO σ-ALGEBRA

B.3.1 DEFINITION

A collection M of subsets of a set X is said to be a σ-algebra in X if M has the following properties:

i. $X \in M$

ii. If $A \in M$, then $A^c \in M$, with A^c the complement of A relative to X

iii. If $A = \overset{\infty}{\underset{n=1}{\cup}} A_n$ and if $A_n \in M$, for $n = 1, 2, \ldots$ then $A \in M$.

B.3.2 MEASURABLE SPACE

If M is a σ-algebra in X, then X is called a measurable space. The members of M are called the measurable sets in X.

If X is a measurable space, Y a topological space and f a mapping of X into Y, then f is said to be measurable provided that $f^{-1}(V)$ is a measurable set in X for every open set V in Y.

Remark: It is probably more satisfactory to use the term "measurable space" to the ordered pair (X, M) rather than to X. Similarly, a topological space is an ordered pair (X, τ).

Examples 1

Examples of topological spaces are metric spaces. Let us now elaborate a little bit on the metric space. Is call metric space, a set X in which a distance function (or metric) ρ is defined with the following properties:

a. $0 \leq \rho(x, y) < \infty$ for all x and y elements of X;

b. $\rho(x, y) = 0$ if and only if all $x = y$;

c. $\rho(x, y) = \rho(y, x)$ for all x and y elements of X;

d. $\rho(x, y) \leq \rho(x, z) + \rho(z, y)$ for all x, y, z elements of X (triangle property).

B.3.3 OPEN BALL

If x is an element of X and $r \geq 0$, the open ball with center at x and radius r is the set $\{y \in X : \rho(x, y) < r\}$.

B.3.4 LOCAL CONTINUITY

The definition of continuity given previously is the one of a global continuity. We will now define the continuity locally:

A mapping f of X into Y is said to be continuous at the point $x_0 \in X$ if to every neighborhood V of $f(x_0)$, it corresponds a neighborhood W of x_0 such that $f(W) \subset V$. A neighborhood of a point x is an open set which contains x.

The following proposition relates to the two definitions of continuity.

Proposition

Let X and Y be two topological spaces; a mapping f of X into Y is continuous if and only if f is continuous at every point of X.

Theorem 1 (Continuity of composite functions)

Let Y and Z be two topological spaces and let $g : Y \to Z$ be continuous

   a. If X is a topological space, with $f : X \to Y$ continuous and $h = gof$, then $h : X \to Z$ is continuous.
   b. If X is a measurable space, with $f : X \to Y$ measurable, and $h = gof$, then $h : X \to Z$ is measurable.

Proof

If V is open in, Z then $g^{-1}(V)$ is open in Y and $h^{-1}(V) = f^{-1}(g^{-1}(V))$.

If f is continuous, it follows that $h^{-1}(V)$ is open, proving a).

If f is measurable, it follows that $h^{-1}(V)$ is also measurable, proving b).

Theorem 2

Let u and v be real measurable functions on a measurable space X; let ϕ be a continuous mapping of the plane into a topological space Y and define $h(x) = \phi(u(x), v(x))$ for $x \in X$; then $h : X \to Y$ is measurable.

Proof

Let $f(x) = (u(x), v(x))$; then vf maps X into the plane. Since, Theorem 1 shows that it is enough to prove the measurability of f.

If R is any rectangle in the plane with sides parallel to the axes, then R is the Cartesian product of two segments I_1 and I_2, and $f^{-1}(R) = u^{-1}(I_1) \cap v^{-1}(I_2)$, which is measurable by an assumption on u and v. Every open set V in the plane is a countable union of such rectangles R_i and $f^{-1}(V)$ is measurable because $f^{-1}(V) = f^{-1}\left(\bigcup_{i=1}^{\infty} R_i \right) = \bigcup_{i=1}^{\infty} f^{-1}(R_i)$.

B.3.6 PROPOSITIONS

Let X be a measurable space. The following propositions are corollaries of Theorems 1 and 2:

   i. If $f = u + iv$, where u and v are real measurable functions on X, then f is complex measurable function on X. (This follows from the previous theorem with $\phi(z) = z$.)

ii. If $f = u + iv$ is a complex measurable function on X, then u, v and $|f|$ are real measurable functions on X. (This follows from Theorem 1, with $g(z) = R(z), \mathrm{Im}(z)$ and $|z|$.)

iii. If f and g are complex measurable functions on X, then so are $f + g$ and fg.

The complex case follows then from (i) and (ii):

iv. If E is a measurable set in X and if $\chi_E(x) = \begin{cases} 1 & \text{for} \quad x \in E \\ 0 & \text{for} \quad x \notin E \end{cases}$

then, $\chi_E(x)$ is a measurable function.

We call $\chi_E(x)$ the characteristic function of the set E. The letter χ is used here as a characteristic function.

v. If f is a complex measurable function on X, there is a complex measurable function α on X such that $|\alpha| = 1$ and $f = \alpha |f|$.

Theorem 3

If $\mathfrak{I}$ is any collection of subsets of X, there exists a smallest σ-algebra M^* in X such that $\mathfrak{I} \subset M^*; M^*$ is sometimes called the σ-algebra generated by $\mathfrak{I}$.

B.3.7 Borel Sets

Let X be a topological space. By Theorem 3, there exists a smallest σ-algebra B in X such that every open set in X belongs to B. The members of B are called the Borel sets of X.

Examples of Borel sets include closed sets, all countable unions of closed sets and all intersections of open sets. Countable unions of closed sets are called F_σ's and countable intersections of open sets are called G_σ's. These notations are due to Hausdorff.

Theorem 4

Suppose M is a σ-algebra in X and Y is a topological space. Let f maps X into Y.

a. If Ω is the collection of all sets $E \subset Y$ such that $f^{-1}(E) \in M$, then Ω is a σ-algebra in Y.

b. If f is measurable and E is a Borel set in Y, then $f^{-1}(E) \in M$.

c. If $Y = [-\infty + \infty]$ and $f^{-1}((\alpha, \infty]) \in M$ for every real α, then f is measurable.

Proof

a. Follows from the relations $f^{-1}(Y) = X$, $f^{-1}(Y - A) = X - f^{-1}(A)$ and $f^{-1}(A_1 \cup A_2 \cup \cdots)$
$= f^{-1}(A_1) \cup f^{-1}(A_2) \cup \cdots$

b. To prove (b) let Ω be as in (a); the measurability of f implies that Ω contains all open sets in Y and since Ω is a σ-algebra, it contains all Borel sets in Y.

c. To prove (c) let Ω be the collection of all $E \subset [-\infty + \infty]$ such that $f^{-1}(E) \in M$. Since Ω is a σ-algebra in $[-\infty + \infty]$ and since $(\alpha, \infty] \in \Omega$ for all real α, the same is true for the set

$$[-\infty, \alpha) = \bigcup_{n=1}^{\infty} \left[-\infty, \alpha - \frac{1}{n} \right] = \bigcup_{n=1}^{\infty} \left[\alpha - \frac{1}{n}, \infty \right]^c \text{ and } (\alpha\beta) = [-\infty, \beta) \cap (\alpha, +\infty].$$

Since every open set in $[-\infty,+\infty]$ is a countable union of segments of the above types Ω contains every open set, so f is measurable.

Theorem 5

If $f_n : X \to [-\infty,+\infty]$ is measurable for $n = 1,2,\cdots$ and $g = \sup_{n\geq 1}f_n, h = \sup_{n\to\infty}f_n$, then g and h are measurable.

Proof

Considering $g^{-1}=\left((\alpha,\infty]\right)=\bigcup_{n=1}^{\infty}f_n^{-1}((\alpha,\infty))$, Theorem 4(c) implies that g is measurable. The same result holds with inf in place of sup and also, since $h = \inf_{k\geq 1}\{\sup_{i\geq k}f_i\}$, then h is measurable.

B.3.8 COROLLARY

i) The limit of every pointwise convergent sequence of complex measurable functions is measurable.
ii) If f and g are measurable, then so are $\max\{f,g\}$ and $\min\{f,g\}$. It is true for the functions $f^+ = \max\{f,0\}$ and $f^- = -\min\{f,0\}$. Note that f^+ is called the positive part of f and f^- is called the negative part of f.

B.4 SIMPLE FUNCTIONS

A function s on a measurable space X whose range consists of only finitely many points in $[0,\infty)$ will be called a simple function.

If $\alpha_1,\ldots,\alpha_n$ are the distinct values of a simple function s, and if $A_i = \{x : s(x) = \alpha_i\}$, then clearly:

$s = \sum_{i=1}^{n}\alpha_i\chi_{A_i}$, where χ_{A_i} is the characteristic function of A_i, as defined in section f, IV.

Note that the simple function s is measurable if and only if each of the sets A_i is measurable.

Theorem 6

Let $f : X \to [0,+\infty]$ be measurable. Then, there exists simple measurable functions s_n on X such that

a. $0 \leq s_1 \leq s_2 \leq s_3 \leq \cdot,f$
b. $s_n(x) \to f(x)$ as $n \to \infty$, for every $x \in X$.

Proof

For the proof, consider $n = 1,2,3,\cdots$ for $1 \leq i \leq n2^n$ and define:

(1) $E_{n,i} = f^{-1}\left[\left(\dfrac{i-1}{2^n},\dfrac{i}{2^n}\right)\right]$ and $F_n = f^{-1}[n,\infty]$ and write:

(2) $S_n = \sum_{i=1}^{n}\dfrac{i-1}{2^n}\chi_{E_{n,i}} + n\chi_{F_n}.$

Theorem 4 (b) shows that $E_{n,i}$ and F_n are measurable sets. We can notice that the functions defined in (2) satisfy the conditions (a) of the Theorem 6.

If x is such that $f(x) < \infty$, then $s_n(x) \geq f(x) - 2^{-n}$ when n is large enough, if $f(x) = \infty$ then $s_n(x) = f(x)$ which proves (b).

B.5   ELEMENTARY PROPERTIES OF MEASURES

B.5.1 DEFINITION

(a) A positive measure is a function μ defined on a σ-algebra, M, whose range is in $[0, \infty]$ which is accountably additive. That is, if the set $\{A_i\}$ is a disjoint countable collection of members of M, then

$$\mu\left(\bigcup_{i=1}^{\infty} A_i\right) = \sum_{i=1}^{\infty} \mu(A_i);$$

we shall assume $\mu(A) < \infty$ for at least an $A \in M$. This helps avoid trivialities.

(b) A measure space is a measurable space which has a positive measure defined on the σ-algebra of its measurable sets.
(c) A complex measure is a complex-valued countable additive function defined on a σ-algebra.

Note: A positive measure can just be called a measure. If $\mu(E) = 0$ for every $E \in M$, then μ is a positive measure. Also, for a complex measure, it is understood that $\mu(E)$ is a complex number for every $E \in M$. The real measures are a subclass of the complex ones.

Theorem 7
Let μ be a positive measure defined on a σ-algebra M. Then:

(a) $\mu(\phi) = 0$;

(b) $\mu(A_1 \cup \cdots \cup A_n) = \mu(A_1) + \cdots + \mu(A_n)$ are pairwise disjoint members of M;

(c) $A \subset B$ implies $\mu(A) \leq \mu(B)$ if $A \in M, \ B \in M$;

(d) $\mu(A_n) \to \mu(A)$ as $n \to \infty$ if $A = \bigcup_{n=1}^{\infty} A_n, A_n \in M$ and $A_1 \subset A_2 \cdot \subset A_3 \subset \cdots$;

(e) $\mu(A_n) \to \mu(A)$ as $n \to \infty$ if $A = \bigcap_{n=1}^{\infty} A_n, A_n \in M$ and $A_1 \supset A_2 \cdot \supset A_3 \supset \cdots, \mu(A_1)$ is finite.

B.6   INTEGRATION OF POSITIVE FUNCTION

Consider here M a σ-algebra in a set X and μ a positive measure on M. We can define as follows.

B.6.1   DEFINITION

Consider s a measurable simple function on X of the form

$$s = \sum \alpha_i \chi_{A_i} \tag{B.6.1}$$

where $\alpha_1, \cdot, \alpha_n$ are distinct values of s. Consider also $E \in M$, then

$$\int_E s d\mu = \sum_{i=1}^{n} \alpha_i \mu\left(A_i \cap E\right) \tag{B.6.2}$$

If $f : X \to [0, +\infty]$ is measurable and $E \in M$, we define

(3) $\int_E f d\mu = \sup \int_E s d\mu$ the supremum being taken over all simple measurable functions s such that $0 \leq s \leq f$.

The left side of equation (3) is called the Lebesgue integral of f over E, with respect to the measure of μ. It is a number in $[0, \infty]$.

B.6.2   CONSEQUENCE

Assume f, g are measurable functions, then:

(a) if $0 \leq f \leq g$, then $\int f d\mu \leq \int_E g d\mu$.

(b) If $A \subset B$ and $f \geq 0$ then $\int_A f d\mu \leq \int_B g d\mu$.

(c) If $f \geq 0$ and c a constant, with $0 \leq c \leq \infty$, then $\int_E c f d\mu = c \int_E f d\mu$.

(d) If $f(x) = 0$ for all $x \in E$, then $\int_E f d\mu = 0$ even if $\mu(E) = \infty$.

(e) If $\mu(E) = 0$, then $\int_E f d\mu = 0$ even if $f(x) = \infty$ for every $x \in E$.

(f) If $f \geq 0$, then $\int_E f d\mu = \int_K \chi_E f d\mu$.

PROPOSITION

Let s and t be measurable simple functions on X. For $E \in M$, define

(1) $\varphi(E) = \int_E s d\mu$;

then φ is a measure on M. Also,

(2) $\int_X (s + t) d\mu = \int_X s d\mu + \int_X t d\mu$.

Theorem 8 (Lebesgue's monotone convergence theorem)
Let $\{f_n\}$ be a sequence of measurable functions on X and suppose that

(a) $0 \leq f_1(x) \leq f_2(x) \leq \cdots \leq \infty$ for every $x \in X$;

(b) $f_n(x) \to f(x)$ as $n \to \infty$, for every $x \in X$.

Then f is measurable, and $\int_k f_n d\mu \to \int_K f d\mu$ as $n \to \infty$.

B.7   INTEGRATION OF COMPLEX FUNCTIONS

Consider as before μ a positive measure on an arbitrary measurable space X.

B.7.1 Definition

We call $L^1(\mu)$ the collection of all complex measurable functions f on X for which $\int_X |f| d\mu < \infty$. It is understood that the measurability of f implies that of $|f|$. The elements of $L^1(\mu)$ are called Lebesgue integrable functions (with respect to μ) or summable functions.

B.7.2 Properties

Consider the complex function $f = u + iv$ where u and v are real measurable functions on X and with $f \in L^1(\mu)$ we define:

(i) $\int_E f d\mu = \int_E u^+ d\mu - \int_E u^- d\mu + i\int_E v^+ d\mu - i\int_E v^- d\mu$ for every measurable set E.

   u^+ and u^- are the positive and negative parts of u as defined before.

(ii) $\int_E f d\mu = \int_E f^+ d\mu - \int_E f^- d\mu$.

   With f a measurable function in $[-\infty, +\infty]$.

(iii) For every f and $g \in L^1(\mu)$ and α, β complex numbers, then $\alpha f + \beta g \in L^1(\mu)$ and

   $\int_X (\alpha f + \beta g) d\mu = \alpha \int_X f d\mu + \beta \int_X g d\mu$.

(iv) For $f \in L^1(\mu), \left| \int_X f d\mu \right| \leq \int_X |f| d\mu$.

B.8 TRIGONOMETRIC IDENTITIES

Here are some useful identities in trigonometry.

$$\cos\left(\frac{\pi}{2} - x\right) = -\sin(x)$$

$$\cos\left(\frac{\pi}{2} + x\right) = -\sin(x)$$

$$\cos(\pi + x) = -\cos(x)$$

$$\cos(\pi - x) = -\cos(x)$$

$$\cos^2(x) + \sin^2(x) = 1$$

$$\cos\left(\frac{\pi}{2} + x\right) = -\sin(x)$$

$$\sin\left(\frac{\pi}{2} + x\right) = \cos(x)$$

$$\sin\left(\frac{\pi}{2} - x\right) = \cos(x)$$

$$\sin(\pi + x) = -\sin(x)$$

$$\sin(\pi - x) = \sin(x)$$

$$\cos(a+b) = \cos(a)\cos(b) - \sin(a)\sin(b)$$

$$\cos(a-b) = \cos(a)\cos(b) + \sin(a)\sin(b)$$

$$\cos(a+b) + \cos(a-b) = 2\cos(a)\cos(b)$$

$$\cos(a+b) - \cos(a-b) = -2\sin(a)\sin(b)$$

$$\cos(2x) = 2\cos^2(x) - 1 = 1 - 2\sin^2(x)$$

Assume now that $a+b = p; a-b = q$;

therefore $a = \frac{p+q}{2}$ and $b = \frac{p-q}{2}$. Also,

$$\cos(p) + \cos(q) = 2\cos\left(\frac{p+q}{2}\right)\cos\left(\frac{p-q}{2}\right)$$

$$\cos(p) - \cos(q) = -2\sin\left(\frac{p+q}{2}\right)\sin\left(\frac{p-q}{2}\right)$$

$$\sin(a+b) = \sin(a)\cos(b) + \cos(a)\sin(b)$$

$$\sin(a-b) = \sin(a)\cos(b) - \cos(a)\sin(b)$$

$$\sin(a+b) + \sin(a-b) = 2\sin(a)\cos(b)$$

$$\sin(a+b) - \sin(a-b) = 2\cos(a)\sin(b)$$

$$\sin(p) + \sin(q) = 2\sin\left(\frac{p+q}{2}\right)\cos\left(\frac{p-q}{2}\right)$$

$$\sin(p) - \sin(q) = 2\cos\left(\frac{p+q}{2}\right)\sin\left(\frac{p-q}{2}\right)$$

$$\sin(2x) = 2\sin(x)\cos(x)$$

Bibliography

Akhiezer, N. I., & Glazman, I. M. (1993). *Theory of linear operators in Hilbert space*. Dover Publications, Inc.

Arfken, G. B. (1970). *Mathematical methods for physicists* (2nd ed.). Academic Press.

Ballentine, L. E. (1990). *Quantum mechanics*. Prentice Hall, Inc.

Byron, F., & Fuller, R. (1969). *Mathematics of classical and quantum physics*. Dover Publication, Inc.

Champeney, D. C. (1973). *Fourier transforms and their physical applications*. Academic Press Inc.

Chaos, A. V. (1986). *Holden*. Manchester University Press.

Cushing, J. (1975). *Applied analytical mathematics for physical scientists*. John Wiley & Sons, Inc.

Goldstein. (1959). *Classical mechanics*. Addison-Wesley Publishing company, Inc.

Goodbody, A. M. (1982). *Cartesian tensors, with applications to mechanics-fluids mechanics and elasticity*. Ellis Horwood Limited.

Goodman, A.W. & Saff, E. B. (1981). *Calculus: Concepts and calculations*. Macmillan Publishing Co., Inc.

Hellwig, G. (1964). *Partial differential equations an introduction* (English ed.). Blaisdell Publishing Company.

Helmberg, G. (2008). *Introduction to spectral theory in Hilbert space*. Dover.

Hogg, R. V., & Craig, A. T. (1970). *Introduction to mathematical statistics* (4th ed.). Macmillan Publishing Co., Inc.

Holl, D. L., Maple, C., & Vinograde, B. (1959). *Introduction to the laplace transform*. Appleton-Century-Crofts, Inc.

Jaeger, J. C., & Starfield, A. M. (1974). *An introduction to applied mathematics*. (2nd ed.). Oxford at the Clarendon Press.

Jeffreys, Sir H., & Swirles, B. (1972). *Methods of mathematical physics* (3rd ed.). Cambridge University Press.

Jones, L. M. (1979). *An introduction to mathematical methods of physics*. The Benjamin, Cummings Publishing Company, Inc.

Larson, R. (2013). *College Algebra* (9th ed.). Cengage learning.

Larson, R. & Falvo, D. (2009). *Elementary linear algebra* (6th ed.). by Houghton Mifflin Harcourt Publishing Company.

Levine, H. (1997). *Partial differential equations* (vol. 6). American Mathematical Society, International Press.

McOwen, R. C. (2003). *Partial differential equations methods and application* (2nd ed.). Prentice Hall.

Morrison, F. A. (2001). *Understanding rheology*. Oxford University Press.

Newbury, N., Newman, M., Ruhl, J., Staggs, S., & Thorsett, S. (1990). *Princeton problems in physics with solution*. Princeton University Press.

Papoulis, A. (1962). *The fourier integral and its applications*. McGraw-Hill Book Company, Inc.

Parshall, K. H., & Rice, A. C. (2002). *Providence, R.I., mathematics unbound: The evolution of an international mathematical research community, 1800–1945*. American Mathematical Society.

Perko, L. (2000). *Differential equations and dynamical systems* (3rd ed.). Springer.

Pinsky, M. A. (1991). *Partial differential equations and boundary-value problems with applications* (2nd ed.). McGraw-Hill, Inc.

Rudin, W. (1966). *Real and complex analysis*. Mc Graw Hill Inc.

Sanchez, D. A., Allen, R. C., Walter, J. R. & Kyner, T. (1988). *Differential equations* (2nd ed.). Addison-Wesley.

Sansone, G. (1959). *Orthogonal functions* (Rev. English ed.). Interscience Publishers.

Schowalter, W. R. (1978). *Mechanics of non-newtonian fluids*. Pergamon Press, Inc.

Smirnov, A. V. (2004, September). *Introduction to tensor calculus*. McGraw-Hill, New York.

Smirnov, V. (1975). *Cours de Mathematiques superieures, Tome 4 1ere partie*. Edition Mir Moscou.

Stakgold, I. (1998). *Green's functions and boundary value problems* (2nd ed.). A Wiley-Interscience publication, John Wiley & Sons, Inc.

Strogatz, S. H. (1994). *Nonlinear dynamics and chaos*. Perseus Books Publishing.

Thangavelu, S. (1993). *Lectures on Hermite and Laguerre expansions*. Princeton University Press.

Vinuesa, J. (1989). *Orthogonal polynomials and their applications: Proceedings of the international congress*. M. Dekker.

Walpole, R. & Myers, R. (1993). *Probability and statistics for engineers and scientists* (5th ed.). Prentice Hall, NJ07632.

Williamson, R. E., Crowell, R. H., & Trotter, H. F. (1968). *Calculus of vector functions* (2nd ed.). Prentice Hall, Inc.

Young, N. (1988). *An introduction to Hilbert space*. Cambrigde University Press.

Index

Printed in the United States
by Baker & Taylor Publisher Services